Fourth Edition

PRINCIPLES OF

...OGY

Fourth Edition

PRINCIPLES OF ECOTOXICOLOGY

C.H. Walker, R.M. Sibly, S.P. Hopkin, D.B. Peakall

CRC Press
Taylor & Francis Group
Boca Raton   London   New York

CRC Press is an imprint of the
Taylor & Francis Group, an **informa** business

CRC Press
Taylor & Francis Group
6000 Broken Sound Parkway NW, Suite 300
Boca Raton, FL 33487-2742

© 2012 by Taylor & Francis Group, LLC
CRC Press is an imprint of Taylor & Francis Group, an Informa business

No claim to original U.S. Government works

Printed and bound by CPI Group (UK) Ltd, Croydon, CR0 4YY
Version Date: 20120206

International Standard Book Number: 978-1-4398-6266-7 (Paperback)

Visit the Taylor & Francis Web site at
http://www.taylorandfrancis.com

and the CRC Press Web site at
http://www.crcpress.com

This fourth edition is dedicated to the memory of

Dr. Steve Hopkin (1956–2006)

who made key contributions to the first two editions.

Contents

Section II   Effects of Pollutants on Individual Organisms

Section III   Effects of Pollutants on Populations and Communities

Preface to Fourth Edition

Since publication of the first edition of *Principles of Ecotoxicology*, both David Peakall and Steve Hopkin have died—serious losses to the international scientific community. Both made key contributions to the first two editions. The last edition was dedicated to the memory of David; we dedicate this edition to the memory of Steve.

The origins of this book lie in the MSc course titled "Ecotoxicology of Natural Populations," first taught at Reading in 1991. Ecotoxicology was then emerging as a distinct subject of interdisciplinary character. The structure of the course reflected this characteristic and was taught by people of widely differing backgrounds ranging from chemistry and biochemistry to population genetics and ecology. Combining the different disciplines in an integrated way was something of a challenge.

The experience of teaching the course persuaded the authors of the need for a textbook that would deal with the basic principles of such a wide-ranging subject. The intention has been to approach ecotoxicology in a broad interdisciplinary way, cutting across traditional subject boundaries. However, the nature of the text is bound to reflect the experiences and interests of the authors.

Since publication of the first edition, important advances occurred in some areas of ecotoxicology and progress was disappointingly slow in others, reflecting the high costs of certain types of research and the difficulty of obtaining support for some areas of study that do not represent the highest priorities of funding agencies. The objective of this fourth edition has been to bring the text up to date and strengthen the treatments of certain topics that we believe are very important now. We have continued to follow our original purpose—emphasizing principles rather than practice. Other works that describe testing protocols and modern statistical techniques for analyzing data from ecotoxicological studies are cited in the text and on lists of further reading at the end of each chapter. Throughout the book, small changes and additions have been made and a new final chapter has been added to discuss prospects for the future development of ecotoxicology.

A theme running through this new edition is how the concepts discussed may contribute to improved methods of environmental risk assessment. To what extent will it be possible to adopt a more mechanistic approach to risk assessment using biomarker assays or comprehensive simulations of animal populations in realistically modelled landscapes? How feasible is it to use results from tests whose endpoints are not toxicity data for laboratory species but changes in populations, communities, or ecosystems? Protocols for ecotoxicity testing are currently subject to much debate and touch not only on scientific issues but on ethical, economic, and political ones as well. In recent years, the suffering of laboratory animals caused by toxicity testing has become an important issue in Western countries.

In producing this fourth edition the authors gratefully acknowledge the feedback, help, and advice from many users of the book, especially Dr. Peter Hodson, Dr. Russ McClain, Dr. Erik Muller, Dr. Marinus L. Otte, Dr. William E. Robinson, and Dr. Denise M. Woodward.

Colin Walker and Richard Sibly

Acknowledgments

Many people have contributed to this book in many ways. Although we cannot acknowledge them all, we would specifically like to mention our MSc students who contributed much in discussion and feedback, and Amanda Callaghan, Peter Calow, Peter Dyte, Mark Fellowes, Valery Forbes, Glen Fox, Andy Hart, Graham Holloway, Tom Hutchinson, Paul Jepson, Laurent Lagadic, Alan McCaffery, Mark Macnair, Steve Maund, Pavel Migula, Diane Nacci, Miroslav Nakonieczny, Ian Newton, Demetris Savva, Ken Simkiss, Nick Sotherton, Nico van Straalen, Pernille Thorbek, Chris Topping, Charles Tyler, Paule Vasseur, and George Warner.

We have made every effort to contact authors and/or copyright holders of works reprinted in *Principles of Ecotoxicology*. This has not been possible in all cases, and we will welcome correspondence from individuals and companies we have been unable to trace.

Authors

Steve Hopkin died in 2006. He was a zoologist who worked on electron microscopy and x-ray analysis for his Ph.D. and later investigated the effects of metals on soil ecology at the University of Bristol. During his time at University of Reading, his teaching and research focused on the roles of essential and nonessential metals in the biology of soil invertebrates.

The late **David Peakall** studied chemistry and commenced his research as a physical chemist. He moved into biochemistry and finally into environmental toxicology. The last move was in keeping with his long-standing interest and active involvement in ornithology. During the last fifteen years of his scientific career, he was chief of the wildlife toxicology division of the Canadian Wildlife Service where he played a major role in studies of the Great Lakes.

Richard Sibly applied a degree in mathematics first in animal behavior and then more widely in ecology, including studies of the population effects of environmental chemicals. His recent work has focused on the new metabolic theory of ecology and developing and using agent-based models (ABMs) in population ecology and environmental risk assessment.

Colin Walker originally qualified as an agricultural chemist. He was responsible for chemical and biochemical studies of environmental pollutants at the Monk's Wood Experimental Station during the mid-1960s when certain effects of organochlorine insecticides were established. This work led to restrictions on the use of cyclodienes and DDT. He subsequently joined the University of Reading where he taught and conducted research on the molecular basis of toxicity with particular reference to ecotoxicology. Now retired, he is currently affiliated with the Department of Biosciences at the University of Exeter where he teaches a course in ecotoxicology.

Introduction

The term *ecotoxicology* was introduced by Truhaut in 1969 and was derived from the words *ecology* and *toxicology*. The introduction of the new term reflected a growing concern about the effects of environmental chemicals on species other than humans. It identified an area of study concerned with the harmful effects of chemicals (toxicology) within the context of ecology. Until now environmental toxicology focused mainly on the harmful effects of environmental chemicals on humans, e.g., the effects of smoke on urban communities. However, environmental toxicology in its widest sense encompasses the effects of chemicals on ecosystems as well. Thus ecotoxicology is a discipline within the wider field of environmental toxicology (see Calow, 1994). In the present text, it is defined as the study of harmful effects of chemicals upon ecosystems and includes effects on individuals and consequent effects at the levels of population and above.

Despite this definition, much early work described as ecotoxicology related little to ecology or toxicology. It was concerned with the detection and determination of chemicals in samples of animals and plants. Seldom could the analytical results be related to effects on individual organisms, let alone effects on populations or communities. Analytical techniques such as gas chromatography, thin layer chromatography, and atomic absorption facilitated the detection of very low concentrations of chemicals in biota. Establishing the biological significance of the results was a more difficult matter! One of the main themes of this text is the problem of progressing from the measurement of concentrations of environmental chemicals to establishing their effects at the levels of the individual, the population, and the community.

New disciplines frequently present problems of terminology, and ecotoxicology is no exception. Several important ecotoxicology terms are used inconsistently in the literature. Their use in the present text will now be explained. Both pollutants and environmental contaminants are chemicals that exist at levels judged to be above those that would normally occur in any component of the environment. This immediately raises the question of what is to be considered normal. For most man-made organic chemicals such as pesticides, the situation is simple—any detectable level is abnormal because the compounds did not exist in the environment until humans released them. Conversely, chemicals such as metals, sulfur dioxide, nitrogen oxides, polycyclic aromatic hydrocarbons (PAHs), and methyl mercury occur naturally and existed in the environment before humans appeared. The variations in concentrations of these chemicals from place to place and from time to time make it difficult to judge their normal ranges.

The distinction sometimes made between pollutants and contaminants raises further difficulties. In the wider literature, the term pollutant implies that the chemical in question causes environmental harm. A contaminant is a chemical present at levels above those that might be judged normal and may or may not have caused environmental damage. In this text, examples of environmental chemicals have been chosen that are widely regarded as pollutants. They have been shown to have caused environmental harm or clearly exhibit the potential to do so at environmentally realistic levels.

By contrast, the contaminant term implies that a chemical is not necessarily harmful at environmentally realistic levels. The difficulties with this distinction are threefold. The first is the general toxicological principle that toxicity is related to dose (Chapter 5). Thus a compound may answer to the description of pollutant in one situation but not in another—a problem mentioned earlier. The second issue is the lack of general agreement about what constitutes environmental harm or damage. Some scientists regard deleterious biochemical changes in an individual organism as harmful; others reserve the term for declines in populations. Third, the effects of measured levels of chemicals in living organisms or in their environments are seldom known, but the term pollutant is applied to them. Judgment of this issue is made more difficult by the possibility of potentiation of toxicity when organisms are exposed to mixtures of environmental chemicals. To minimize these terminology, problems of "pollutant" will refer to environmental chemicals that exceed normal background levels and can cause harm. It would be attractive to reserve the term for particular chemicals in situations in which they have been shown to cause harm, but due to problems of measurement, this usage would be too restrictive. Harm encompasses the biochemical or physiological changes that adversely affect individual organisms' birth, growth, or mortality rates. Such changes would necessarily produce population declines if other processes (e.g., density dependence) did not compensate (Chapter 12).

Whether a contaminant is a pollutant therefore depends on its level in the environment, on the organism considered, and whether the organism is harmed. Thus a compound may answer to the description of pollutant for one organism but not for another. Because of the problems in demonstrating harmful effects in the field, the terms pollutant and contaminant will, to a large extent, be used synonymously because it can seldom be said that contaminants have no potential to cause environmental harm in any situation. We will use environmental chemical to describe any chemical that occurs naturally in the environment without judging whether it should be regarded as a pollutant or as a contaminant.

Another word that has been used inconsistently in the literature is the term biomarker. Here, biomarkers are defined as biological responses to environmental chemicals at the individual level or below, demonstrating departure from normal status. Biomarker responses may be at the molecular, cellular, or whole-organism level. Some workers would regard population responses (changes in number or gene frequency) as biomarkers. However, as the latter tend to be much longer term than the former, it may be unwise to use the same term for both. In the present text, the term biomarker will be restricted to biological responses at the level of the whole organism or below. An important thing to emphasize about biomarkers is that they represent measurements of effects, which can be related to the presence of particular levels of environmental chemicals; they provide a means of interpreting environmental levels of pollutants in biological terms.

Finally, the organic pollutants considered in this text are examples of xenobiotics (foreign compounds). They play no part in the normal biochemistry of living organisms. Xenobiotics will be discussed in Chapter 5.

Although ecotoxicology is a relatively new discipline, it is worth emphasizing that chemical warfare has been waged in the natural environment since early in the evolutionary history of Planet Earth. Both plants and animals produce chemical weapons. Plants produce metabolites that are toxic to animals that graze upon them. In turn, animals developed enzyme systems (e.g., forms of cytochrome P450) that can metabolize and thereby detoxify such compounds. This phenomenon has been called a co-evolutionary arms race (Harborne, 1993).

It is worth recalling that a number of naturally occurring compounds exert insecticidal actions and this property has been harnessed commercially by marketing these compounds as insecticides—or using them as models for the development of novel insecticides. Examples include natural pyrethrins and nicotine, both of which have been used as insecticides and as models for novel commercially developed insecticides. Conversely, the detoxifying enzymes present in insects can protect against commercially developed insecticides. Indeed, the continued and indiscriminate use of insecticides leads to the emergence of resistant strains of insects via a process that might be described as unnatural selection. Whichever term is used, the study of the molecular basis of the development of resistance provides a superb example of the operation of the principle of natural selection—a phenomenon of great interest to Darwinians.

These considerations extend beyond the responses to chemical weapons produced by animals and plants during the so-called co-evolutionary arms race. Other naturally occurring chemicals including methyl mercury, methyl arsenical compounds, polycyclic aromatic hydrocarbons, and certain heavy metals exert selective pressures on living organisms. These chemicals originate from the weathering of rocks, natural events such as volcanic activity, and the metabolic actions of microorganisms. Thus, ecotoxicology can be viewed against the background of the effects of a wide range of naturally occurring toxic compounds acting on living organisms during the course of evolution.

Approach and Organization of This Book

An exciting feature of ecotoxicology is that it represents a molecules-to-ecosystems approach that relates to the genes-to-physiologies approach originally identified by Clarke (1975) and extensively developed in North America in the 1980s (Feder et al., 1987). Moreover, it analyzes experimental manipulations on the largest of scales (although the experiments were not designed as such). Thus metal pollution, acid rain, and the applications of pesticides have affected whole ecosystems, sometimes with dramatic consequences for their populations. In ecotoxicology, the ecosystem response is studied at all levels. Initially (Figure 0.1), the molecular structures of pollutants and their properties and environmental fate are considered (Section I of this book).

Ecophysiologists generally analyze the impact of pollutants on an organism's growth rates, birth rates, and death rates; indeed, as explained above, pollutants can adversely affect these vital rates. This makes it desirable to understand how adverse effects on vital rates have implications for populations (Chapters 12 and 13). Consequently it is, in principle, possible to evaluate pollutants quantitatively in terms of their population effects. This emphasis on vital rates as crucial intervening variables, linking physiological effects to population effects, is a particular feature of this book. The approach is continued in Chapter 13 to consider whether and how quickly resistant genes increase in populations. The rate at which resistant genes increase is measured by the growth rate of the population of resistant genes. The population growth rate of resistant genes is a measure of their Darwinian fitness. Although this is not the conventional population-genetic measure of fitness, it is particularly useful in ecotoxicology because it (alone) shows explicitly how the fitness of resistant genes depends on the effects those genes have on their carriers. To summarize, the approach taken in this book allows linkages to be made between the different levels of organization shown in Figure 0.1, from molecules to physiologies to populations, right through to ecosystems. This is the underlying basis for the biomarker strategy, which seeks to measure sequences of responses to pollutants from the molecular level to the level of ecosystems (Chapters 10, 15, and 16). The use of biomarkers in biomonitoring is described in Chapter 11. These four chapters are placed at the end of their respective sections of the book. They represent the practical realization of theoretical aspects described in earlier chapters.

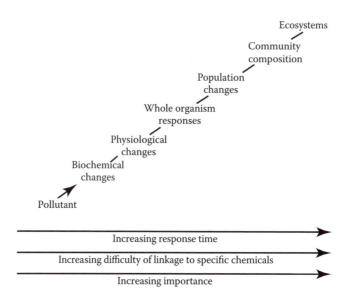

FIGURE 0.1
Schematic relationship of linkages between responses at different organizational levels.

The text is divided into three parts. Section I describes major classes of organic and inorganic pollutants, their entry into the environment, and their movement, storage, and transformation within the environment. Thus, this part bears a certain resemblance to toxicokinetics in classic toxicology, which deals with the uptake, distribution, metabolism, and excretion of xenobiotics by living organisms (Chapter 5). The difference is in complexity. Ecotoxicology deals with movements of pollutants in air, water, soils, and sediments and through food chains, with chemical transformation and biotransformation.

Section II deals with the effects of pollutants on living organisms and resembles toxicodynamics in classic toxicology. The difference is again one of complexity. Toxicodynamics focuses on interactions of xenobiotics and their sites of action; ecotoxicology covers a wide range of effects on individual organisms at differing organizational levels (molecular, cellular, and whole animal). Toxicity data obtained in laboratories are used for the purposes of risk assessment. Effects of pollutants are studied in the laboratory, an approach that can lead to the development of biomarker assays (Chapter 10). The use of biomarker assays in biomonitoring is discussed in Chapter 11. The chapter also considers effects at the population level, thereby looking ahead to the final part of the text.

Section III addresses the questions of greatest interest to ecologists. What effects do pollutants have at the levels of population, community, and whole ecosystem? This takes the discussion into the disciplines of population biology and population genetics. Classic toxicology is concerned with chemical toxicity to individuals. Ecotoxicologists are interested in effects at the level of population, community, and whole ecosystem. Effects at the population level may be changes in numbers of individuals (Chapter 12), changes in gene frequency (as in resistance; Chapter 13), or changes in ecosystem function (e.g., soil nitrification; Chapter 14). Effects may be sublethal (e.g., on physiology or behavior) rather than lethal. They may be indirect (e.g., decline in predator numbers because of direct chemical toxicity may lead to an increase in numbers of its prey). It is often difficult to establish effects of pollutants on natural populations. However, the development of appropriate biomarker assays can help resolve this problem.

Applications and Conclusions

The principal areas of application of ecotoxicology are the biomonitoring of environmental pollution (including the use of bioassays and biomarkers), the investigation of pollution problems, the conducting of field trials (particularly of pesticides), the study of the development of resistance, and finally, risk assessments of environmental chemicals—an area of growing importance that receives particular emphasis in this fourth edition of *Principles of Ecotoxicology*.

Risk assessment is required to establish whether novel chemicals are safe to use. In particular, there is need to show they will not harm populations of non-target organisms. Current practice is largely based on laboratory testing of chemicals on a small number of organisms chosen in part for practicality (e.g., they complete their life cycles quickly in laboratory conditions) and in part as being representative of other organisms with similar lifestyles. These tests have yielded much valuable data and we refer to the results of such tests at appropriate places in the text. We do not describe in detail the methods currently in use in risk assessment, which vary to some extent between states. Our aim instead has been to provide a basis for understanding what happens to chemicals in the real world, where they go and how they ultimately degrade, and how they affect the individuals and populations that encounter them. Thus we hope our book will provide a solid foundation for all those contemplating a career in the important and interesting field of risk assessment.

A companion volume to this book covers the mechanistic aspects of ecotoxicology in greater depth and detail than this text. It is titled *Organic Pollutants: An Ecotoxicological Perspective*, 2nd Edition by C. H. Walker (2009). In the chapters that deal with individual groups of pollutants, the companion volume is structured in a similar way to the present text as shown in the following table.

Divisions in This Text	Corresponding Sections in Chapters 5–12 of *Organic Pollutants*
Section I: Pollutants and their fates in ecosystems	Chemical properties, metabolism, and environmental fate
Section II: Effects of pollutants on individual organisms	Toxicity of pollutants(s)
Section III: Effects of pollutants on populations and communities	Ecological effects

We hope that our book illustrates the truly interdisciplinary character of ecotoxicology. The study of the harmful effects of chemicals on ecosystems draws on the knowledge and skills of ecologists, physiologists, biochemists, toxicologists, chemists, meteorologists, soil scientists, and others. It is nevertheless a discipline with a distinct character. In addition to the important applied aspects that address current public concerns, it has firm roots in basic science. Chemical warfare is nearly as old as life itself and the evolution of detoxification mechanisms by animals to avoid the toxic effects of xenobiotics produced by plants is paralleled by the recent development of resistance by pests to pesticides made by humans.

Section I

Pollutants and Their Fate in Ecosystems

1

Major Classes of Pollutants

Many different chemicals are regarded as pollutants, ranging from simple inorganic ions to complex organic molecules. In this chapter, representatives will be identified of all the major classes of pollutants, and their properties and occurrence will be briefly reviewed. These pollutants will be used as examples throughout the text. Their fate in the living environment will be the subject of the remainder of Section I. Their effects upon individuals and ecosystems will be considered in Sections II and III, respectively.

1.1 Inorganic Ions

1.1.1 Metals

A metal is defined by chemists as an element that has a characteristic lustrous appearance, is a good conductor of electricity, and generally enters chemical reactions as a cation. Although metals are usually considered pollutants, it is important to recognize that they are natural substances.

With the exception of radioisotopes produced by man-made nuclear reactions (bombs and reactors), all metals have been present on the earth since its formation. There are a few examples of localized metal pollution resulting from natural weathering of ore bodies (e.g., Hågvar and Abrahamsen, 1990). However, in most cases, metals become pollutants after human activity, mainly through mining and smelting, releases them from the rocks in which they were deposited during volcanic activity or subsequent erosion, and relocates them where they can cause environmental damage. In April 1998, for example, a major spill of metal-rich mining waste near Doñana National Park in southwest Spain caused one of the worst environmental disasters in western Europe (Pastor et al., 2004).

The extent to which human activity contributes to global cycles of metals can be described by the anthropogenic enrichment factor (AEF; Table 1.1). The table clearly shows that human activity is responsible for most global movement of cadmium, lead, zinc, and mercury, but is relatively unimportant in the cycling of manganese. The very high AEF for lead is due mostly to the widespread use and subsequent release of lead-based additives to gasoline. For most radioactive isotopes, the AEF is 100%.

Elements considered metals are identified within the periodic system that classifies all elements (Figure 1.1). Groups of elements sharing similar chemical properties are listed in vertical columns. The first two columns show elements that readily lose one or two outer electrons to yield monovalent cations (column 1) or divalent cations (column 2). Among these are many of the most common metals found in surface waters and soils in their stable ionic forms (Na^+, K^+, Mg^{2+}, and Ca^{2+}). The next 10 columns list the transition elements that are regarded as more complex than the alkali and alkali earth elements that constitute the first two groups. Moving from left to right through the three main series of transition

TABLE 1.1

Anthropogenic Enrichment Factors (AEFs) for Total Global Annual
Emissions of Cadmium, Lead, Zinc, Manganese, and Mercury in
the 1980s

Metal	Anthropogenic Sources (A)	Natural Sources	Total (T)	AEF (A/T) (%)
Cadmium (Cd)	8	1	9	89
Lead (Pb)	300	10	310	97
Zinc (Zn)	130	50	180	72
Manganese (Mn)	40	300	340	12
Mercury (Hg)	100	50	150	66

Note: All values = 10^6 kg year^{-1} determined from various sources.

elements, the nuclei become larger and the outer electrons show less tendency to escape (i.e., form cations) than the elements listed in columns 1 and 2. Consequently, these elements have a tendency to share electrons with other elements, leading to the formation of covalent bonds and complex ions (e.g., by copper, iron, cobalt, or nickel). Some of the larger atoms tend to retain electrons and remain in the elemental state (e.g., silver and gold, the so-called noble metals).

Other characteristics of iron, copper, and certain other transition elements are variable in valency and participation in electron transfer reactions. Electron transfer reactions involving oxygen can lead to the production of toxic oxyradicals. This toxicity mechanism is now known to be of considerable importance for both animals and plants. Some oxyradicals such as the superoxide anion ($\cdot O_2^-$) and the hydroxyl radical ($\cdot OH$) can cause serious cellular damage.

Elements in the remaining vertical groups as we move from left to right exhibit less tendency to form cations. The elements progress from metals to metalloids, the latter showing characteristics of both metals and nonmetals, until the nonmetal elements (C, N, O, P, S, Cl, Br, etc.) are reached. The final vertical column contains the very stable inert gases that

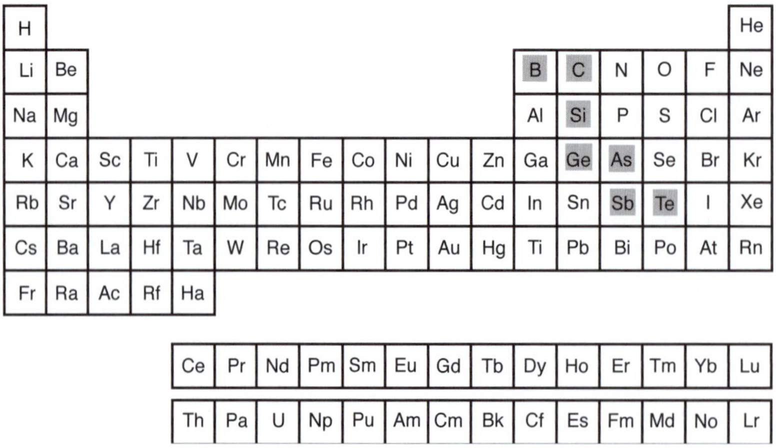

FIGURE 1.1
Periodic table of the elements. Those considered metals are surrounded by bold lines. Metalloids (with properties of metals and nonmetals) are shaded. (*Source:* Hopkin, S.P. (1989). *Ecophysiology of Metals in Terrestrial Invertebrates.* Elsevier Applied Science Publishers. With permission.)

exhibit almost no chemical reactivity. The two horizontal boxes below the main periodic classification list the rare elements of the lanthanide and actinide series that are metallic in character.

The tendency to form covalent bonds shown by metalloids and also by metals close to them in the periodic classification table has two important toxicological consequences. First, these elements can bind covalently to organic groups, thereby forming lipophilic compounds and ions. Some of these compounds such as tetraalkyl lead, tributyl tin oxide, methyl mercury salts, and methylated forms of arsenic are highly toxic. Because of their lipophilicity, their distribution within animals and plants and their toxic actions usually differ from activities of simple ionic forms of the same elements. Organometallic compounds are discussed in Section 1.3. Second, these elements can exert toxic effects by binding to nonmetallic constituents of cellular macromolecules, e.g., the binding of copper, mercury, lead, and arsenic to sulfydryl groups of proteins.

The term *heavy metals* has been used extensively to describe metals that act as environmental pollutants. For a metal to be considered heavy, it must have a density relative to water of greater than 5. However, the term has been replaced by a classification scheme that considers chemistry rather than relative density (Nieboer and Richardson, 1980, Table 1.2). This approach is more logical because some metals are not heavy but can act as important environmental pollutants. Aluminum, for example, is a metal with a relative density of only 1.5. However, it is an extremely important pollutant in acidified lakes, where it becomes soluble and is toxic to fauna. The gills of fish are particularly susceptible to aluminum poisoning. Aluminum may be deposited in the human brain and has been implicated in Alzheimer's disease.

Metals are nonbiodegradable. Unlike some organic pesticides, metals cannot be broken down into less harmful components. Detoxification by organisms consists of hiding active metal ions within a protein such as metallothionein (binding covalently to sulfur) or depositing them in insoluble forms in intracellular granules for long-term storage or excretion in the feces (see Chapter 8).

TABLE 1.2

Separation of Some Essential and Nonessential Metal Ions of Importance as Pollutants into Class A (Oxygen Seeking), Class B (Sulfur or Nitrogen Seeking), and Borderline Elements

Class A	Borderline	Class B
Calcium	Zinc	Cadmium
Magnesium	Lead	Copper
Manganese	Iron	Mercury
Potassium	Chromium	Silver
Strontium	Cobalt	
Sodium	Nickel	
	Arsenic	
	Vanadium	

Note: Based on classification scheme of Nieboer, E. and Richardson, D.H.S. (1980). *Environmental Pollution*, 1B, 3–26. The distinction is important in determining rates of transport across cell membranes and sites of intracellular storage in metal binding proteins and metal containing granules.

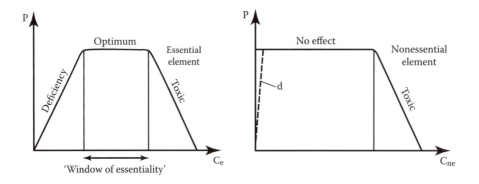

FIGURE 1.2
Relationships between performance (growth, fecundity, survival; P) and concentrations of essential (Ce) or non-essential (C_{ne}) elements in the diets of animals. Possible deficiency effects at ultratrace levels (d) of apparently nonessential elements may be discovered as the sensitivities of analytical techniques are improved.

Essential elements all have windows of essentiality within which dietary concentrations in animals and soil concentrations in plants must be maintained if the organisms are to grow and reproduce normally (Figure 1.2). In addition to carbon, hydrogen, oxygen, and nitrogen, all animals need the seven major mineral elements (calcium, phosphorus, potassium, magnesium, sodium, chlorine, and sulfur) to maintain ionic balance and as integral parts of amino acids, nucleic acids, and structural compounds. Thirteen other so-called trace elements (iron, iodine, copper, manganese, zinc, cobalt, molybdenum, selenium, chromium, nickel, vanadium, silicon, and arsenic) are also required. Zinc, for example, is an essential component of at least 150 enzymes; copper is essential for the normal function of cytochrome oxidase; and iron is a component of hemoglobin, the oxygen-carrying pigment in red blood cells. Boron is required only by plants. A few other elements such as lithium, aluminum, fluorine, and tin are essential at ultratrace levels. The window of essentiality for some elements is very narrow. Selenium, for example, was considered a dangerous toxin for a long time until its importance to the glutathione peroxidase enzyme was discovered. The level determines whether an element acts as a poison.

In addition to their toxicities above certain levels, nonessential metals such as mercury and cadmium may also affect organisms by inducing deficiencies of essential elements through competition at active sites in biologically important molecules (Table 1.3; see Chapter 7). Such antagonism also occurs between essential elements. A concentration of only 5 μg Mo (molybdenum) g^{-1} in a cattle diet is sufficient to reduce copper intake by 75%, which often leads to symptoms of copper deficiency.

1.1.2 Anions

Certain inorganic pollutants are not particularly toxic but cause environmental problems because they are used in such large quantities. These include anions such as nitrates and phosphates. Nitrate fertilizers are used extensively in agriculture. During crop growth, most of the fertilizer applied is absorbed by plant roots. However, when growth ceases, nitrate released during the decomposition of dead plant material passes through the soil and may enrich adjacent water courses. The increase in available nitrogen may cause blooms in algal populations. This effect is called eutrophication and eventually leads to oxygen starvation as microorganisms break down the dead algal tissues (Nosengo, 2003).

TABLE 1.3

Levels of Activities of Carbonic Anhydrase
Expressed Relative to Levels of Zinc for
Different Metals Substituted in Proteins

Metal	Normal Activity (% Hydration of CO_2)
Zinc	100
Cobalt	56
Nickel	5
Cadmium	4
Manganese	4
Copper	1
Mercury	0.05

Source: Coleman, J.E. (1967). *Nature*, 214, 193–194.
With permission.

The safe limit for nitrates in drinking water in the United Kingdom has been set at 50 parts per million (ppm). Human health problems may arise if young babies ingest bottled milk made with nitrate-contaminated water. During their first few months of life, human infants have anaerobic stomachs. The nitrates are converted to nitrites in this oxygen-poor environment. The nitrites bind to hemoglobin and reduce its capacity to carry oxygen, causing an infant to develop blue-baby syndrome or methemoglobinemia. The problem does not arise with breast-fed babies.

In regions of intensive agriculture, the 100 ppm level is exceeded in water extracted from rivers or in bore holes where nitrates have leached down to the aquifers. The problem can be solved by removing the nitrate chemically at a water treatment plant or by diluting the contaminated water with water from a relatively nitrate-free source. The long-term solution is of course to reduce nitrate use. This is now achieved in so-called exclusion zones around sources of water for human consumption.

Similar problems of eutrophication can also arise with phosphates used as fertilizers. Washing powders constitute additional sources of phosphates. These materials have been made less resistant to breakdown in recent years through cooperation between soap manufacturers and water-treatment companies. In the 1950s and 1960s, it was common to see huge build-ups of foam below weirs and waterfalls downstream of the outfalls of sewage treatment works.

1.2 Organic Pollutants

The great majority of compounds that contain carbon are described as organic; the few exceptions are simple molecules such as CO_2 and CO. Carbon has the ability to form a bewildering diversity of complex organic compounds, many of which provide the basic fabrics of living organisms. The reason for this is the tendency of carbon atoms to form stable bonds with one another, thereby creating molecules in the forms of rings and extended chains. Carbon can also form stable bonds with hydrogen, oxygen, and nitrogen atoms.

Molecules built of carbon alone (e.g., graphite and diamond) or of carbon and hydrogen (hydrocarbons) have very little polarity and consequently low water solubility. Polar molecules have associated electrical charges; nonpolar molecules have few or none. Molecules with strong charges are described as highly polar; molecules of low charge have low polarity. Polar compounds tend to be water soluble because the charges on them are attracted to opposite charges on water molecules. For example, a positive charge on an organic molecule will be attached to a negative charge on a water molecule.

Carbon compounds tend to be more polar and more chemically reactive when they contain functional groups such as OH, HCO, and NO_2. In these examples, the oxygen atom attracts electrons away from neighboring carbon atoms, thereby creating a charge imbalance on the molecule. Molecules of high polarity tend to enter into chemical and biochemical reactions more readily than do molecules of low polarity.

The behavior of organic compounds is dependent upon their molecular structure—molecular size, molecular shape, and the presence of functional groups being important determinants of metabolic fate and toxicity. Thus it is important to know the formulae of pollutants in order to understand or predict what happens to them in the living environment. The principles operating here are illustrated by examples given in Chapters 5 and 7. Readers with a limited knowledge of chemistry are referred to the text of Manahan (1994), which contains two useful concise chapters on basic principles.

The pollutants that will be described here are predominantly man-made (anthropogenic) compounds that have appeared in the natural environment only during the last century. This is a very short time in evolutionary terms, and there has been only limited opportunity for the evolution of protective mechanisms against their toxic effects (e.g., detoxication by enzymes) beyond preexisting mechanisms that operate against natural xenobiotics (see Box 1.1). In this respect, they differ from inorganic pollutants and from those naturally occurring xenobiotics that have substantial toxicity (e.g., nicotine, pyrethrins, and rotenone are compounds produced by plants that are highly toxic to certain species of insect). Aromatic hydrocarbons represent a special case. They have been generated by the combustion of organic matter since the appearance of higher plants on earth (e.g., as a result of forest fires started by volcanic lava). Like metals that are mined, their environmental levels increase substantially as a consequence of human activity (as with the combustion of coal or gasoline to produce aromatic hydrocarbons).

1.2.1 Hydrocarbons

Hydrocarbons are compounds composed only of carbon and hydrogen. Some hydrocarbons of low molecular weight (e.g., methane, ethane, and ethylene) exist as gases at normal temperature and pressure. However, most hydrocarbons are liquids or solids. They are of low polarity (electrical charge) and consequently have low water solubility, but they are highly soluble in oils and in most organic solvents. They are not very soluble in polar organic solvents such as methanol and ethanol.

Hydrocarbons fall into two classes, (1) alkanes, alkenes, and alkynes, and (2) aromatic hydrocarbons (Figure 1.3). Aromatic hydrocarbons are distinguished by the presence of one or more benzene rings in their structures. Benzene rings are six-membered carbon structures that are unsaturated—not all available carbon valences are taken by linkages to hydrogen. In fact, benzene rings have delocalized electrons that can move freely over the entire ring system and do not remain in the immediate vicinity of any single atom. Other hydrocarbons lack this feature. They vary greatly in molecular size and may be fully saturated (e.g., hexane and octane) or unsaturated.

BOX 1.1 CHEMICAL WARFARE

Chemical warfare has been in progress for a very long period on the evolutionary scale, and the employment of chemical weapons by humans is, from this point of view, a very recent event. In higher plants, for example, enzymes have evolved that can synthesize secondary compounds with high toxicity for the animals that feed upon them, thereby giving protection against grazing. In response, detoxication mechanisms (e.g., detoxifying enzymes) have evolved in mammals; they give protection against these plant toxins. Taken together, these two related phenomena have been termed a coevolutionary arms race (Ehrlich and Raven, 1964; Harborne, 1993), during the course of which many toxic natural xenobiotics have emerged, together with defense mechanisms against them. The capacity to synthesize chemical weapons has also evolved in mammals, both weapons of defense and others that are used for attack. Microorganisms produce antibiotics that are toxic to other microorganisms that compete with them.

Human use of pesticides, biocides, and chemical warfare agents should be seen against this background (Walker, 2001). Humans are newcomers to this game. Indeed, early pesticides were frequently natural products. Examples include pyrethrin, nicotine, and rotenone insecticides. Also, the design and subsequent synthesis of novel pesticides has sometimes been based on the structures of toxic natural xenobiotics. Examples include the neonicotinoid insecticides related to nicotine, anticoagulant rodenticides related to coumarol, pyrethrins related to pyrethrum, and carbamate insecticides related to physostigmine (see Section 1.2 and Walker, 2001).

In view of the foregoing, it is hardly surprising that most novel pesticides are readily degraded by detoxifying enzymes (e.g., monooxygenases and/or esterases) that were present in target species long before the discoveries of the compounds. It is generally believed that such enzymes evolved during the course of a coevolutionary arms race.

Also worth noting is the popular belief in the safety of organic produce—food containing relatively few or no synthetic organic chemicals such as pesticides and pharmaceuticals. In fact, many natural products are known to be highly toxic and contain botulinum toxin, tetradotoxin, aflatoxin, and ergot alkaloids (see Trewavas, 2004). All these examples have caused food poisoning, so an "organic" label alone does not guarantee safety! At least pesticides and pharmaceuticals have been subject to rigorous testing and the resulting restrictions on their use consider human health risks.

Unsaturated hydrocarbons contain carbon–carbon double bonds (e.g., ethylene) or carbon–carbon triple bonds (e.g., acetylene). Saturated hydrocarbons are called alkanes (Figure 1.3). Unsaturated hydrocarbons with carbon–carbon double bonds are alkenes. Unsaturated hydrocarbons with carbon–carbon triple bonds are alkynes. They may exist as single chains, branched chains, or rings. The properties of these two groups of hydrocarbons will now be discussed.

The properties of nonaromatic hydrocarbons depend upon molecular weight and degree of unsaturation. Alkanes are essentially stable and unreactive and have the general formula C_nH_{2n+2}. The first four members of the series exist as gases (n < 4). Where n = 5 to 17, they are liquids at normal temperature and pressure. Where n = 18 or more, they are solids. Alkenes and alkynes are more chemically reactive because they contain carbon–carbon double or triple bonds. As with alkanes, the lower members of the series are gases; the higher ones are liquids or solids.

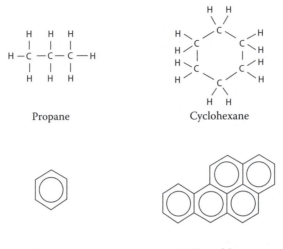

FIGURE 1.3

Hydrocarbons consist only of hydrogen and carbon. They exhibit low polarity and thus low solubility in water and high solubility in oils and organic solvents. Propane and cyclohexane are examples of alkanes. Benzene, and benzo(a)pyrene are aromatic compounds that contain six-membered carbon (benzene) rings with delocalized electrons. The benzene ring is represented as a hexagon (six-membered carbon frame) surrounding a circle (cloud of delocalized electrons). Aromatic hydrocarbons undergo certain characteristic biotransformations influenced by the delocalized electrons (Section 5.1.5).

Aromatic hydrocarbons exist as liquids or solids—none has a boiling point below 80°C at normal atmospheric pressure. They are more reactive than alkanes and are susceptible to chemical and biochemical transformation. Many polycyclic aromatic hydrocarbons (PAHs) are planar (flat) molecules consisting of three or more six-membered (benzene) rings directly linked together.

The major sources of hydrocarbons are deposits of petroleum and natural gas in the upper strata of the Earth's crust. These fossil fuels originate from the remains of plants and animals of earlier geological times (notably the Carboniferous period). Although non-aromatic hydrocarbons predominate in these deposits, crude oils also contain significant amounts of PAHs. PAHs are also formed through the incomplete combustion of organic materials and are thus generated when coal, oil, and gasoline are burned, when trees or houses burn, and when people smoke cigarettes. Major sources of hydrocarbon pollution are spills of crude oils (e.g., tanker disasters) and the combustion of fossil fuels (notably the use of brown coal in Eastern Europe).

1.2.2 Polychlorinated Biphenyls (PCBs)

These are commercial mixtures of related compounds (congeners) that are useful for their physical properties. They are stable, unreactive, viscous liquids of low volatility used as hydraulic fluids, coolants, insulation fluids in transformers, and plasticizers in paints. Among 209 possible PCB congeners, some 120 are present in commercial products such as Aroclor 1254, Aroclor 1260, and Clophen A60. The last two digits in the numbers refer to the percentage of chlorine in the PCB mixture. The larger the number, the greater the proportion of more highly chlorinated PCBs in a mixture.

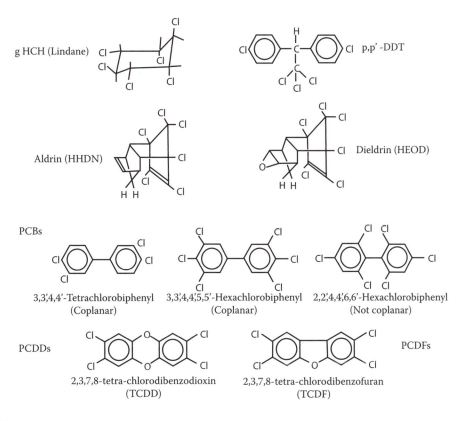

FIGURE 1.4
Organohalogens are organic compounds containing halogen atoms (fluorine, chlorine, bromine, or iodine). The examples shown are organochlorine compounds, although organofluorine compounds (chlorfluorocarbons) and organobromine compounds (polybrominated biphenyls) are also environmental pollutants. The compounds shown here are stable solids of low polarity and water solubility. They are not found in nature; they are often only slowly metabolized and consequently are persistent in living organisms (Chapter 5).

PCB mixtures have very low solubilities in water and high solubilities in oils and organic solvents of low polarity. The water solubilities of Aroclor 1254 and Aroclor 1260 are only 21 μg l^{-1} and 2.7 μg l^{-1}, respectively.

Individual congeners of PCB vary in their stereochemistries depending on the positions of substitution of chlorine atoms. Where there is no substitution in the ortho positions, the two benzene rings tend to remain in the same plane (coplanar PCBs); 3,3′,4,4′-tetrachloro-biphenyl is an example of a coplanar PCB (Figure 1.4). By contrast, substitution of two, three, or four ortho positions with chlorine leads to the movement of the rings out of plane because of the interactions of adjacent chlorines in different rings (chlorine atoms are bulky). The molecular conformation is no longer coplanar and becomes more globular.

PCBs were once used for many purposes—dielectric fluids, heat transformers, lubricants, and vacuum pump fluids, as plasticizers in paints, and in carbonless copy paper. In many countries, PCB use is banned or severely restricted. Major sources of pollution are (or have been) manufacturing wastes and careless disposal or dumping (Waid, 1985 and 1987).

1.2.3 Polychlorinated Benzodioxins (PCDDs)

The best known member of this group of compounds is 2,3,7,8-tetrachlorodibenzodioxin (2,3,7,8-TCDD; Figure 1.4), usually simply called dioxin. This compound is extremely toxic to mammals (LD_{50} = 10 to 200 µg kg l^{-1} in rats and mice). PCDDs are flat molecules formed by the linking of two benzene rings by two oxygen bridges with varying substitutions of chlorine on the available ring positions. PCDDs include 75 possible congeners. PCDDs are chemically stable and exhibit very low water solubilities (below 1 µg l^{-1} at 20°C) and limited solubility in most organic solvents, even though they are lipophilic (Box 1.2).

PCDDs are not produced commercially; they are unwanted by-products generated during the synthesis of other compounds. They are also formed during the combustion of PCBs and by the interaction of chlorophenols during disposal of industrial wastes. In general, they are formed when chlorophenols interact. PCDD residues have been detected throughout the environment (especially in aquatic environments), albeit at low concentrations, e.g., in fish and fish-eating birds (Elliott et al., 2001).

1.2.4 Polychlorinated Dibenzofurans (PCDFs)

These compounds are similar to PCDDs both in structure (Figure 1.4) and origin. Many congeners exist. PCDFs arise as unwanted by-products—they are not synthesized intentionally. They have not, however, received as much attention as PCDDs and do not appear to have raised such serious environmental problems.

1.2.5 Polybrominated Biphenyls (PBBs)

Mixtures of polybrominated biphenyls have been marketed as fire retardants (e.g., "Firemaster"). These mixtures bear a general resemblance to PCB mixtures and are lipophilic, stable, and unreactive. As with PCBs, some congeners are very persistent in living

---

BOX 1.2   PCDDS

The problem of environmental pollution by PCCDs is best illustrated by three well publicized examples involving 2,3,7,8-TCDD.

1. Dioxin (2,3,7,8-TCDD) is a contaminant of commercial herbicides such as 2,4-D and 2,4,5-T. A number of investigations have focused on pollution caused by these preparations, most notoriously, the spraying by the US Air Force of the defoliant herbicide Agent Orange in Vietnamese jungles from 1960 to 1969 (see Section 1.2.11).
2. The release of PCDDs by combustion furnaces used to dispose of PCB wastes at the Rechem plant in Scotland in the 1980s. PCDD residues were detected in soils and cattle in the surrounding area. The PCDD releases indicated incorrect operation of the furnace.
3. The release of PCDDs into the air from the Seveso chemical plant in northern Italy in 1976. The PCDDs formed a chemical cloud that contained trichlorophenols. Many exposed people developed a skin condition known as chloracne. However, no fatalities or serious toxic effects were attributed to PCDDs.

organisms and have long biological half-lives. In one incident in the United States, a PBB mixture was accidentally fed to cattle, appearing as substantial residues in meat products consumed by humans in Wisconsin and neighboring states (Carter, 1976).

1.2.6 Organochlorine Insecticides

Organochlorine insecticides constitute a relatively large group of insecticides with considerable diversity of structure, properties, and uses (Brooks, 1974). The three major types are DDT and related compounds, the chlorinated cyclodiene insecticides (aldrin and dieldrin), and hexachlorocyclohexanes (HCHs) such as lindane (Figure 1.4).

Organochlorine insecticides are stable solids of limited vapor pressure, very low water solubility, and high lipophilicity. Some are highly persistent in their original form or as stable metabolites. All the examples given here act as nerve poisons (see Chapter 7). Commercial DDT contains 70% to 80% of the insecticidal isomer p,p′-DDT. Related insecticides include rhothane (DDD) and methoxychlor. The insecticidal properties of DDT were discovered by Paul Müller of the firm of Ciba-Geigy in 1939. DDT was used, mainly for vector control during the Second World War and widely used thereafter for the control of agricultural pests, vectors of disease (e.g., malarial mosquitoes), ectoparasites of farm animals, and insects in domestic and industrial structures. Because of its low solubility in water (<1 mg 1^{-1}), DDT has been formulated as an emulsifiable concentrate for application as a spray. (Emulsifiable concentrates are solutions of pesticides in organic liquids; when added to water, they form creamy emulsions that can be sprayed on crops.)

DDT has an acute oral LD_{50} of 113 to 450 mg kg^{-1} and is considered only moderately toxic to vertebrates (Chapter 6). However, it has been shown to cause eggshell thinning in certain sensitive species of birds at very low doses through the action of its stable metabolite p,p′-DDE (see Chapters 7, 12, and 16).

In addition to p,p′-DDT, some 20% of the commercial insecticide is o,p′-DDT, which is more readily biodegradable than p,p′-DDT and exhibits very low toxicity to insects and vertebrates. However, it also exerts estrogenic activity, for example, in rats, and has, with other compounds of the DDT group, been implicated in cases of alleged endocrine disruption in the natural environment (see Chapter 7).

Kelthane (dicofol) is another pesticide related in structure to DDT, which has been marketed as an acaricide. It has shown only weak insecticidal activity, is of limited persistence, and evidence indicates that it may act as an endocrine disruptor (see Chapter 7).

The chlorinated cyclodiene insecticides were introduced after DDT (very widely during the 1950s) and some of them created serious environmental problems because they (or their stable metabolites) were highly toxic to vertebrates and exhibited marked biological persistence. Aldrin, dieldrin, and heptachlor are examples of cyclodiene insecticides showing this undesirable combination of properties (acute oral LD_{50} to rats is 40 to 60 mg kg^{-1}). Chlordane is a similar insecticide, but has lower vertebrate toxicity than the other examples cited. Endrin and, to a lesser extent, endosulfan are cyclodiene insecticides of very high vertebrate toxicity but only limited biological persistence. In general, the cyclodienes resemble DDT in that they are stable lipophilic solids of very low water solubility but differ from it in their modes of action (Chapter 7). Endosulfan is an exception; it exhibits appreciable water solubility.

The cyclodienes were introduced into Western countries during the 1950s and were used in diverse formulations for many purposes. Because of their insolubility in water, emulsifiable concentrate and wettable powder formulations were normally used for spraying to control certain crop pests and vectors of disease (e.g., tsetse fly). They were also used in dips and sprays to control ectoparasites of livestock and as seed dressings for cereals and other crops. The use of aldrin, dieldrin, and heptachlor for the as seed dressings produced very serious ecological consequences that will be discussed in some detail in Chapters 12 and 16.

By the 1990s, the use of DDT and its relatives and the cyclodiene insecticides for most purposes had been banned on the ground of perceived human health risks or hazards to the environment. However, some of these compounds continue to be used on a limited scale in some countries. For example, DDT is used in certain developing countries to control vectors such as malarial mosquitoes that represent greater hazards to people than the side effects of the chemical. Although these compounds are little used today, their ecotoxicology is discussed in detail in this text for two reasons. First, the marked persistence of compounds such as dieldrin and p,p´-DDE has ensured that significant residues are still present in once heavily contaminated soils and/or sediments and will only slowly disappear over the decades to come. These residues continue to be released slowly into aquatic and terrestrial food chains and can reach significant concentrations in animals at higher trophic levels. The second reason is that they have been studied in considerable depth and detail—probably more than other types of organic pollutants. A great deal has been learned about the ecological hazards associated with them and, in the present text, they serve as useful examples and models for persistent lipophilic pollutants more generally. Further, their persistence is associated with a potential to cause neurotoxic effects and associated behavioral effects at sublethal levels (see Section 15.13, Example B).

Hexachlorocyclohexane (HCH) has been marketed as a crude mixture of isomers (BHC) but more commonly as a refined product containing mainly the γ isomers known as γ-HCH, γ-BHC, or lindane (Figure 1.4). γ-HCH has similar properties to other organochlorine insecticides but is somewhat more polar and water soluble (7 mg l^{-1}). Emulsifiable concentrates of HCH have been used to control agricultural pests and parasites of farm animals. It has also been used as an insecticidal dressing on cereal seeds. HCH is only moderately toxic to rats (LD_{50} is 60 to 250 mg kg^{-1}).

1.2.7 Organophosphorous Insecticides (OPs)

During the Second World War, interest developed in organophosphorous compounds that act as nerve poisons (neurotoxins) by inhibiting the enzyme acetylcholinesterase (AChE) enzyme (Chapter 7). These compounds were produced for two main uses—as insecticides and as nerve gases for chemical warfare (Ballantyne and Marrs, 1992). They are organic esters of phosphorus acids (Figure 1.5). Today, a large number of organophosphorous compounds are marketed as insecticides, and most follow the basic formula shown in Figure 1.5.

Most organophosphorous insecticides (OPs) are liquids of lipophilic character and some volatility; a few are solids. They are generally less stable than organochlorine insecticides and more readily broken down by chemical or biochemical agents (Eto, 1974; Fest and Schmidt, 1982). Thus they tend to be relatively short-lived when free in the environment and the environmental hazards they present are largely associated with short-term (acute) toxicity. OPs and other neurotoxic insecticides raise concerns about their behavioral effects in animals at sublethal doses. They are more polar and water soluble than the main types of organochlorine insecticides. Their water solubility is highly variable, with some compounds (e.g., dimethoate) showing appreciable solubility. The active forms of some OPs are sufficiently water soluble to act as effective systemic insecticides, reaching high

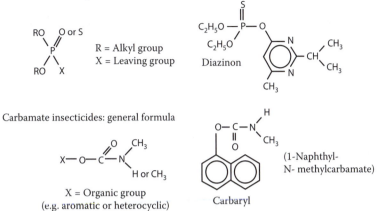

FIGURE 1.5
Organophosphorous and carbamate insecticides are toxic to insects because they inhibit the acetylcholinesterase enzyme (Chapter 7). They vary in their polarity and water solubility and are generally more reactive and less stable and persistent than the organochlorine insecticides. The leaving group of the organophosphorous compounds breaks away from the rest of the molecule when hydrolysis occurs (Chapter 5).

enough concentrations in the phloem structures of plants to poison sap-feeding insects (cf. organochlorine compounds and pyrethroids).

The formulation of organophosphorous compounds is important in determining the environmental hazard that they present. Many are formulated as emulsifiable concentrates for spraying. Others are incorporated into seed dressings or granular formulations. Granular formulations are required for the most toxic OPs (e.g., disyston and phorate) because they are safer to handle than emulsifiable concentrates and other types of formulations. The insecticide is locked within the granules and released slowly into the environment.

In many countries, OPs are still applied to crops as sprays, granules, seed dressings, and root dips to control ectoparasites of farm and domestic animals (sheep dipping) and sometimes to control internal parasites such as the ox warble fly. Other uses include control of certain vertebrate pests such as quelea birds in parts of Africa, locusts, stored product pests such as beetles, insect vectors of disease such as mosquitoes, and salmon parasites at fish farms (Hites et al., 2003).

1.2.8 Carbamate Insecticides

Carbamate insecticides are derivatives of carbamic acid developed later than organochlorine compounds (OCs) and OPs (Figure 1.5; Kuhr and Dorough, 1977). Like OPs, however, they inhibit acetylcholinesterase. Carbamates are usually solids but may be liquids. They vary greatly in water solubility. Like OPs, they are readily degradable by chemical and biochemical agents and do not usually raise problems of persistence. They present short-term toxicity hazards. Some (aldicarb and carbofuran) act as systemic insecticides. A few such as methiocarb are used as molluscicides to control slugs and snails. It is important to distinguish between the insecticidal carbamates and the herbicidal carbamates (propham, chlorpropham) that exhibit only low toxicity to animals.

Carbamate insecticides are formulated similarly to OPs. The most toxic carbamates (aldicarb and carbofuran) are available only as granules. They are used principally to control insect pests of agricultural and horticultural crops, although they also have some use as nematicides and molluscicides.

1.2.9 Pyrethroid Insecticides

Naturally occurring pyrethrin insecticides found in the flowering heads of *Chrysanthemum* species provided models for the development of synthetic pyrethroids. The synthetic chemicals are generally more stable chemically and biochemically than natural pyrethrins (Leahey, 1985).

Pyrethroids are solids of very low water solubility that act as neurotoxins similarly to DDT (see Chapter 7). They are esters formed between an organic acid (usually chrysanthemic acid) and an organic base (see Figure 1.6). Although pyrethroids are more stable than pyrethrins, they are readily biodegradable and have short biological half-lives. They can, however, bind to particles in soils and sediments and show some persistence in these locations. With their low water solubilities, they do not show significant systemic properties and are not used as systemic insecticides.

The hazards that they present relate mainly to short-term toxicity. However, it should be emphasized that they are highly selective between insects on the one hand and mammals and birds on the other. The main environmental concerns relate to their toxicity to fish and nontarget invertebrates. Other concerns relate to their effects on animal behavior at sublethal doses (Section 8.4.2).

Pyrethroids are formulated mainly as emulsifiable concentrates for spraying. They are used to control a wide range of insect pests of agricultural and horticultural crops worldwide, and are currently used to control insect vectors of disease such as tsetse flies in parts of Africa.

1.2.10 Neonicotinoids

Another group of neurotoxic insecticides introduced since the pyrethroids became widely available are the neonicotinoids. These are structurally similar to the natural compound nicotine. Like nicotine, they can interact with nicotinic receptors located on the cholinergic synapses of animals. However, they are less polar than the natural compound and show more toxicity to insects than to vertebrates. One example of a neonicotinoid is imidacloprid shown in Figure 1.7. Other examples include thiacloprid, clothianidin, acetamiprid and thiamethoxam (Jeschke and Nauen, 2008). They show moderate water solubility and have systemic properties. They have been very successful in controlling sucking insects such as aphids.

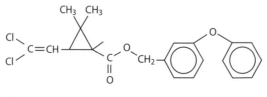

Permethrin

FIGURE 1.6
Pyrethroid insecticides have low polarity and limited water solubility. They are related in structure to the natural pyrethrins that are also toxic to insects.

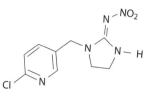

FIGURE 1.7
Imidacloprid is an example of a neonicotinoid insecticide that structurally resembles nicotine.

Neonicotinoids have become widely used in recent years. Questions have been asked about their possible effects on bees and other pollinating insects. As with other neuro-toxic insecticides, evidence indicates that sublethal doses can cause behavioral effects (see Section 8.4.2).

1.2.11 Phenoxy Herbicides (Plant Growth Regulators)

Phenoxy herbicides constitute the single most important group of herbicides. Familiar examples are 2,4-D, MCPA, CMPP, 2,4-DB, and 2,4,5-T (for general formulae, see Figure 1.8). They act by disturbing growth processes in a manner similar to that of the natural plant growth regulator indoleacetic acid. They are derivatives of phenoxyalkane carboxylic acids. When formulated as alkali salts, they are highly water soluble; when formulated as simple esters, they are lipophilic and reveal low water solubility.

Most phenoxy herbicides are readily biodegradable and so are not strongly persistent in living organisms or in soil. They are selectively toxic between monocotyledonous and dicotyledonous plants. Their principal use is to control dicot weeds in monocot crops such as cereals and grasses. They cause two types of environmental hazards. The first is unwanted phytotoxicity as a consequence of spray or vapor drift. The second issue is that formulations of certain herbicides of this type have sometimes been contaminated with the highly toxic TCDD (Agent Orange), a formulation containing 2,4-D and 2,4,5-T used as a defoliant in Vietnam (Stellman et al., 2003; see Section 1.2.3).

Water-soluble salts containing, for example, Na^+ and K^+ are formulated as aqueous solutions (concentrates), whereas lipophilic esters are formulated as emulsifiable concentrates. Some of the esters have caused unwanted phytotoxicity due to their volatilities (vapor drift).

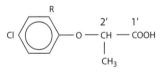

R = CH₃ or Cl

Phenoxyacetic acids (general formula)

FIGURE 1.8
Phenoxyalkanoic acid herbicides. The example is a general formula for phenoxyacetic acids that include 2,4-D and MCPA. Others are phenoxypropionic acids such as CMPP and phenoxybutyric acids such as 2,4-DB). All of them regulate plant growth and resemble indoleacetic acid, a natural growth regulator.

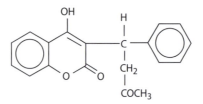

FIGURE 1.9
Warfarin and related compounds such as diphenacoum and brodifacoum, are anticoagulant rodenticides. They are complex molecules bearing some structural resemblance to vitamin K. Their toxic actions arise from competition with vitamin K in the liver (vitamin K antagonism).

1.2.12 Anticoagulant Rodenticides

For many years, warfarin (Figure 1.9) has been used as a rodenticide. It is a lipophilic molecule of low water solubility that acts as an antagonist to vitamin K (Chapter 7). More recently, because wild rodents developed resistance to warfarin, a number of structurally related second-generation anticoagulant rodenticides (sometimes called super warfarins) have been marketed. These include diphenacoum, bromadiolone, brodifacoum, and flocoumafen. Their general properties are similar to those of warfarin, but they are more toxic to mammals and birds and are markedly persistent in the livers of vertebrates. Thus they may be transferred from rodents to the vertebrate predators and scavengers that feed upon them. Owls in the United Kingdom, for example, have been found to contain their residues. Rodenticides are usually incorporated into bait that is placed where it will be taken by wild rodents.

1.2.13 Detergents

Detergents are organic compounds that have both polar and nonpolar characteristics. They tend to exist at phase boundaries where they are associated with polar and nonpolar media. Figure 1.10 illustrates examples. Detergents are of three types: anionic, cationic, and nonionic. The first two have permanent negative or positive charges attached to nonpolar (hydrophobic) C–C chains. Nonionic detergents have no permanent charges; rather they have a number of atoms that are weakly electropositive and electronegative due to the electron-attracting power of oxygen atoms.

Detergents are used in domestic and industrial activities. The major entry point into water is from sewage plants into surface waters. They are also used in pesticide formulations and for dispersing oil spills at sea. The degradation of alkylphenol polyethoxylates (nonionic detergents) can lead to the formation of alkylphenols (particularly nonylphenols), that act as endocrine disruptors (see Chapter 10).

1.2.14 Chlorophenols

A number of polychlorinated phenols (PCPs) occur as environmental pollutants (Figure 1.11). A major source is pulp mill effluent. PCPs are formed by the chemical action of chlorine (used as a bleaching agent) on phenolic substances present in wood pulp (Södergren, 1991). Pentachlorophenol is used as a wood preservative and is a major source of pollution. Chlorinated phenols have acidic properties and are water soluble, chemically reactive, and of limited persistence. The tendency of some PCPs to interact and form PCDDs is covered in Section 1.2.3.

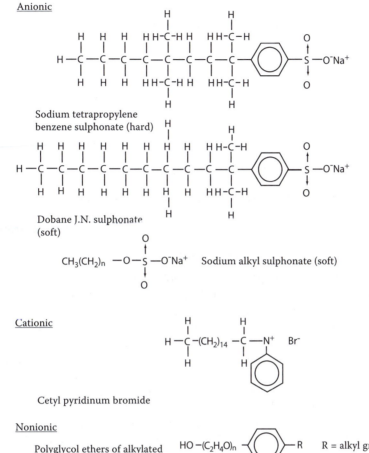

Anionic

Sodium tetrapropylene benzene sulphonate (hard)

Dobane J.N. sulphonate (soft)

CH₃(CH₂)ₙ —O—S—O⁻Na⁺   Sodium alkyl sulphonate (soft)

Cationic

Cetyl pyridinum bromide

Nonionic

Polyglycol ethers of alkylated phenols, e.g., lissapol N stergene $HO-(C_2H_4O)_n$ ⬡ -R   R = alkyl group

FIGURE 1.10
Detergent molecules contain both polar and nonpolar elements. They may have permanent negative charges (anionic detergents), permanent positive charges (cationic detergents), or a collection of small positive and negative charges over their structures (nonionic detergents).

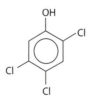

2,4,5-Trichlorophenol

FIGURE 1.11
Chlorinated phenols have acidic properties; they release H⁺ ions when they dissolve in water. They can interact to form dioxin (see Section 1.2.3).

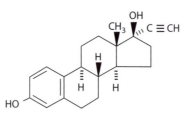

FIGURE 1.12
17A-Ethinylestradiol (EE2) is a potent synthetic estrogen bearing a structural resemblance to estradiol.

1.2.15 Ethinylestradiol (EE2)

EE2 is a potent estrogen, widely used in oral contraceptives. Its structure closely resembles that of the natural hormone estradiol (Figure 1.12). It shows some persistence in fish and surface waters. Its persistence in fish is partly due to the tendency of its glucuronide conjugate to undergo enterohepatic circulation resulting from the action of the B-glucuronidase enzyme in the gut (see Section 5.1.6). The glucuronide appears relatively unstable in the environment. The concern is that EE2, acting in concert with other environmental estrogens present in surface waters, may cause the feminization and consequent infertility of male fish. The possible ecological consequences are discussed in Section 15.2.

1.2.16 Pharmaceuticals

With recent rapid advances in medical science and the associated growth of the pharmaceutical industry, it is not surprising that traces of drugs are sometimes detected in surface waters. These are referred to as Pharmaceuticals and Personal Care Products as Pollutants (PPCPs) in the United States. Where drugs are widely used they are likely to appear as contaminants of sewage outflows. Indeed, the case of EE2 discussed above is an example of this. On the available evidence, levels of drugs detected in surface waters appear relatively low and unlikely to cause ecotoxicity. For example, a study of the beta blockers propranolol, metoprol, and nadolol suggested that environmental levels were not high enough to exert significant toxic effects on aquatic invertebrates (Huggett et al., 2002). However, little work has been done in this area and we see no grounds for complacency. The great number of PPCPs and complex mixtures of environmental chemicals in surface waters lead to the question of possible potentiation of toxicity (see Chapter 9), despite very low concentrations. Additional questions concern antibiotics and drugs used to treat farm animals or fish that may contribute to the pollution of fresh waters.

From 1997 to 2006, diclofenac, a non-steroidal anti-inflammatory drug, became a source of environmental concern because of its veterinary use; it has been widely used in India because it is relatively persistent in cattle. Vultures feeding on dead cattle have been poisoned by diclofenac and population declines of these birds are attributed to it (Green et al., 2006; Senacha et al., 2008; Section 17.1). Diclofenac is another example of a lipophilic organochlorine compound showing marked persistence and causing secondary poisoning (cf. Sections 1.2.2, 1.2.3, and 1.2.6).

1.3 Organometallic Compounds

Some metal ions are so insoluble that they are relatively nontoxic to animals if ingested. Liquid mercury, for example, can be swallowed in small amounts by humans and produce little long-term effect. Until the end of the nineteenth century, drinking liquid mercury was recommended as a cure for constipation! The low toxicity of tin is demonstrated by its use as a lining in food containers.

Nevertheless, the toxicities of several metals are greatly enhanced if they are bound deliberately or accidentally to organic ligands. This action changes their lipophilicity and consequently their distribution and behavior in the environment and within individual organisms. In chemical industries, mercury, tin, and lead are modified in this way to synthesize pesticides and other commercial products.

In addition to industrial synthesis of organometallic compounds (intentional or not), such compounds are sometimes generated in the environment by natural processes, for example, in soils and sediments. In particular, mercury and arsenic (a metalloid) are methylated in this way. Methyl mercury has been detected in canned tuna of considerable age—produced well before methyl mercury was synthesized industrially. A more recent concern is the natural synthesis of methyl arsenic compounds in sea foods (see Craig, 1986; Walker, 2009).

Organomercury compounds were used widely as antifungal seed dressings in the United Kingdom as late as 1993. Organolead compounds have been used extensively to control caterpillars on fruit crops, and also as "antiknock" agents in petrol. Organotin compounds, particularly tributyl tin, are extremely toxic. Their main use is for preventing boring animals from destroying timber and as components of antifouling paints applied to the outer surfaces of boats and fish cages to inhibit marine organisms. When these substances leach into the environment, they can affect nontarget organisms. Tributyl tin, for example, has devastated populations of *Nucella lapillus* (dog whelks) near boating areas in many countries (see Sections 13.6.4 and 16.3). Methyl arsenic compounds were once used as herbicides; dimethyl arsenic acid was used as a defoliant during the Vietnam war.

A tragic example of the effects of organomercury compounds occurred in Minamata Bay, Japan, in the 1950s. Metallic mercury released from a paper factory on the shores of the bay was methylated by bacteria in the sediments and formed methyl mercury (Kudo et al., 1980). Mercury in its methylated form is much more bioavailable than liquid mercury. The compound moved rapidly along the food chain until it reached high concentrations in fish. The local people relied heavily on locally caught fish and were thus vulnerable to mercury poisoning. About 100 people died and many suffered severe disabilities. Such incidents are most severe when a local population is highly dependent on a single food source; they are rare in more developed regions where food comes from more diverse sources. Similar problems continue to occur in the Amazon Basin. Huge quantities of mercury are dumped into the river as by-products of gold refining. Evidence indicate the mercury is becoming methylated and is transported into food chains (Pfeiffer et al., 1989 and 1993).

1.4 Radioactive Isotopes

1.4.1 Introduction

Since the development of nuclear energy and atomic weapons, ongoing debate has surrounded the safety of low levels of radioactivity in the environment. We are all exposed to background radiation from cosmic rays and the natural decay of radioactive isotopes. Some consider such exposure beneficial as it promotes natural DNA repair mechanisms and constitutes a type of immunization. Others claim that there is no safe level of radiation. The contributions of various sources to total natural background radiation depends largely on local geology. One of the most important sources is radon gas that may reach dangerous levels in poorly ventilated houses, especially if they are sited on igneous rocks (Mose et al., 1992).

Three factors determine whether radioactive isotopes are harmful to organisms: (1) the nature and intensity of the radioactive decay based on the mass and energy of the particles produced; (2) the half-life of the isotope; and (3) the biochemistry of the radioactive element. Radioactive isotopes of essential elements will follow the same biochemical pathways as their stable forms and accumulate in certain organs. Furthermore, some nonessential radioactive elements may be biochemical analogues of essential elements and follow similar routes in living tissues.

1.4.2 Natures and Intensities of Radioactive Decay Products

When an atom of a radioactive substance decays, it can produce one of four types of particles: alpha (α), beta (β), gamma (γ), or neutron. It can subsequently decay one or more times until the atom is stable. The intensity of a radioactive substance is measured in the SI unit called a becquerel (Bq) and represents the number of atoms that disintegrate per second. Formerly, radioactivity was measured in curies (Ci) and was equal to the number of disintegrations per second of 1 g of radium: 1 Ci = 3.7×10^{10} Bq and 1 Bq = 2.7×10^{-11} Ci.

An alpha particle is a positively charged helium nucleus consisting of two protons and two neutrons. Alpha particles are relatively massive compared with other radioactive emissions. Although they travel only a few centimeters in air, and only a few millimeters at most in biological tissues, their large mass makes them very damaging if they collide with cells, especially if inhaled into the lungs of vertebrates.

A beta particle is a negatively charged electron. Beta particles have more penetrating ability than alpha particles, with a range of a few meters in air. Sources of beta radiation can be shielded by thin layers of glass or Perspex. However, the small mass means beta particles cause less tissue damage than alpha particles.

Gamma rays are quanta of electromagnetic radiation. They are highly penetrating and can pass through several centimeters of lead. As a rule, they cause damage similar to that caused by beta particles.

Neutrons carry no charges and are liberated only when certain elements are bombarded with alpha or gamma rays. They react with other elements only by direct collision. Production of neutrons is the basis for nuclear fission in a reactor. They can pass through several centimeters of lead.

In biological terms, if we were to measure the radioactivity of a substance only in becquerels, we would get little information about its effects on tissues because a becquerel measures only frequency, not the nature of the nuclear disintegration. Two other SI units used to measure radiation are the gray and the sievert.

The gray (Gy) is equal to the amount of radiation that causes 1 kg of tissue to absorb 1 joule of energy. However, different kinds of radiation producing the same amount of energy impose different amounts of damage on living tissues. This may be difficult to understand and a simple analogy may be helpful. If a heavyweight boxer tapped your chin gently with his fist 100 times and imparted x joules of energy each time, you would find the action irritating, but your jaw would not incur long-term damage. However, if the boxer were to impart 100 times x joules of energy in a single punch, unconsciousness and a broken jaw would result. The boxer would have imparted the same amount of energy to your chin, but the rate at which it was applied determined the amount of damage. Thus another SI measurement, the sievert (Sv), is needed. The sievert accounts for the different ways the same amount of energy can be imparted to tissues. The "safe" annual radiation exposure of the general public is usually cited as 5 mSv. A dose of 20 Sv (received by several hundred workers involved in the Chernobyl clean-up; Edwards, 1994) is equal to 20 Gy of beta or gamma emissions or only 1 Gy of alpha particles. Thus alpha particles have about 20 times the effect of beta or gamma radiation for the same number of grays.

1.4.3 Half-Lives

Another important consideration is the half-life of a radioactive isotope—the time required for half the atoms of a radioactive isotope to decay. The decay curve follows an exponential decline. For example, ^{32}P is a radioactive isotope of phosphorus used extensively in molecular biology and plant physiology experiments. It has a half-life of 14 days. If we started on day 0 with 1 g of ^{32}P, on day 14 we would have 0.5 g of ^{32}P and 0.5 g of ^{32}S (stable sulfur). On day 28, we would have 0.25 g of ^{32}P and 0.75 g of ^{32}S, and so on.

After the passage of 10 half-lives, the radioactivity of an isotope is no longer significantly different from background levels and is considered "safe." Thus ^{32}P can be discarded with normal refuse if it is held in shielded conditions for 140 days before disposal.

The method of disposal of radioactive waste is dictated by the half-life of the longest-lived isotope. Thus if short-lived isotopes can be separated from long-lived ones before disposal, the volume of waste that must undergo long-term storage can be greatly reduced. Storage is easier in a laboratory or hospital where different isotopes can be separately maintained. Low-level waste is usually placed in concrete and buried in surface trenches. However, medium- and high-level radioactive waste from nuclear reactors contains a complex cocktail of highly radioactive isotopes, some with very long half-lives. The only option in this case is secure long-term underground storage. In the United Kingdom, the long-term goal is to bury the waste deep underground, although at the time of writing the exact locations remain the subjects of much political debate. Table 1.4 lists examples of the wide range of half-lives of different isotopes.

1.4.4 Biochemistry

Some radioactive isotopes are especially dangerous because they follow the same biochemical pathways in the body as stable elements. For example, more than 80% of the total iodine in the human body is contained in the thyroid gland where it forms an essential component of the thyroxin growth hormone. If radioactive iodine is ingested, it becomes

TABLE 1.4

Half-Lives of Some Radioactive Isotopes

Isotope	Half-Life
Plutonium-241	13 years
Plutonium-239	24,000 years
Iodine-131	8 days
Iodine-129	160×10^6 years
Strontium-90	28 years
Cesium-137	30 years

concentrated in the thyroid gland and thyroid cancer may result. Evidence indicates an increase in thyroid cancers among people living in the vicinity of Chernobyl after the disaster (Kazakov et al., 1992).

The strontium-90 radioactive isotope (Table 1.4) follows the same pathways as calcium in humans. Most people retain amounts of this isotope in their bones caused by its global dispersion after atmospheric atom bomb tests in the 1950s and 1960s. Cesium-137 follows potassium pathways and was a particular problem in areas such as northwest England and Scandinavia that experienced Chernobyl fallout (Crout et al., 1991; Simkiss, 1993; Åhman and Åhman, 1994). In areas where vegetation strongly recycles nutrients, it may take many decades before Chernobyl contamination declines to background levels.

1.5 Gaseous Pollutants

The most important gaseous pollutants are ozone (O_3) and oxides of carbon, nitrogen, and sulfur. On a global level, the main concern about ozone is its reduced concentration in the upper atmosphere. The well-publicized ozone hole over the Antarctic (and now detected at high latitudes in the northern hemisphere) arose from the degradative effects of CFCs on ozone molecules (see Chapters 3 and 4). CFCs are released from aerosol containers, from the coolants in domestic refrigerators, and from foam packaging. The ozone layer absorbs ultraviolet light, so that one hazard associated with its thinning is an increase in the rates of skin cancer. Another environmental impact is that the increased radiation may decrease photosynthesis of phytoplankton in the Antarctic.

At a local level, ozone is produced in photochemical smogs. The oxides of nitrogen from car exhausts (NO and NO_2, sometimes known as NO_x or NOX gases) and fumes from other fossil fuel consumptions react with moisture under the action of sunlight (Bower et al., 1994). High concentrations of ozone in the air irritate respiratory epithelia of animals and can directly affect the growth rates of some plants (Mehlhorn et al., 1991). Tobacco is particularly sensitive to ozone-induced damage (Heggestad, 1991). Ozone is considered a major factor in forest die-back in Germany (Postel, 1984).

Levels of carbon dioxide rarely reach toxic concentrations, except in very confined places. The effects of CO_2 on global warming will be discussed in Chapter 14.

Sulfur dioxide is a product of volcanoes and fossil fuel burning. It dissolves in water droplets, forming sulfurous and sulfuric acids that then fall as acid rain. Rain with a low pH may directly damage the leaves and roots of plants. Furthermore, nutrients may be washed from acidified soils as the hydrogen ions displace essential elements from soil particles. Plants growing on such soils may become deficient in one or more trace elements. A deficiency of magnesium, for example, is thought to be one cause of forest die-back in Germany. Acid rain may also increase the mobility of metal pollutants in soils if the pH drops sufficiently low (see Figure 4.3). The effects of acid rain are much greater on soils and lakes with low buffering capacities. One reason Scandinavia has been so badly affected by acid rain is the poor buffering capacity of the soil developed from the granite bedrock that underlies much of the region (Martinson et al., 2005). The effects of acid rain on ecosystems are discussed in Section 14.3.1.

1.6 Nanoparticles

Into the present millenium there has been a considerable growth of interest in the ecotoxicology of nanoparticles. For a recent review of this topic see Scown, van Aerle, and Tyler [2010] This is a wide ranging topic, still in its infancy, that can only be dealt with in bare outline here. There are many questions about the ecotoxicological significance of nanoparticles but, as yet, very few definitive answers.

Nanoparticles have been defined as particles with one dimension which is less than 100×10^{-6} mm [i. e., less than 1×10^{-7} m]. By comparison, colloids have been defined as particles having one dimension in the range 10^{-6} to 1×10^{-3} mm [i. e., $10^{-6} - 10^{-9}$ m]. Thus, colloidal particles of smaller size fall into the same size range as nanoparticles. Particles of colloidal or nanoparticulate size occur naturally and are found in surface waters, soils and the atmosphere. Much of the current interest in this topic concerns engineered nanoparticles [ENPs] which occur in a large number of industrial products. A detailed discussion of this topic lies beyond the scope of this book. Suffice it to say that products containing nanoparticles include certain paints, sprays, surface coatings, cosmetics, pesticide formulations, and preparations for the bioremediation of soils. The materials constituting ENPs are diverse. Perhaps the most widely studied nanoparticle is composed of titanium oxide [Ti O_2]. Others that have attracted interest are made of silver, aluminium oxide, cerium oxide, zinc oxide, or carbon [carbon nanotubes]. In particles of this type there are, inevitably, questions about what influence their presence in particulate form may have on their uptake by and their toxicity to animals.

Oberdorster [2001] found large quantities of nanoparticles in air, but only a very small proportion of these were ENPs. There was a very high background level of other nanoparticles. Much work has been done already on particles of colloidal size in soils and in surface waters and sediments. Indeed, there has been great interest in the clays and humic materials of soils, which are, of course, important places of storage for plant nutrients such as K+, NH_{4+} and Mg_{2+}. The extractable organic material in soils has been fractionated into so called 'humic' and 'fulvic' acids, which are, themselves, complex mixtures of colloidal material. These soil colloids are important because they bind both metal ions and organic pollutants (see Section 5.2.1). Their marked binding capacities are related to their large surface area/volume ratios which are characteristic of particles of small size. Many lipophilic molecules are strongly adsorbed by them. Also, they can become strongly bound to ENPs to form complex associations [Yang et al., 2009].

There are many questions now about the ecotoxicology of complex associations between ENPs and naturally occurring nanoparticles. How does existence in these complex associations affect the uptake of pesticides and/or heavy metals by animals which exist in surface waters or in soils? If uptake of such pollutants is increased, what effect does this have on ecotoxicity? These and many other questions have yet to be resolved.

1.7 Summary

In this chapter, major pollutants were classified into five groups: inorganic ions, organic pollutants, organometallic compounds, radioactive isotopes, and gases. Their structures, properties, and occurrence were described briefly. These pollutants will be used as examples throughout the subsequent text. It is important to relate the properties of individual compounds to their environmental fates (Chapters 2 through 5), their toxicities to individual organisms (Chapters 6 through 11), and their effects on populations, communities, and ecosystems (Chapters 12 through 15). The influences of properties such as polarity, vapor pressure, partition coefficient, and molecular stability on the movement and distribution of environmental chemicals are covered in Chapter 3. Chemical properties have been used to develop models for predicting environmental fates (Chapters 3 and 5) and toxicities (Chapter 6) of chemicals. Nanoparticles and their possible ecotoxicological significance were also discussed briefly.

Further Reading

Åhman, B. and Åhman, G. (1994). A very interesting study of the consequences of Chernobyl fallout in reindeer and the local human population that relies on them.

Bunce, N. (1991). *Environmental Chemistry.* An account of the pollutants discussed here but contains little about pesticides and radionuclides.

Craig, P.J., Ed. (1986). Organometallic Compounds in the Environment; Principles and Reactions. Longman Publishing. A valuable source of information about this group of pollutants; it covers the important question of their generation within the natural environment.

Crosby, D.G. (1998). *Environmental Toxicology and Chemistry.* Provides useful background on many organic pollutants.

Edwards, T. (1994). Excellent article that captures the impact of Chernobyl in the usual National Geographic style.

Guthrie, F.E. and Perry, J.J., Eds. (1980). *Introduction to Environmental Toxicology.* Multiauthor work giving in depth information about the most important organic and inorganic pollutants.

Hassall, K.A. (1990). *The Biochemistry and Uses of Pesticides,* 2nd ed. Readable account of the chemical properties of major types of pesticides.

Jukes, T. (1985). Interesting discussion of the consequences of the narrow window of essentiality for selenium.

Manahan, S.E. (1994). *Environmental Chemistry,* 6th ed. Detailed and comprehensive text covering inorganic and organic chemicals. Two useful chapters on basic principles for those with limited backgrounds in chemistry.

Merian, E., Ed. (1991). *Metals and Their Compounds in the Environment.* Good information about metals in this comprehensive text of 1438 pages.

Nieboer, E. and Richardson, D.H.S. (1980). Influential paper on the chemistry of metal ions that had a major impact on studies of metal toxicity.

Paasivirta, J. (1991). *Chemical Ecotoxicology.* Brief overview of most of the important organic and inorganic pollutants.

Robertson, L.W. and Hansen, L.G. (2001). PCBs. Recent Advances in Environmental Toxicology and Health Effects. Symposium on PCB Toxicology.

Schwarzenbach, R.P., Gschwend, P.M., and Imboden, D.M. (1993). *Environmental Organic Chemistry.* Authoritative work providing considerable detail on the chemistries of major organic pollutants; clear treatment of physicochemical aspects and reaction mechanisms.

Scown, T.M., van Aerle, R., and Tyler C.R. (2010). Do engineered nanoparticles pose a significant threat in the aquatic environment? Critical Reviews in *Toxicology,* 10, 653–670. Wide ranging review of ENPs in the aquatic environment; discusses different types of nanoparticles and their effects on aquatic organisms.

Walker, C.H. (2009). *Organic Pollutants: An Ecotoxicological Perspective,* 2nd ed. Taylor & Francis: London. Comprehensive review of these important environmental chemicals.

2

Routes By Which Pollutants Enter Ecosystems

Pollutants may enter ecosystems as the consequence of human activity in the following ways:

1. Unintended release from human activities (nuclear accidents, mining operations, shipwrecks, and fires)
2. Disposal of wastes (sewage and industrial effluents)
3. Deliberate application of biocides (e.g., pest and vector control)

Some of the chemicals so released can also reach unusually high levels locally as a result of natural processes such as weathering of rocks (metals and inorganic anions) and volcanic activity with associated forest fires (SO_2, CO_2, and aromatic hydrocarbons). As noted in the introduction, it is difficult to define what actually constitutes pollution. Some authorities prefer to restrict the terms *pollution* and *pollutant* to the consequences of human activities. However, in some cases it is impossible to determine the relative contributions of human and natural processes that affect the general environment.

2.1 Entry into Surface Waters

The discharge of sewage into surface waters represents a major source of pollutants globally (Table 2.1). Domestic wastes are discharged mainly into sewage systems. Industrial wastes are discharged into sewage systems or directly into surface waters.

The quality of the sewage discharged into surface waters depends on the quality of the raw sewage received by the plants and the treatment of the sewage that takes place there (Benn and McAuliffe, 1975). Urine, feces, paper, soap, and synthetic detergents are important constituents of domestic waste. Industrial wastes vary and their quality depends on the nature of the specific operation. A variety of treatments may be carried out at sewage works to improve the quality of sewage before its discharge into surface waters.

During primary treatment, sewage is passed through a sedimentation tank where it is retained for several hours (Figure 2.1). At this stage, primary sludge settles out. After this, during secondary treatment, biological oxidation and flocculation of most of the remaining organic matter takes place. Typically, this is carried out by an activated sludge process or by biological filtration. Conversion of ammonia to nitrites and nitrates by microorganisms occurs. Detergents are removed by biological oxidation. Much of the organic matter entering sewage works is converted into sludge and disposed of by spreading on land as a fertilizer or by dumping on land or at sea. The sewage effluent resulting from secondary treatments may be subjected to further treatment to remove

TABLE 2.1

Major Routes of Entry to Surface Waters

Route	Major Pollutants	Comments
Sewage outfalls	Wide range of organic and inorganic pollutants from commercial and domestic sources; detergents generally present	Highly variable; dependent on inputs to sewage plants and types of sewage treatments
Outfalls from commercial premises	Dependent on commercial activity; wide range of pollutants from chemical industry; metals from mining; pulp mills are important sources in some areas	Concentration of pollutants in effluents must stay below statutory limits
Outfalls of nuclear power stations	Radionuclides	Subject to regular monitoring and close control in most countries
Runoff from land	Various pollutants dumped on land surface; pesticides	Generally uncontrolled and difficult to measure
From the air	(i) Precipitation with rain or snow	May be transported over great distances
	(ii) Direct application of biocides	Control of pests, parasites, disease vectors, and aquatic weeds
	(iii) Accidental contamination by sprays or dusts	Aerial spraying a potential problem
Dumping at sea	Raw sewage; radiochemicals and toxic wastes in sealed containers dumped in deep ocean	Concern about release from containers that may degrade over long term
Release from oil rigs and terminals	Hydrocarbons	May be accidental or result of war (e.g., Gulf War in Kuwait)
Shipwrecks	Hydrocarbons and other organic pollutants	Wrecks of oil tankers

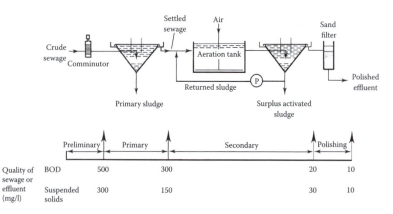

FIGURE 2.1

Conventional treatment of sewage by the activated sludge process. The top of diagram illustrates typical stages in sewage treatment. The lower figure indicates the quality of sewage at different stages of treatment. (*Source:* Benn, F.R. and McAuliffe, C.A., Eds. (1975). *Chemistry and Pollution.* Macmillan Publishing. With permission.)

constituents such as phosphate, nitrate, silicates, and borates, depending on the quality of final effluent required.

Important properties of sewage are levels of suspended solids, chemical oxygen demand (COD), and biochemical oxygen demand (BOD). COD measures the amount of oxygen required to achieve a complete chemical oxidation of one liter of a sewage sample. BOD measures the amount of dissolved oxygen used by microorganisms to oxidize the organic matter in one liter of a sewage sample. Sewage discharged into surface waters should have COD and BOD values below required limits. If the values are too high, the organic content of the sewage is too high for discharge into receiving waters, and a discharge could cause substantial reduction in the oxygen level of the water and create serious consequences for aquatic organisms.

In practice, the quality of sewage effluent varies enormously from country to country and from place to place. Because of the high cost of sewage treatment, no more is done than the situation requires, even in developed countries. Full advantage is taken of the capacity of receiving waters to achieve degradation of sewage components. Sewage is a rich source of organic and inorganic pollutants, prominent among them detergents, which are extensively used in both domestic and industrial premises. These detergents have given rise to serious pollution problems. For further discussion of this question, see Benn and McAuliffe (1975).

The types of pollutants in industrial effluents depend largely on the industrial processes that are being followed. Metals are associated with mining and smelting operations, chlorophenols and fungicides with pulp mills, insecticides with moth-proofing factories, a variety of organic chemicals with the chemical industry, and radionuclides with atomic power stations. In developed countries, there are close controls over the permitted levels of release of chemicals in industrial effluents. Offshore industrial activities, such as oil extraction and manganese nodule extraction, lead to the direct discharge of pollutants to the sea.

In addition to direct discharge, pollutants are sometimes dumped into surface waters at considerable distances from where they are produced. Such dumping is largely restricted to the sea. Sometimes sludge from sewage works is transported far out to sea and dumped. Radioactive wastes and chemical weapons in sealed containers have been dumped at sea. This practice raises questions about releases over the long term because these containers will eventually disintegrate. It is usual to dump dangerous wastes where the sea is deep to minimize risks of contaminating surface layers.

Another source of pollutants is the release of oil from tankers, most dramatically by shipwrecks that discharge large quantities of oils over one area in a short time. However, it is important to put such incidents as these—and the disastrous release of oil during the Gulf War—into a wider perspective. The total input of petroleum hydrocarbons into the marine environment has been estimated at 3.2 million tons per year. Although oil tanker disasters can cause great damage, the quantities they release are far less than spills from normal tanker operations and discharges from industrial and municipal waste.

Biocides are sometimes deliberately applied to surface waters to control invertebrates or plants. Herbicides are used to control aquatic weeds in lakes and watercourses. Insecticides are applied to control fish parasites at freshwater and marine fish farms and to control pests in watercress beds. Tributyl tin fungicides have been incorporated into antifouling boat paints and this has led to marine pollution.

Most of the examples given thus far resulted from deliberate actions that are, at least in theory, carefully controlled. However, many cases of accidental pollution occur which are not under human control. Pollutants present in air may enter surface waters as a consequence of precipitation of dust or droplets or with rain or snow or simply as a result of partition from air into water. Pollutants present on land surfaces, for example metals or pesticides, may be washed into rivers, streams, and oceans by heavy rainfall. They may be in the free or particulate state or attached to soil or mineral particles. Aerial spraying of pesticides creates risks of spray drift into surface waters. Some pesticides are extremely toxic to aquatic organisms and spray droplets should never directly contaminate waters.

Release of pollutants into moving surface water is followed by dilution and degradation. Consequently, biological effects are most likely to be seen at or near points of release. Where pollutants enter rivers, a biological gradient may occur downstream from the outfall. Sensitive organisms may be absent near the outfall but reappear downstream. In fast-flowing rivers, the dilution effect is marked, and pollutants are unlikely to reach high concentrations at a reasonable distance below the outfall.

Because of their size and the action of currents, oceans can effectively dilute incoming pollutants. Of greater concern are lakes and small inland seas subject to pollution transported by rivers and other routes. Because these water bodies have no effective outlets, pollutants build up as water evaporates, sometimes with serious consequences. Much depends upon the rates of degradation or precipitation that remove pollutants from the water. The pollution of the Great Lakes of North America provides a good example and will be discussed in Chapter 16.

2.2 Contamination of Land

As with pollution of surface waters, contamination of land may or may not be deliberate. Deliberate contamination may involve waste disposal or the control of animals, plants, or microorganisms with biocides. Accidental contamination may be the result of short-term or long-term aerial transport, flooding by rivers or seas, or collisions of tankers and trucks carrying toxic chemicals (Table 2.2).

Waste dumping at landfill sites is a widespread practice. Indeed, many old sites can be considered ecological time bombs. In addition to the issues of disposal of domestic and industrial wastes, toxic wastes require special handling. Of particular concern are radioactive wastes from nuclear power stations. Stringent regulations require safe disposal to minimize contamination of the land and neighboring surface waters. One practice is to embed the disposed radioactive material in concrete.

The use of sewage sludge as fertilizer on agricultural land constitutes another source of pollution. Heavy metals, nitrates, phosphates, and detergents are all added to soil as sludge. Land is also contaminated by aerially transported materials. Smoke and dust from chimneys carrying a variety of organic and inorganic pollutants can fall on neighboring land. Gases such as sulfur dioxide, nitrogen oxides, and hydrogen fluoride released from chimneys damage vegetation near industrial operations.

Thus pollution of land surfaces may occur in the immediate vicinity of domestic and industrial premises that cause air pollution. Additionally, as with surface waters, pollutants reach land after traveling considerable distances. They are carried by rain or snow, in solution or suspension, or with associated dust particles.

Land is sometimes also contaminated by pollutants after flooding. Considerable areas of land surfaces are treated with biocides to control vertebrate and invertebrate pests, plant diseases, weeds, and disease vectors. This is important in areas of intensive agriculture, where a variety of different pesticides are applied over the course of a farming year. Pesticides are applied as different formulations—sprays, granules, dusts, and seed

TABLE 2.2

Major Routes of Land Contamination

Route	Major Pollutants	Comments
Waste dumping, (e.g., rubbish dumps, landfills, industrial dumps)	Wide range of pollutants	Some industrial dumps are high in specific pollutants such as oil, metal ore, PCBs
Pesticide applications to agricultural land and forests	Insecticides, rodenticides, herbicides, and fungicides as sprays, dusts, seed dressings, etc.	Most countries strictly regulate application of pesticides
Control of insect disease vectors	Insecticides	Major pollution of large areas resulting from measures for controlling malarial mosquitoes and tsetse flies
Application of sewage to agricultural land	Heavy metals, nitrates, detergents	
Flooding by rivers or seas	Various pollutants including those associated with sewage	
Precipitation from air as dust or droplets or in rain or snow	Pollutants associated with soot and dust, acid rain, pesticides	Transport may be over short distances (spray drift, soot and dust from chimneys) or long distances (dropped in rain and snow)

dressings. The manner of application and the nature of the formulation influence distribution in crops and soils. Spray drift is a potential problem outside target sites during pesticide application, particularly via aerial spraying. The extent of drift depends on the wind strength and direction at the time of the operation. Pesticides can also move through soil to contaminate groundwater, especially where the soil profile has cracks that allow rapid percolation of water. A field experiment conducted by the Ministry of Agriculture, Fisheries and Food (MAFF) at Rosemaund, England, demonstrated that a variety of pesticides find their ways into drains and watercourses after heavy rain.

Pesticides are also applied over large areas for purposes other than agriculture. Insecticides are used extensively in Africa to control tsetse fly and locust swarms. Aerial spraying of insecticides has taken place over forests in Canada to control pests (see Chapter 16) and over nesting colonies of quelea (bird pests) in parts of Africa.

2.3 Discharge into Atmosphere

Pollutants enter the atmosphere in a gaseous state, as droplets or particles, or in association with droplets or particles. In the gaseous state, they may be transported over considerable distances by air masses. Particles and droplets are more likely to move only relatively short distances before falling to the ground. However, they can undergo long-distance transport when they are small in diameter.

Residential and industrial chimneys are important sources of atmospheric pollution (Table 2.3). Carbon dioxide (CO_2), sulfur dioxide (SO_2), oxides of nitrogen (NO_x), hydrogen fluoride, and chlorofluorocarbons (CFCs) are examples of gases released in this way.

TABLE 2.3

Major Points of Entry into Atmosphere

Route	Major Pollutants	Comments
Domestic chimneys	Many organic compounds including hydrocarbons associated with smoke particles or as vapors; SO_2, CO_2, NO_x, other gases	Level of pollution depends on quality of fuel burned and clean-up of flue gases
Chimneys of industrial plants, power stations, etc.	Same as domestic chimneys along with other pollutants caused by activities at sites; radiochemicals from nuclear power stations*	Procedures for cleaning up effluent gases of hazardous substances are important
Internal combustion and jet engines	CO_2, (NO_x) hydrocarbons and other organic pollutants; lead compounds (mainly inorganic but some organic) where leaded fuel is used	Level of pollution depends on engine designs and exhaust system; growing use of lead-free gasoline is restricting lead pollution
Pesticide applications	Insecticides, fungicides, and herbicides	Volatile pesticides enter air as vapors; droplets of pesticide spray and pesticide dust formulations reach atmosphere
Refrigerators	Chlorofluoromethanes	
Aerosols	Chlorofluoromethanes such as CF_2Cl_2, $CFCl_3$	Many countries now strictly control use as propellants

* For further details, see text.

The combustion of fuels releases CO_2, SO_2, NO_x and a variety of organic compounds (e.g., PAHs) produced by incomplete combustion. The level of pollution depends on the quality of the fuel and the extent to which flue gases are cleaned up. Some forms of coal (e.g., brown coal from Silesia) are high in sulfur and can cause very serious SO_2 pollution.

Many organic compounds released from chimneys are present in smoke particles. The subsequent movement of pollutants is dependent on atmospheric conditions and on the height and form of the chimney releasing them. Under clear and warm conditions, pollutants are diluted quickly by the mixing with air. As the earth's surface is warmed by sunlight, hot air near a chimney rises, producing convection currents and carrying pollutants. Cold, clean air will flow in to replace it. With a side wind, airborne pollutants will be carried away from the initial point of release and diluted further. The process may be reversed at night as the air cools. If no wind is present, a layer of mist or fog may form, trapping a well of cold air beneath it (Figure 2.2). In the morning, the sun will be unable to penetrate the layer of mist or fog, preventing warming of the air and dispersal of pollutants that become trapped in the area of the chimney that emitted them. Thus, the dispersal of air pollutants is favored by warm, dry conditions and steady side winds. The dispersal is more effective from high chimneys than from low chimneys. In general, the higher the point of release, the greater the height that pollutants will reach in the atmosphere and the greater the distance they are likely to travel (Chapter 3).

The internal combustion engine is another important source of air pollution. Along with vehicles on roads, airplane and ship engines pollute the air and seas. Jet engines create significant air pollution. During the operation of an internal combustion engine, chemical reactions generate substances not originally present in the gasoline and air mixture delivered by the carburetor. Carbon monoxide and nitrogen oxides are released along with organic molecules that are the products of incomplete combustion including PAHs, aldehydes, ketones, and the normal constituents of gasoline. Gasoline is also a source of

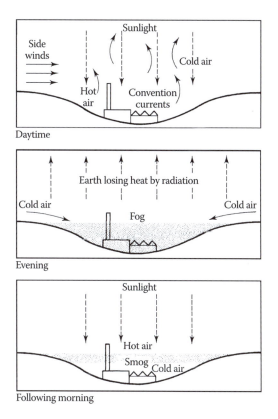

Daytime

Evening

Following morning

FIGURE 2.2
Inversion effects in air pollution. (*Source:* Benn, F.R. and McAuliffe, C.A., Eds. (1975). *Chemistry and Pollution.* Macmillan Publishing. With permission.)

organolead and inorganic lead compounds that arise when tetraalkyl lead is used in gasolines as an antiknock agent to control semiexplosive burning during engine operation. There is now a strong movement toward use of lead-free gasolines to reduce the emissions of lead in various forms from car exhausts. The control of emissions from internal combustion engines is a complex subject that will be considered only briefly here. In a vehicle without pollution controls, effectively all the carbon monoxide, nitrogen oxides, inorganic lead compounds, and about 65% hydrocarbons are released from the exhaust. Evaporation from the fuel tank and carburetor accounts for a significant loss of hydrocarbons and volatile lead tetraalkyl. Finally, substantial loss of hydrocarbons occurs from leakage around the pistons and into the crankcase—known as crankcase blowby.

Design improvements in modern engines reduce all these sources of pollution. Exhaust emissions decreased substantially after the installation of catalytic converters and filters into exhaust systems to remove nitrogen oxides, carbon monoxide, and hydrocarbons. However, catalytic converters are rapidly poisoned by tetraalkyl lead. This problem hastened the phasing out of leaded gasolines. Further improvements in exhaust emission were made by optimizing engine performance by improving air:fuel ratios, ignition timing, and cylinder design. Crankcase blowby has been reduced by recycling blowby gases via the carburetor system. Finally, evaporative losses from the carburetor and fuel tank have been reduced by improvements in design.

Considerable improvements in controlling pollutant releases by internal combustion engines have been effected in recent years. However, motor vehicle use continues to increase and thus vehicles remain major sources of air pollution. The extent to which improvements have been made varies by country. The current standards in Western Europe, North America, and Australia are not common in other parts of the world. Considerable concern also relates to releases of particulates from diesel engines.

Air pollution also arises from pesticide use. The application of pesticides as sprays or as dusts is not efficient; substantial proportions of pesticides are wasted. Aerosol droplets, dust particles with adhering pesticides, and pesticides in the gaseous state pass into the air. The problem worsens when pesticides are applied aerially. Climatic factors influence the extent to which pesticides contaminate the atmosphere. Strong side winds tend to move them away from original areas of application, with the risk that neighboring areas downwind will be contaminated. Volatilization is most rapid at high air temperatures. Thus pesticides show a greater tendency to volatilize into the air under tropical conditions than under temperate conditions. This point needs to be borne in mind when attempting to extrapolate from field studies performed in the temperate zone to predict pesticide fates under tropical conditions.

Another factor of importance is droplet size. Very small spray droplets produced during low-volume spraying fall more slowly to the ground than large droplets because their sedimentation velocity is slower, and they can travel relatively long distances before reaching the ground. In general, environmental factors such as wind speed, temperature, and humidity must be considered when planning spray operations to maximize the amount of pesticide reaching its target and to minimize air pollution.

Radiochemical pollution of the air from explosions of atomic devices on or above land surfaces was a problem for many years. By international agreement, however, these explosions were discontinued, but concern remains over accidental releases from establishments such as power stations and atomic research stations that handle nuclear materials. The seriousness of the problem was clearly illustrated by the Chernobyl accident of 1986 in the Ukraine, when a nuclear reactor caught fire and caused widespread air pollution with radionuclides (see Chapter 1). Half the reactor contents were dispersed.

Low molecular-weight halogenated hydrocarbons such as chlorofluorohydrocarbons (CFCs) used as propellants and in refrigerators and chlorinated compounds (CH_2Cl_2) used for dry cleaning represent another important group of air pollutants. These volatile substances can escape into the air during normal use and after waste disposal. CFCs can reach zones of the upper atmosphere where they can cause serious damage to the ozone layer (ozonosphere). See Chapter 3.

Many pollutants found in air exist in the same forms that were originally released from the land or water surface. Some pollutants are generated by chemical reactions within the atmosphere. This is the case with the photochemical smog that creates problems in Los Angeles and other cities that combine large numbers of vehicles with high levels of solar radiation. When no wind is present, nitrogen dioxide and organic compounds released from car exhausts along with oxygen produce a complex series of reactions. The products include ozone and organic compounds such as peroxyacetyl nitrate (an eye irritant).

TABLE 2.4

Gaseous Pollutants Released Globally (Tons per Year)

Pollutant	Anthropogenic Sources	Natural Sources[a]
CO_2	6,000,000,000	100,000,000,000
SO_x	100,000,000	50,000,000
NO_x	68,000,000	20,000,000
CFCs	1,100,000	0

Sources: Tolba, M.K. (1992). *Saving Our Planet.* London, Chapman & Hall; UNEP (1993). Environmental Data Report 1993–94. Oxford, Blackwell. With permission.

[a] Considerable uncertainty surrounds natural sources data.

2.4 Quantification of Release of Pollutants

Legislation intended to control pollution focuses on the amounts of pollutants that may enter the environment and the rates at which they may be released. International agreements are necessary to control environmental inputs of major pollutants. In the case of carbon dioxide, the United Nations Conference on Environment and Development at Rio de Janeiro in June 1992 recommended reducing carbon dioxide emissions to 1990 levels by the year 2000. Even this modest goal was not supported by many countries, including the US—the largest producer of CO_2.

A more recent conference in Kyoto, Japan, in 1997 recommended an 8% decrease from 1990 levels of six greenhouse gases before 2012. Although definite targets have been proposed by the European Union, the agreement has not been ratified by the US The cautious approach of the US and the opposition of the oil-producing countries have made the outcome uncertain. More has been achieved in agreements about CFCs. Industrialized countries agreed to phase them out by the year 2000 (a 10-year time lag was allowed to developing countries). This date was later advanced to 1996. The European Economic Community (EEC) has been active in negotiating reductions of SO_2 and NO_x to limit the effects of acid rain (Table 2.4).

In ecotoxicity testing of new industrial chemicals, the protocols are influenced by the amount of chemical produced per annum, because this indicates the possible scale of a potential pollution problem. Under the current regulations of the European Commission, if production is at a low level, only a minimal base set of tests is usually required. However, when production exceeds certain thresholds, additional testing is required (Walker et al., 1991a). Knowledge of the rates and release patterns of pollutants is necessary when modeling their environmental fates (see Chapter 3).

2.5 Summary

In this chapter, the major routes by which pollutants enter surface waters, land surfaces, and the atmosphere have been identified. A distinction is drawn between deliberate and regulated releases (e.g., application of biocides to control pests or vectors of disease) and accidental unregulated release (e.g., nuclear accidents, shipwrecks, and fires). Some pollutants are released by natural processes in addition to the activities of humans, thus making the sources of problems difficult to determine. Examples include the release of metals due to the weathering of rock and the production of SO_2, CO_2, and aromatic hydrocarbons via volcanic activity and associated forest fires.

The statutory regulation of pollution depends on definitions of the amounts of pollutants that may be released into the environment and on the rates at which releases may occur by specified routes. Thus permitted rates of release may be defined for chemicals in sewage or factory effluents and car exhausts. Permitted rates of application are defined for pesticides used in agriculture. The releases of pollutants such as CO_2 and CFCs into the atmosphere are matters of international concern, and attempts have been made to define the permitted rates of release by individual countries in the longer term.

Further Reading

Benn, F.R. and McAuliffe, C.A., Eds. (1975). *Chemistry and Pollution*. Useful chapters on sewage treatment and major sources of air pollution.

Butler, J.D. (1979). *Air Pollution Chemistry*. Authoritative work on air pollution.

Clark, R.B. (1992). *Marine Pollution*, 3rd ed. Standard text explaining marine pollution.

Manahan, S.E. (1994). *Environmental Chemistry*, 6th ed. Comprehensive text discussing releases of pollutants.

Salomons, W., Bayne, B.L., Duursma, E.K. et al., Eds. (1988). *Pollution of the North Sea: An Assessment*. Specialized chapters on pollution of North Sea.

3

Long-Range Movements and Global Transport of Pollutants

Pollutants are capable of movement over considerable distances. They can be carried over boundaries between countries, thereby raising political as well as environmental problems, because even the unintended export of environmental chemicals tends not to be welcomed by countries that receive them. Scandinavian countries, for example, have objected to the deposition of aerially transported sulfur dioxide (SO_2) originating from the British Isles. For the most part, transport over large distances is a consequence of the mass movement of air or water. However, movement may also be by diffusion that may be rapid in air but less rapid in water. Movement by diffusion may be localized or spread over large distances especially in air.

This chapter discusses the long-range transport of pollutants in surface waters and air and their global distribution via the different compartments of the environment: air, water, land surface, and biota. Transport is governed largely by abiotic physical processes such as mass movements of air and water and diffusion. Other, generally local movements that depend on biotic factors will be described elsewhere. Thus movements in food chains, in migrating animals and birds, and in soil will be considered in Chapters 4 and 5. This chapter concludes with a discussion of models to describe or predict the environmental distributions of chemicals.

3.1 Factors Determining Movements and Distributions of Pollutants

For convenience, the environment can be divided into four distinct yet interconnected compartments or phases: the air (atmosphere), surface waters (hydrosphere), land surface (principally soil or the lithosphere), and living organisms (biosphere). As noted above, movements dependent on biotic factors, for example, along food chains, will not be considered in this simple analysis. Biota are represented by the exposed surfaces of animals and plants, e.g., the integuments of insects and the cuticles of plants. The movements of chemicals within water and air and across the interphases between compartments are determined by physical processes; movement depends on the properties of the chemicals and the properties of the environmental compartments. The principal factors involved will now be reviewed before we describe their roles in determining the environmental fates of chemicals and their incorporation into descriptive and predictive models.

3.1.1 Polarity and Water Solubility

Water is an example of a polar liquid. The oxygen atom strongly attracts electrons away from the two hydrogen atoms (see Section 1.1.1), with the consequence that oxygen develops

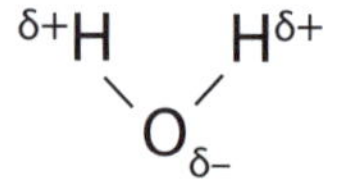

$$\delta+H \qquad H^{\delta+}$$
$$\diagdown O \diagup$$
$$O_{\delta-}$$

FIGURE 3.1
Water molecule.

a partial negative charge, and the hydrogens develop partial positive charges. The positive and negative charges are separated, and the molecule is said to be polar. By contrast, little charge separation occurs in nonpolar compounds such as nonaromatic hydrocarbons.

Opposite charges tend to attract one another, and the positive charges on hydrogen atoms of water molecules (Figure 3.1) are attracted to the negative charges on the oxygen atoms of their neighbors, thereby forming weak hydrogen bonds. Thus water molecules tend to form aggregates in which each molecule is surrounded by four others. Cations and anions have affinity for those parts of water molecules that bear the opposite charge (thus cations are attracted by the partial negative charge on oxygen). This leads to disruption of the water aggregates and the ions go into solution.

In general, many inorganic salts and polar organic compounds (e.g., simple alcohols and amines) have appreciable water solubility, whereas nonpolar organic liquids and solids have virtually none. Solubility depends on the strength of charge on the solute. Among inorganic salts, those formed by alkali and alkali earth metals, the first two series of the periodic classification, readily release their ions, whereas those of metals on the right-hand side of the table (e.g., lead, mercury, and tin) have a greater tendency to form covalent bonds rather than ionic ones and are accordingly of lower water solubility. Among organic compounds, the presence of polarizing atoms such as oxygen and nitrogen in the molecular structure tends to increase charge separation and consequently water solubility. Figure 3.2 shows examples of the water solubilities of pollutants.

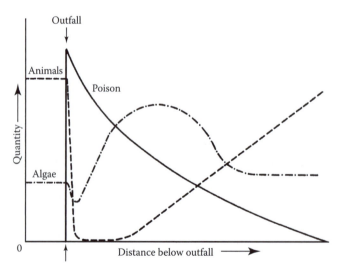

FIGURE 3.2
Effects of poisonous effluent on a river. Quantity (vertical axis) is the concentration of chemical or number of individuals per unit volume in water. (*Source:* Hynes, H.B.N. (1960). *The Biology of Polluted Waters.* Liverpool University Press. With permission.)

An important consequence of polarity in biochemistry is the so-called hydrophobic effect (Tanford, 1980). The tendency of water molecules to form aggregates actively excludes nonpolar (hydrophobic) substances such as lipids and hydrocarbons. Hence the attractive forces between water molecules contribute to the formation of phospholipid bilayers and phase boundaries between lipids and water in living cells. An important aspect of this water–lipid interface is the movement of lipophilic pollutants into and through membranes and the toxicological consequences thereof. This will be discussed in Chapter 5.

3.1.2 Partition Coefficients

Nonpolar liquids such as octanol, hexane, and olive oil are immiscible with water. If a nonpolar liquid is mixed with water, two phases will separate, and the less dense of the two liquids will rise to the top. Solutes partition between the two phases, and when equilibrium is reached, the ratio of the concentrations in the two phases is the partition coefficient. Thus, in the case of octanol and water, the relationship is

$$K_{ow} = \frac{\text{Conc. in octanol}}{\text{Conc. in water}}$$

K_{ow} is the octanol–water partition coefficient that has a high value for substances of low polarity and so provides an index of hydrophobicity. The value should be constant for any solute at equilibrium distribution between two defined immiscible liquids at a particular temperature. K_{ow} values are used to predict environmental distributions (Section 3.4) and bioconcentrations of environmental chemicals (Section 5.1.8). Some examples of K_{ow} values of environmental chemicals are given in Table 3.1.

TABLE 3.1

Properties of Pollutants

Compound	Water Solubility at 20–25°C (mg/l)	Log K_{ow}	Vapor Pressure (torr)
Sodium chloride	2.6×10^5		
Calcium chloride ($6H_2O$)	4.3×10^5		
Mercuric chloride	6.9×10^4		6.0×10^{-3}
Lead chloride	6.4×10^3		
Malathion	145	2.36	1.0×10^{-5}
Carbaryl	40		5.0×10^{-3}
Cypermethrin	<0.2		3.9×10^{-12}
p,p´-DDT	1.2×10^{-3}	6.96	1.9×10^{-7}
Dieldrin	<0.2		1.8×10^{-7}
2,2´, 4,4´, 5,5´-HCB	5.5×10^{-3}	6.57	8.1×10^{-7}
2,3,7,8-TCDD	8.0×10^{-6}	6.53	1.5×10^{-9}
Benzo(a)pyrene	4.0×10^{-3}	5.97	5.5×10^{-9}
Tributyl tin chloride	9.7×10^3	3.70	7.5×10^{-9}
Methyl mercuric chloride			8.5×10^{-3}

3.1.3 Vapor Pressure

The tendency for a liquid or solid to volatilize is expressed by its vapor pressure, defined as the pressure exerted by the vapor of a substance on its own solid or liquid surface at equilibrium. It may be expressed in millimeters of mercury (also known as torr) or as a fraction of normal atmospheric pressure, which is 760 torr. Vapor pressure increases with rising temperature, as surface molecules increase in kinetic energy. When the vapor pressures of liquids reach atmospheric pressure, they boil. Solids also exert a vapor pressure and some solids vaporize without melting (sublimation). Table 3.1 shows examples of vapor pressures of environmental chemicals.

3.1.4 Partition between Environmental Compartments

Just as chemicals partition between immiscible liquids, they also partition between compartments of the environment—between air and water, air and soil, etc. Again, the distribution between different phases at equilibrium can be described by what are in effect partition coefficients, although they are usually known by other terms. Henry's constant, for example, relates to the distribution of a volatile chemical between air and water. A particular situation occurs at the interfaces of air with solid and water with solid, found in soil (see Section 5.2.1). Here, chemicals may be adsorbed at the solid surface rather than being absorbed into the solid matrix. The movement of a substance from one compartment to another is driven by its escaping tendency or fugacity. The construction of models of environmental fate based on the concept of fugacity using distribution coefficients will be described in Section 3.4.

3.1.5 Molecular Stability and Recalcitrant Molecules

The time of residence of a chemical in the environment and consequently the distance it can travel depend on its molecular stability. Environmental compounds are broken down by both chemical and biochemical processes. Common methods of chemical transformation are hydrolysis (e.g., esters such as organophosphorous and carbamate insecticides), oxidation, and photodegradation (many types of chemicals). The stability of a chemical is important, but environmental factors such as temperature, level of solar radiation, nature of adsorbing surface, and pH influence the rate at which chemical degradation occurs. Many organic pollutants are readily biotransformed by the actions of enzyme systems (see Section 5.1.5). However, very large differences between groups and species mean that compounds that are readily metabolized by one species may be highly persistent in others. Also, recalcitrant molecules are highly resistant to both chemical and biochemical transformation and have long half-lives in biota, soils, sediments, and water. It is now clear that a number of polyhalogenated compounds have this characteristic. Examples include p,p'-DDE, dieldrin, some PCBs, and dioxins (e.g., TCDD). The environmental problems associated with polyhalogenated compounds are recurring themes throughout this text.

Although degradability is regarded as a desirable characteristic of an environmental chemical because it limits persistence, movement, and biomagnification, it is necessary to strike a cautionary note. Some transformations actually lead to increased toxicity. Many carcinogens, for example, undergo metabolic activation within living organisms (see Chapter 5). The properties of metabolites and the products of chemical transformation must be considered.

3.2 Transport in Water

The pollutants present in surface waters exist in diverse states. They may be in solution or in suspension. Suspended material may be in the form of droplets (e.g., oil) or particles, and pollutants may be dissolved in droplets or absorbed by solid particles. All these forms can be transported by water over considerable distances. Particulate materials can fall to the bottoms of surface waters, e.g., where the rivers enter the sea their rate of flow is checked and the coarser transported particles fall to the bottom with estuarine deposits. Liquid droplets may rise to the surface or may be carried by particles down to the sediment, depending on their density. With oil pollution, both events occur—light oil rises to the surface and heavy oil residues go into sediments.

In rivers, pollutants are transported over varying distances depending on factors such as their stability and physical states of the pollutants and the speed of flow of the river. The distance traveled is likely to be greatest where stable compounds are in solution in a fast flowing river. In general, the concentration of a pollutant continually falls with increasing distance below an outfall, and this may be reflected in the changing compositions of the fauna and flora (Figure 3.2). The importance of long-distance transport of pollutants by rivers was clearly demonstrated when the Rhine River became polluted with the insecticide endosulfan in 1969. The initial release was evidently in the middle section of the river near Frankfurt, but the transported compound was detected by Dutch scientists working near the Rhine estuary some 500 km downstream.

After pollutants reach lakes or oceans, they may be transported by currents. The major oceans of the world are traversed by surface currents, so it is possible for pollutants to move from one continent to another. These currents are wind driven, and they move roughly at right angles away from the direction of prevailing winds. In both the Atlantic and the Pacific oceans, large circular patterns of currents (gyres) cover most of the surface area. The movement is clockwise in the northern hemisphere and counterclockwise in the southern hemisphere. The Gulf Stream of the North Atlantic is part of a clockwise gyre system and brings warm water to the shores of the British Isles.

The density of sea water is an important factor. Water may increase in density as the result of a fall in temperature or an increase in salt concentration (e.g., because of evaporation). When water masses increase in density, they move toward the bottom of the ocean. Downward movements are countered by upward movements of advecting water from the lower levels of the ocean. Deep water circulation in the oceans of the world is illustrated in Figure 3.3.

It is sometimes assumed that oceans are so large that dilution will quickly reduce pollutant concentrations to such low levels that they no longer constitute a problem. One shortcoming of this argument is that the distribution of pollutants in oceans is far from uniform. The movement of particulate matter with currents and its subsequent precipitation ensure unequal distribution. This problem is readily seen with the precipitation of sediment in estuaries as mentioned earlier. Inshore waters tend to have substantially higher levels of pollution than the open sea.

When persistent pollutants enter marine food chains, they can be moved over large distances by migrating animals and birds. Some fish, whales, and fish-eating sea birds migrate thousands of miles, taking pollutants with them. This can transfer pollutants from one ecosystem to another, e.g., where contaminated fish or birds migrate over large distances and are then eaten by predators.

FIGURE 3.3
Deep water circulation in the oceans. Thick lines = major bottom currents; thin lines = return flows. (*Source:* Turekian, K.K. (1976). *Oceans,* 2nd ed. Prentice-Hall. With permission.)

3.3 Transport in Air

Traces of persistent pollutants such as organochlorine insecticides and PCBs have been detected in snow and in animals living in polar regions, far removed from any point of release. This clearly illustrates that pollutants can be transported very large distances by the movement of air masses. The main sink of CFCs is in the stratosphere, indicating the importance of vertical movement through the troposphere for small, stable, and volatile molecules. The translocation of pollutants over large distances depends on their physical state and on movements of air masses.

Consider the physical states of pollutants. Some air pollutants are in a gaseous state. Examples include CO, NO_X, SO_2, HF, and small volatile halogenated molecules such as CFCs, trichloroethylene (CCl_2CHCl), and carbon tetrachloride (CCl_4). These may move through the air via mass transport and diffusion. The question of mass transport will be covered shortly. Diffusion of gases is of two types. The first is diffusion along a concentration gradient that proceeds at rates determined by Fick's law of diffusion:

$$\theta = -D(C_2 - D_1)$$

where θ is the rate of diffusion, D is the diffusion constant, and $(C_2 - C_1)$ is the concentration gradient. Thus net diffusion occurs in a direction that will tend to remove the gradient. The steeper the concentration gradient, the faster the rate of movement. The second type is thermal diffusion. In situations where a thermal gradient exists, hot molecules of high velocity move faster than cold molecules of low velocity.

Pollutants also exist as droplets or particles or more commonly in association with droplets and particles. Particles of dust or soot or droplets of water can be of complex composition and contain a range of polluting substances. Pollutants may also be incorporated

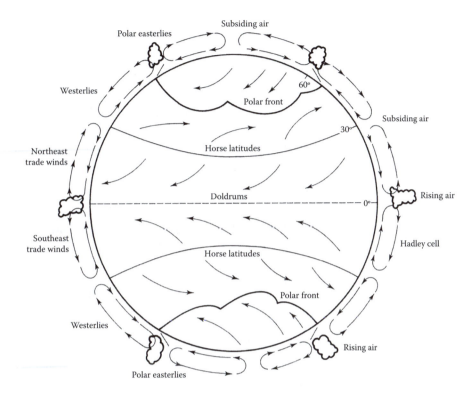

FIGURE 3.4
Idealized global circulation of air (*Source:* Lutgens, F.K. and Tarbuck, E.J. (1992). *The Atmosphere: An Introduction to Meteorology.* Prentice-Hall. With permission.)

into rain droplets during the course of precipitation (wash-out) or during the formation of droplets in clouds (rain-out). Soluble gases such as SO_2 and NO_X tend to dissolve in rain droplets. Also, rain may bring down dust particles present in the air.

Air movements on the global scale are relatively complex. First, the layer of air close to the Earth's surface (up to about 4 km high) is subject to particular turbulence and localized air flow. Pollutants released within this zone are likely to return to ground quickly, traveling only relatively short distances. On the other hand, pollutants that reach greater heights may be transported over considerable distances by circulating air masses.

The part of the atmosphere relevant to this discussion extends some 35 km above the Earth's surface. This is the lower atmosphere that accounts for about 99% of the total air mass. It is divided into the troposphere (first 10 or 11 km) and the stratosphere that lies above it. The troposphere is characterized by strong vertical mixing—individual molecules can move through the entire height in a matter of days. Little vertical mixing occurs in the stratosphere above it. The boundary between these two layers is the tropopause. Within the stratosphere lies a band of relatively high ozone concentration called the ozone layer or ozonosphere.

Within the troposphere are regular patterns of air circulation characteristic of different climatic zones (Figure 3.4).

Both the northern and southern hemispheres are divided into three circulation zones. First, at the equator, sharply rising currents of hot air on both sides draw in flows of cooler air from the north and south, respectively. In the upper part of the troposphere, the rising air cools and then moves northward or southward. This poleward air flow begins to sub-side and move toward the Earth's surface at 20 to 35 degrees latitude. Some of this air will then move in a surface flow toward the equator, thus completing the cycle.

The surface winds resulting from this circulation are termed the northeast and south-east trade winds in their respective hemispheres. These terms illustrate the point that the air flow is not directly on a north-to-south axis—it also has a west-to-east component due to the influence of the Earth's rotation on air movement in the troposphere. Between approximately 30 and 60 degrees latitude, this circulation pattern in both hemispheres undergoes a reversal—the surface flow is poleward and not toward the equator. Finally, in the polar regions beyond 60 degrees latitude, air flow is again reversed, as shown in Figure 3.4.

It is clear from this description that air pollutants, including those associated with small particles and droplets, may be transported over large distances when they enter the main air circulation a few kilometers above the Earth's surface. It follows that the release of pol-lutants some distance above Earth, e.g., from airplanes or from high chimneys, may lead to long-distance transport, changing pollution from a local problem to a global issue.

Pollutants may reach land in rain or snow or may be transferred to land surfaces by dry deposition at distances far removed from their original points of release. Dry deposition involves the direct absorption of gaseous components of air into surface waters or land surfaces. Findings of significant levels of organochlorine compounds in the Arctic have been attributed to a cold condensation effect. This movement of molecules from one phase of the environment to another must be considered when constructing models that attempt to predict the distributions of pollutants through different compartments.

This discussion has been concerned with the movements of pollutants released into the atmosphere. Brief mention should also be made of molecules generated by chemical reac-tions in the atmosphere. As mentioned earlier, some pollutants are generated as a result of the interactions of chemicals released from internal combustion engines, especially under conditions that give rise to photochemical smog. Ozone is generated from molecu-lar oxygen in the stratosphere under the influence of solar radiation. Most ozone produc-tion occurs in the equatorial zone, after which it diffuses toward the polar regions, giving rise to a more or less continuous ozone layer in the stratosphere. In the 1980s, a hole in the ozone layer above the South Pole was discovered. This was subsequently attributed to the destruction of ozone when it interacts with CFCs—volatile pollutants that can diffuse into the stratosphere.

3.4 Models for Environmental Distribution of Chemicals

Models have been constructed that attempt to describe the movements and distributions of chemicals among different compartments of the environment in mathematical terms (descriptive models). They may also be used to predict the movements and distributions of environmental chemicals (predictive models; Jorgensen, 1991). Broadly speaking, the mod-els are thermodynamic or kinetic. Thermodynamic models do not include a time dimen-sion; they are concerned with the distribution that will be found when thermodynamic equilibrium is reached. Kinetic models, on the other hand, are concerned with the *rates* at which processes of transfer or transformation occur; that is, they include a time factor.

For the purposes of modeling, the environment can be divided into physically distinct compartments separated by phase boundaries (e.g. air–water, water–gas). At the simplest level, the distribution of a chemical between two phases at equilibrium is described by a partition coefficient. This represents a simple thermodynamic model in which no time factor is involved. An example of this is the octanol–water coefficient (K_{ow}) described in Section 3.1.

True equilibrium states are usually found in closed systems in which the molecule(s) under consideration do not enter or leave the system. This is not typical of the natural environment where the systems under consideration are usually open and pollutants enter, leave, and undergo chemical transformation and/or biotransformation. Here, the partition coefficients can describe the distribution of a pollutant between two phases if the system is in a steady state in which the concentrations of a pollutant in the phases under consideration are constant and do not change over time. Such would be the case with the water of a river carrying a constant concentration of a lipophilic pollutant over a sediment high in organic matter. Some pollutant would partition into the sediment from the water, but an equivalent quantity would partition from the sediment into the water. Thus the concentration of pollutant in sediment and its ambient water would remain constant. In practice, the distinction between equilibrium and steady state is not very important in the present context because, in both cases, partition coefficients effectively describe the distribution of a pollutant between two adjacent compartments.

In attempting to describe the distribution of chemicals through several compartments over relatively large areas, some success has been achieved by the use of fugacity models. These are, again, thermodynamic; they are based on physicochemical properties of chemicals that determine distribution and on environmental variables such as temperature, pH, quality and quantity of light and water, and air movement. The environmental variables are complex and predictable only to a very limited degree. For this reason, fugacity models have had only limited success when used predictively. However, they have been useful as evaluative models that describe the environmental distributions of pollutants under defined conditions (Calamari and Vighi, 1992). One virtue of this approach is that it imposes a ranking order upon a group of chemicals based on their tendencies to move into air, sediments, and other environment compartments.

Fugacity models to describe the distribution of environmental chemicals were first introduced by Mackay (1991); see also Bacci, (1993). The underlying principle is that fugacity is a measure of the tendency of a molecule to escape from one phase or compartment into another. It is measured by the same dimensions as pressure. When considering the distribution of a chemical through several adjoining phases, equilibrium is reached when the chemical has the same fugacity in all phases. In any one phase:

$$f = \frac{C}{Z}$$

where C is the concentration of a chemical in the phase, Z is the fugacity capacity constant, and f is the fugacity. Consider now a two-phase system in equilibrium,

$$f_1 = f_2$$

where f_1 and f_2 are fugacities in phase 1 and phase 2, respectively. Thus:

$$\frac{C_1}{Z_1} = \frac{C_2}{Z_2}$$

or

$$\frac{C_1}{Z_1} = \frac{C_2}{Z_2} = K_{12}$$

where K_{12} is the partition coefficient for the chemical between phases 1 and 2, C_1 and C_2 are concentrations in phases 1 and 2, and Z_1 and Z_2 are fugacity capacity constants in phases 1 and 2. Thus the partition coefficient is the ratio of the fugacity capacity constant for the two compartments.

The distribution of a gas between air and water provides an example of how the model works. Here the fugacity in air corresponds to the partial pressure of the gas (P_a). This is a measure of the escaping tendency. The distribution of a gaseous pollutant between air and water is described by Henry's constant (H):

$$\frac{\text{conc. of the gas in water } (W)}{\text{partial pressure of gas } (P)} = H$$

or

$$\frac{W}{H} = P$$

Fugacity capacity constants (Z values) can be calculated for different compartments of the environment. The higher the Z values, the higher the expected concentrations in the compartment in question.

Fugacity models represent the environment as a number of compartments of known volume in which an equilibrium can be established. These compartments are described as "units of the world" and include air, lake water, soil, sediment, and biota. Biota are animals and plants. An example of the distribution of environmental chemicals as described by a fugacity model is shown in Figure 3.5.

These models are, at best, only of limited predictive value at present. The inability to predict various environmental parameters such as temperature and wind speed, and the fact that chemicals are seldom at equilibrium or in a steady state limit their effectiveness.

We have discussed thermodynamic models for systems at equilibrium or in a steady state or that approximate one of these situations. Kinetic models have also been used and these are concerned with the rates at which processes of transfer or transformation occur. At the simplest level, the rate of transfer from one compartment to another follows first-order kinetics and is described by the equation:

$$r = -kC$$

where r is the rate of transfer, k is the rate constant, and C is the concentration of the chemical in the phase (compartment) from which it is escaping. Consider a pollutant that moves

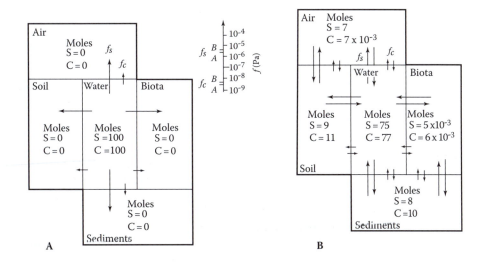

FIGURE 3.5
States of the compartments at two different times. In A, the two insecticides are present only in water. In B, they have moved from water into all the neighboring compartments to achieve equilibrium. The arrows indicate direction of movement in accordance with fugacity values. Thus A shows only movement away from the water compartment. B shows equal movements in all directions of all phase boundaries because the system is in equilibrium. The number of moles (gram molecular weights) is indicated for the two insecticides. The major difference between the two compounds in distribution is the greater tendency for sulfotep to escape from water to air. It has a higher vapor pressure (measure of fugacity). S = sulfotep. C = chlorfenvinphos. f_s and f_c = fugacities of sulfotep and chlorfenvinphos, respectively. Long arrows indicate f_s; short ones indicate f_c. (*Source:* Calamari, D. and Vighi, M.F. (1992). In *Methods to Assess Adverse Organisms (SCOPE)*. John Wiley & Sons. With permission.)

between compartments A and B by diffusion; a constant concentration will be reached after a certain time when equilibrium is reached. Then:

$$r_{A \to B} = r_{B \to A}$$

but

$$r_{A \to B} = k_{AB} C_A$$

and

$$r_{B \to A} = k_{BA} C_B$$

where k_{AB} and k_{BA} are rate constants for movement from A→B and B→A, respectively, and C_A and C_B refer to concentration in compartments A and B, respectively. It follows from this that:

$$\frac{C_A}{C_B} = K_{AB} = \frac{k_{BA}}{k_{AB}}$$

where K is the partition coefficient between compartment A and compartment B. In other words, the partition coefficient is the ratio of the rate constants at equilibrium. Much more complicated kinetic equations than these have been developed, but they lie outside the scope of this book.

To summarize, kinetic models can be used for environmental modeling. In theory, they have an advantage over thermodynamic ones: they can be used to describe the distribution of chemicals between compartments under conditions far removed from equilibrium or steady state. However, this approach has yet to be successfully developed.

3.5 Summary

This chapter has dealt with the long-range movements and global transport of chemicals after their release into the environment. More localized movements involving biotic factors, for example, along food chains or in soil, are described in Chapter 5. The movement of chemicals depends both on their own properties and upon environmental factors. Polarity, partition coefficients, vapor pressure, and molecular stability are all properties of chemicals that can influence their movement and distribution in the environment. Temperature, wind speed, circulation of air masses and movements of surface waters are also critical environmental factors.

In surface waters, pollutants are transported in a dissolved or particulate state by rivers and may later accumulate in lakes or in the estuaries to which they run. Ocean currents can transport pollutants over large distances. In the air, chemicals may exist in the vapor state or within particles or droplets. In the vapor state, movement by diffusion can be very important, as in the case of the movement of CFCs into the ozone layer. Also, pollutants can be transported over large distances by circulating air masses, leading to deposition on water or land surfaces far removed from their original points of release.

There is a growing interest in the development of evaluative and predictive models for the distribution of chemicals through the different compartments of the environment. Fugacity models are useful to determine the escaping tendencies of chemicals from one environmental compartment to another.

Further Reading

Bacci, E. (1993). *Ecotoxicology of Organic Contaminants*. Describes models for the environmental distribution of pollutants.

Calamari, D. and Vighi, M.F. (1992). *Role of evaluative models to assess exposure to pesticides*. Describes use of fugacity models.

Crosby, D.G. (1998). *Environmental Toxicology and Chemistry*.

Dix, H.M. (1981). *Environmental Pollution: Atmosphere, Land, Water and Noise*. Wide-ranging account of the distributions and movements of pollutants in air and water.

Mackay, D. (1991). *Multimedia Environmental Models: The Fugacity Approach*.

Schwarzenbach, R.P., Gschwend, P.M., and Imboden, D.M. (1993). *Environmental Organic Chemistry*.

Turekian, K.K. (1976). *Oceans*, 2nd ed. Describes major ocean currents.

Wayne, R.P. (1991). *Chemistry of Atmospheres*, 2nd ed. Authoritative and readable account of movements in the atmosphere and photochemical reactions.

4

The Fate of Metals and Radioactive Isotopes in Contaminated Ecosystems

4.1 Introduction

Four factors control the fates of inorganic pollutants in contaminated ecosystems. These are (i) localization, (ii) persistence, (iii) bioconcentration and bioaccumulation factors, and (iv) bioavailability.

4.1.1 Localization

A pollutant is toxic when its concentration exceeds a threshold value in a particular environmental compartment. The ultimate compartment is the whole planet, but compartments can be individual organisms or structures as small as single cells or even organelles within cells (see Figure 8.1).

It has been claimed that the solution to pollution is dilution. Tall chimneys operate on the safe dilution approach to discharges to the environment. For example, pollution from the nickel smelting works at Sudbury, Canada, caused severe ecological disruption to the surrounding countryside. The solution was to increase the height of the chimney so that metal particulates were carried farther from the factory. Although the total pollutants discharged remain the same, the concentration locally has been markedly reduced, so that plants have begun to recolonize the vicinity of the factory; however, the solution added to the acid rain of eastern North America.

At the other end of the scale, at the cellular level, organisms may compartmentalize potential toxins in insoluble deposits to prevent interference with essential biochemical reactions in the cytoplasm. For example, the epithelium of the midguts of most invertebrates contains metal-rich granules that act as intracellular sites of storage detoxification (Vijver et al., 2004; Figures 8.4 through 8.6).

4.1.2 Persistence

Metals are nonbiodegradable and do not break down in the environment. Metals that enter soils or sediments have long residence times before they are eluted to other compartments. Furthermore, formation and degradation of specific compounds such as methyl mercury can occur.

Radioactive isotopes of metals decay exponentially and persistence is dictated by the half-lives of the individual isotopes (Table 1.4). Highly radioactive waste from nuclear reactors is extremely persistent as it contains isotopes with half-lives of many thousands or even millions of years.

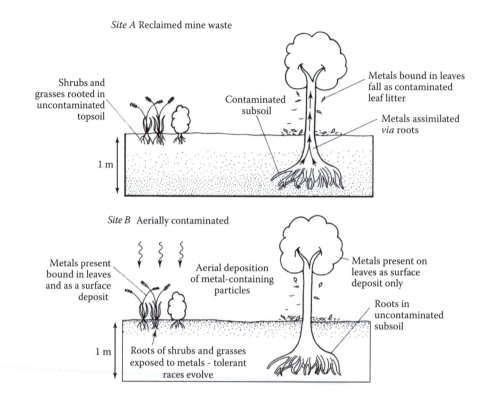

FIGURE 4.1

Comparison of distribution of metals in (A) disused mine site rehabilitated by application of uncontaminated topsoil and (B) site subject to aerial contamination. (From Hopkin, S.P. (1989). *Ecophysiology of Metals in Terrestrial Invertebrates.* Elsevier Applied Science Publishers. With permission.)

4.1.3 Bioconcentration and Bioaccumulation Factors

Some inorganic pollutants are assimilated by organisms to a greater extent than others. This is reflected in the bioconcentration factor (BCF) expressed by

$$BCF = \frac{\text{conc. of the chemical in the organism}}{\text{conc. in the ambient environment}}$$

For the definition of BCF, see Section 5.1.8.

The ambient environment of a terrestrial organism is usually the soil. For an aquatic organism, it is usually water or sediment. The extent of long-term bioaccumulation of inorganic chemicals depends on the rate of excretion (see Chapter 5). Bioaccumulation of cadmium in animals is high relative to most other metals; it is assimilated rapidly and excreted slowly (for examples, see Chapter 11). If an organism exhibits a high bioconcentration factor for a particular substance, the cause may be its biochemistry. For example, animals with calcareous skeletons, exoskeletons, and shells take up lead and/or strontium to a greater extent than those without those structures because lead and strontium follow similar biochemical pathways to calcium for which the organisms have evolved high assimilation efficiencies.

4.1.4 Bioavailability

Another reason for a high bioconcentration factor may be that a certain substance is more bioavailable than one with a low bioconcentration factor. Methylated mercury is taken up more readily than the unmethylated form (Wolfe et al., 1998). pH has a marked effect on the solubilities of metals in soils and water. If the pH declines (for example, because of acid deposition), some metals become more soluble than others and hence more bioavailable. Aluminum is highly insoluble at normal to slightly acidic pH, but below about pH 4.5 its solubility increases dramatically and it becomes the most important cause of fish kills in acidified lakes (see Chapter 14).

4.1.5 Cocktails of Inorganic Pollutants

The subjects of synergism and antagonism between pollutants are dealt with extensively in Chapter 10. However, worth discussing is one aspect of mixtures of pollutants that is often overlooked, i.e., the relationship between the relative toxicities of pollutants to organisms and their relative concentrations in the field (see discussion of risk assessment in Chapter 6). For example, negative effects of cadmium contamination in the diets of soil invertebrates on survival, growth, and reproduction can be detected at about one-tenth of the concentration (by weight) at which they occur with additions of zinc to the diet. Consequently, if the diet concentration of zinc is 10 times that of cadmium, toxicity will be due to both metals equally. However, in regions contaminated by both metals from past mining or smelting activities, zinc is almost always present at about 50 times the concentration of cadmium in soils or on vegetation. Thus zinc is responsible for toxic effects in primary consumers in these situations (Hopkin and Spurgeon, 2000). Nevertheless, because cadmium has a higher bioconcentration factor in most primary consumers than zinc, predators at the next stage in the food chain may be exposed to a zinc–cadmium ratio below 10 and hence be poisoned by cadmium rather than zinc.

4.2 Terrestrial Ecosystems

4.2.1 Introduction

In terrestrial ecosystems, the soil may be contaminated with metals and radioactive isotopes as a result of previous industrial, mining or other activity, or the contamination may be due to deposition from above (Figure 4.1). Airborne contamination may result from agricultural practices such as the application of metal-containing pesticides or metal-contaminated sewage sludge, or as wet or dry deposition from smelting activity, lead-containing car exhausts, atmospheric nuclear weapons testing, or accidents such as Chernobyl.

4.2.2 Metals

Most geological deposits of metals exposed on the surface because of weathering were worked out in previous centuries. Old and more recent mining activities have left a legacy of contaminated sites in which concentrations of metal can be extremely high. Because of the long residence times of metals, mines that have been disused for many years may have very sparse vegetation cover (Figure 4.2A). The plants that manage to survive are often

(A)

(B)

FIGURE 4.2
(A) Parys Mountain, Anglesey, north Wales. During the early nineteenth century, this was the largest copper mine in the world. Mining ceased about 100 years ago, but recolonization by vegetation has been slow because of the very high concentrations of copper in surface soils. (B) Rehabilitation of mining waste at a disused copper mine in the Gusum area in Sweden. The spoil tip is being capped with an impermeable layer before landscaping with a 2-m layer of topsoil on which trees will be planted. (Photographs courtesy of Steve Hopkin.)

metal-tolerant strains that are generically distinct from their non-tolerant ancestors (see Chapter 13).

Rehabilitation of such areas is difficult. The most widely used method at present is capping the contaminated deposit with an impermeable layer, then covering the layer with topsoil on which trees can be planted (Figure 4.2B). Rain falling on the soil flows over the impermeable layer to the edges of the deposit rather than through the metal-contaminated material. This approach greatly reduces the flow of metal-contaminated liquid to groundwater.

Soils may be contaminated at the surface from several sources. Before the development of synthetic organic chemicals, metal-containing pesticides were widely used. In the nineteenth century, it was standard practice to spray Bordeaux mixture in gardens and on crops to control pests, particularly on grape vines, as the name of the mixture suggests. Bordeaux mixture contains copper and is still widely used in the tropics as a fungicide (Lepp and Dickinson, 1994). Arsenic, lead, and chromium were used also, and it is still possible to detect elevated levels of these metals in garden soils of old houses.

One method of disposal of sewage sludge is to spread the waste on agricultural fields (the source of the sewage farm term). However, because drains that supply sewage treatment works also take industrial waste, concentrations of metals in sludge can be very high. The high organic matter content of sewage has a powerful binding capacity for metals that leach very slowly down the soil profile. The number of applications of sludge that can be made into farmland is restricted by the build-up of metals in soils (Alloway and Jackson, 1991).

One of the major sources of metal contamination of soils is the combustion of lead-containing gasoline. In the United Kingdom, for example, leaded gasoline contained about 0.4 g/l until 1985, when the maximum permitted concentration was reduced to 0.15 g/l (gasoline in less developed countries still contains as much as 3 g/l of lead). Leaded gasoline was banned in the European Union after January 1, 2000. The extensive use of lead in the past led to widespread contamination of urban soils (Culbard et al., 1988). Long-range transport of the metal has occurred, and elevated levels of lead can be detected in isolated regions far from industrial activity, such as Greenland (Rosman et al., 1993).

The long residence time of lead in soils means that surface layers will remain contaminated with lead for several hundred years to come. However, the reduction in emissions to the atmosphere has been mirrored by a decline in surface deposition, which in the United Kingdom fell rapidly after 1985 (Jones et al., 1991). Lead concentrations in the air of major cities are now less than one-quarter of their values in the early 1980s. Clear evidence of the role of cars in lead contamination appeared after the collapse of the Berlin Wall. The influx of cars from eastern Germany that ran on leaded gasoline resulted in an increase in the lead content of the moss *Polytrichum formosum*, which was monitored throughout the political change (Markert and Weckert, 1994).

One of the major modern uses of lead is in shotgun pellets. Increasing evidence indicates that lead from this source can be mobilized in the environment and assimilated, in particular by birds (Scheuhammer and Templeton, 1998). Humans can ingest lead if they eat birds that contain the pellets (Johansen et al., 2004). Concern over accumulation in the gizzards of waterfowl and subsequent lead poisoning led to the banning of the use of lead-containing fishing weights and shotgun pellets in many wetland areas.

In contrast to pollution from cars, deposition of metals from smelting activity tends to be fairly localized. One of the best studied sites in the world is the region surrounding a lead, zinc, and cadmium primary smelting works at Avonmouth near Bristol, southwest England (see Hopkin, 1989, and Martin and Bullock, 1994, for detailed descriptions of the area). In the close vicinity of the factory, concentrations of lead, zinc, and cadmium in surface soils are at least two orders of magnitude higher than normal background levels. Significantly elevated levels of cadmium in soils can be detected up to 30 km downwind of the plant. The main effect of this heavy aerial deposition of metals is a reduction in the decomposition rate of dead vegetation that accumulates on the surface as a thick layer. Organisms such as earthworms, woodlice, and millipedes that are responsible for

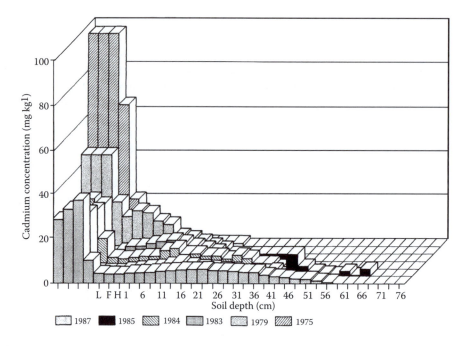

FIGURE 4.3

Concentrations of cadmium (mg/kg dry weight) in Hallen Wood soil profiles from 1975 through 1987. Hallen Wood is 3 km northeast of a primary cadmium, lead, and zinc smelting works at Avonmouth, southwest England. Each value for the mineral soil represents an analysis of a block of soil collected at depths of 0 to 1 cm and then at 2.5-cm intervals to the final depth. The profiles show two main features (i) a reduction over time in concentration in the litter (L), primary (F), and secondary (H) layers; (ii) a progressive wave of cadmium moving down the profile. The increased mobility of cadmium arose from increased acid deposition in the woodland following construction of a tall chimney at a sulfuric acid plant at the smelting works in the mid-1970s. (*Source:* Martin, M.H. and Bullock, R.J. (1994). In *Toxic Metals in Soil–Plant Systems.* John Wiley & Sons With permission.)

the initial fragmentation of leaf litter are absent as a result of the metal contamination of their diets.

The mobility of metals in soils is determined largely by the clay content, the amount of organic matter, and the pH. In general, the higher the clay and/or organic matter content and pH, the more firmly bound are the metals and the longer their residence time in soil. One of the effects of acid deposition in Europe has been forest die-back, due at least partly to nutrient deficiency (mainly magnesium). Essential elements become more mobile in acidified soils and are leached to lower soil layers where the roots of the trees cannot penetrate (Berggren et al., 1990).

In Avonmouth soils, the metals exhibit the classic profile of decreased concentration with depth (Figure 4.3). However, in 1976, a taller chimney was built to vent the sulfuric acid plant on the site, and the pH of soils downwind decreased because of higher acid deposition. The mobility of metals increased and a progressive wave of metals passed down through the soil profile (Figure 4.3).

4.2.3 Radioactivity

Contamination of soils with radioactive material is a relatively recent phenomenon, as most of the elements involved did not exist naturally before the development of nuclear weapons and reactors. Some regions of the world where bombs were tested, such as the Australian and Nevada deserts, are still heavily contaminated. Any clean-up will have to involve the removal of the surface soil, but that creates the problem of disposing of the radioactive material removed.

The production of nuclear energy has a good safety record relative to other methods of energy production. However, it has caused a number of well publicized examples of environmental contamination. Perhaps the best known is the accident at the Chernobyl complex in April 1986 when one of the reactors caught fire, eventually releasing half its contents to the atmosphere (Edwards, 1994). Most of Europe was affected to some extent by the fallout of the radioactivity, which was most severe to the northwest of the site. In Byelorussia (on which 70% of the total fallout landed; Lukashev, 1993), large areas of the country were heavily contaminated and restrictions on certain agricultural practices were introduced (Figure 4.4).

The effects of the Chernobyl fallout outside the former Soviet Union were most persistent in Scandinavia and upland areas of northwest Europe. The vegetation in these nutrient-poor regions is adapted to retain and recycle essential elements. Metal pollutants deposited from the atmosphere pass down through the soil profile extremely slowly. In Cumbria in northwest England, sheep on upland hill farms became contaminated with radioactive cesium for several years and could not be sold for human consumption (the maximum permissible level was 1000 Bq/kg). Lambs were moved to lowland pastures before slaughter where their radiocesium burden was rapidly lost via feces (Crout et al., 1991). The cesium passed much more rapidly down the soil profile in lowland fields than on the hills.

In Sweden, one of the most contaminated areas is occupied by the Saami community, where ^{137}Cs levels in reindeer reached a mean of more than 40,000 Bq/kg (Åhman and Åhman, 1994). Since the disaster, the radioactivity of the reindeer has declined slowly and exhibited a marked seasonal fluctuation (Figure 4.5). This is correlated with the change in diet from summer to winter. During summer, reindeer feed mainly on grass, herbs, and leaves that have low radiocesium contents. In winter, lichens are important parts of their diet (up to 60%); their radiocesium content was more than 10 times that of vascular plants from the same region. This illustrates the importance of regular monitoring of biota after a pollution incident, as the movement and pathways of transport may be more complicated than at first thought.

4.3 Aquatic Systems

The ultimate sink for metals is the ocean. However, because of the massive dilution of contaminants that occurs, it is difficult to prove that metals in the open sea exert significant effects on biota. Indeed, evidence indicates that some metals such as iron are limiting-nutrients for phytoplankton in the open ocean (Coale et al., 1996). Estuaries tell a different story, however, and many are grossly polluted, particularly those fed by rivers that pass through regions of heavy industry and mining activity (Figure 4.6; Grant and Middleton, 1990; Bryan and Langston, 1992). When polluted fresh water reaches the sea, the flow rate slows, suspended sediments settle on the bottom, and dissolved metals are precipitated (see Chapter 3). Even if the discharges to rivers are cleaned up, the estuaries they feed may continue to be affected for many years because of remobilization of past sediment contamination. Such an effect occurred in Minamata Bay in Japan, where sediments were heavily contaminated with mercury (Figure 4.7). A new quay was built in the 1970s to allow much

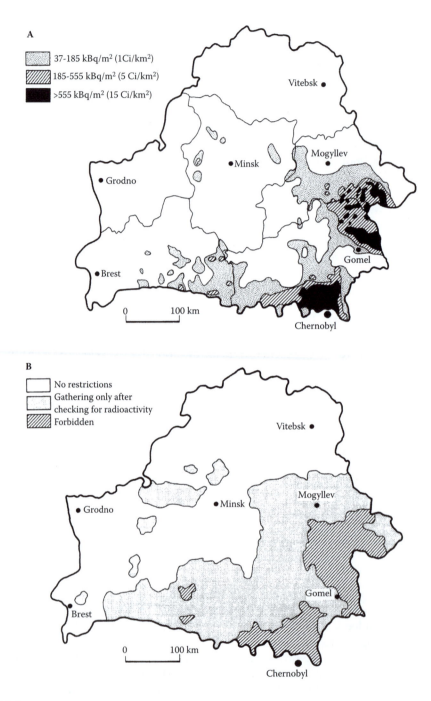

FIGURE 4.4
(A) Distribution of radioactive cesium in surface soils of Byelorussia after the Chernobyl accident. (B) Areas of Byelorussia where mushroom gathering is restricted. (*Source:* Lukashev, V.K. (1993). *Applied Geochemistry* 8, 419–436. With permission.)

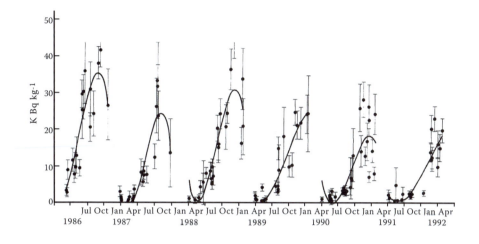

FIGURE 4.5
Activity concentrations of ^{137}Cs in reindeer from the Saami community, Vilhelmina Norra, Sweden, from 1986 to 1992. Mean ± standard deviation from separate slaughter occasions (n = 10 to 825 animals). (*Source:* Åhman, B. and Åhman, G. (1994). *Health Physics* 66, 503–512. With permission.)

larger ships to dock. The actions of their propellers remobilized mercury-contaminated sediment that could be detected much farther out to sea than previously (Kudo et al., 1980).

In water, the solubility of metals is strongly dependent on pH. Streams draining mining areas are often very acidic and contain high concentrations of dissolved metals but little aquatic life. However, as a stream becomes diluted with uncontaminated water farther downstream, the pH rises and metals are precipitated onto the bed. This is the case with the stream that drains Parys Mountain (Figure 4.2A) where heavy deposits of iron and copper coat the submerged rocks, giving the bed of the stream a bizarre orange–brown coloration.

Acid deposition may be stored in snow and released as a sudden pulse of acidity during a spring thaw (Borg et al., 1989). A resulting decline in pH of as much as one unit causes a sudden increase in the levels of soluble metals in lakes and streams (Figure 4.8).

The deliberate release of radioactive waste into the aquatic environment is much more tightly controlled now than it was in the past. Until the 1980s, concrete drums containing radioactive materials were routinely dumped in the ocean until the practice was banned. A bathosphere inspected some of these drums on the floor of the ocean in the early 1980s and they appeared to be intact (Sibuet et al., 1985). However, their long-term integrity must remain in doubt.

The major source of radioactive contamination of the seas in Europe has been effluent from the Sellafield nuclear reprocessing plant in Cumbria, northwest England. Porpoises resident in the Irish Sea contain significantly higher concentrations of ^{137}Cs and ^{40}K than in other European waters (Berrow et al., 1998). During the 1970s, discharges were high. Indeed, by far the most significant current problem is remobilization of this earlier contamination. Present day discharges are very low. Much of the radioactivity is present in sediments deposited in the past and are covered with more recent material. Thus by careful analysis of the layers, it is possible to date the strata and compare their radioactivity with discharge data for a specific year (Mackenzie and Scott, 1993; Mackenzie et al., 1994). Figure 4.9 shows such an approach.

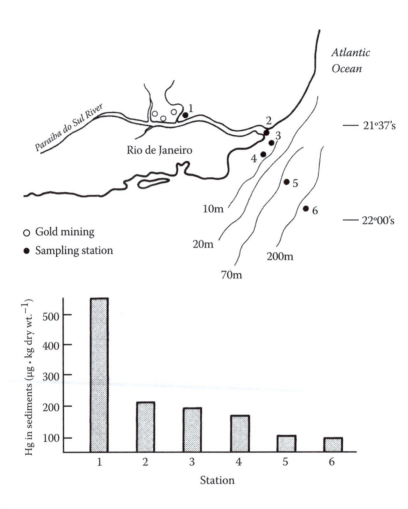

FIGURE 4.6
Mercury (Hg) distribution in sediments of the Paraiba do Sul river, estuary, and adjacent continental shelf, Rio de Janeiro State, southeast Brazil. (*Source:* Pfeiffer, W.C. et al. (1989). *Science of the Total Environment* 87/88, 233–240. With permission.)

4.4 Summary

Metals are nonbiodegradable pollutants, several of which have become widespread in the environment through industrial activities such as mining and smelting. In aerially contaminated soils, they tend to persist for many years in the surface layers. In aquatic systems, metals may become locked in bottom sediments where they may remain for many years. However, if the pH falls, metal solubility increases and metals become more mobile. One of the knock-on effects of acid rain is the transport of metals to lower levels in the soil profile where they may damage deep-rooted plants and contaminate groundwater. Environmental contamination with radioisotopes is a relatively recent phenomenon. Fallout from the Chernobyl disaster was deposited over most of Europe. ^{137}Cs accumulated in a wide range of organisms because of its tendency to follow the same biochemical pathways as essential potassium. Although present levels of release from the Sellafield

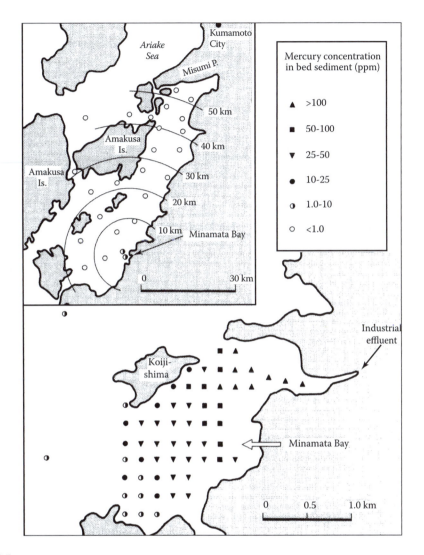

FIGURE 4.7
Mercury concentrations in bed sediments of the Yatsushiro Sea (1975) and Minimata Bay (1973) in Japan. (*Source:* Kudo, A., Miyahara, S., and Miller, D.R. (1980). *Progress in Water Technology* 12, 509–524. With permission.)

reprocessing plant are low, concern continues about the remobilization of radioisotopes in sediments deposited in the 1970s, when discharge rates were high.

Further Reading

Alloway, B.J. and Jackson, A.P. (1991). A study of metal contamination of sewage sludge.

Borg, H., Andersson, P., and Johansson, K. (1989). Paper that clearly demonstrates the impact on metal solubility of the seasonal dip in pH in Swedish lakes.

Coale, K.H., Johnson, K.S., Fitzwater, S.E., Gordon, R.M. et al. (1996). Evidence for iron deficiency of phytoplankton in the open ocean.

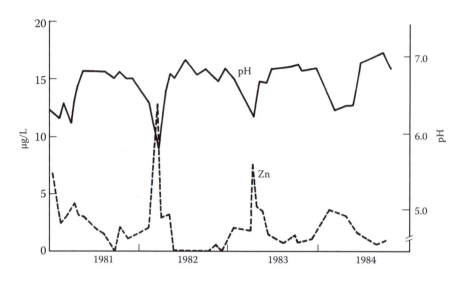

FIGURE 4.8
Variations in outlet water pH and concentrations of zinc in Lake, Holmeshultasjön, Sweden over time. Note the close relationship between pulses of lower pH in the lake and increased levels of dissolved zinc. (*Source:* Borg, H., Andersson, P., and Johansson, K. (1989). *Science of the Total Environment* 87/88, 241–253. With permission.)

Crout, N.M.J., Beresford, H.A., and Howard, B.J. (1991). The reasons behind the very long residence times of cesium in sheep in northeast England and measures taken to solve the problem.

Hopkin, S.P. (1989). *Ecophysiology of Metals in Terrestrial Invertebrates.* Contains general introductory chapters on metals and a detailed account of studies around the Avonmouth metal smelting works (see also Martin and Bullock, 1994).

Jagoe, C.H., Dallas, C.E., Chesser, R.K. et al. (1998). Demonstrates the persistent severe radioactive contamination of fish near Chernobyl.

Pastor, N., Baos, R., Lopez-Lazaro, M. et al. (2004). Interesting account of effects of pollution on the Doñana area (southwest Spain) after the 1998 mining waste spill.

Simkiss, K. (1993). Summary of several relevant papers on radioactivity in the United Kingdom after Chernobyl.

Wolfe, M.F., Schwarzbach, S., and Sulaiman, R.A. (1998). Comprehensive review of distribution and effects of mercury in the environment.

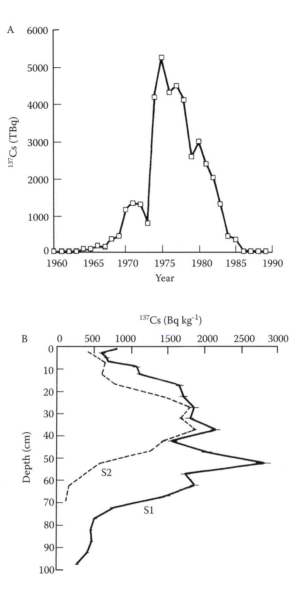

FIGURE 4.9
(A) Variations in annual quantities of ¹³⁷Cs discharged from the Sellafield nuclear fuel reprocessing plant over time. (B) ¹³⁷Cs concentration profiles for Solway Firth salt marsh sediment sections S1 and S2, north of Sellafield. (*Source:* Mackenzie, A.B. et al. (1994). *Journal of Environmental Radioactivity* 23, 36–69. With permission.)

5

Fates of Organic Pollutants in Individuals and in Ecosystems

The organic pollutants discussed in this book are examples of xenobiotics. A xenobiotic is defined here as a compound that is foreign to an organism—it does not play a role in the organism's normal biochemistry. By this definition, a chemical that is normal to one organism may be foreign to another. Thus xenobiotics may be naturally occurring as well as man-made (anthropogenic) and must have existed early in the evolutionary history of this planet.

From an evolutionary view, the role of naturally occurring xenobiotics as chemical warfare agents is of considerable interest. For example, evidence indicates that animals evolved detoxification mechanisms to acquire protection against toxic xenobiotics produced by plants (Walker, 2009; Chapter 1). Most of the organic pollutants cited in Chapter 1 are man-made xenobiotics; they do not occur in nature. It is, however, important to remember that naturally occurring xenobiotics, for example, pyrethrins, nicotine, and various mycotoxins, are subject to the same toxicokinetic processes as man-made pollutants.

Toxicokinetics has relevance to ecotoxicology because it aids the understanding and prediction of the behaviors of organic pollutants within living organisms. However, the fate of a chemical in an entire ecosystem is far more complex and involves movement through soils, surface waters, and air and transfers along food chains. Toxicokinetic models are valuable for predicting the fates of chemicals in individual organisms, but more elaborate models are required to predict fates in entire ecosystems. As discussed earlier, some success has been achieved in predicting the distribution of chemicals through major compartments of the environment (Chapter 3). However, prediction of the distribution of a chemical through the organisms that constitute an ecosystem is another matter. Some systems are far too complicated to lend themselves to this kind of predictive modeling.

We will not describe the general principles of toxicokinetics as they apply to lipophilic xenobiotics in individual organisms. The emphasis will be on animals, with some references to plants. Toxicokinetic models will be briefly discussed, particularly their use for predicting bioconcentration and bioaccumulation. Finally, movements in terrestrial and aquatic ecosystems will be considered separately.

5.1 Fate within Individual Organisms

5.1.1 General Model

Figure 5.1 represents the fate of a xenobiotic in an individual organism. The integrated illustration shows the movements, interactions, and biotransformations that occur after an organism is exposed to a xenobiotic. This section focuses on the situation in animals before drawing attention to special features of plants. It should be stressed that this highly

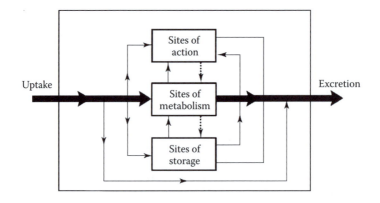

FIGURE 5.1
General model describing fates of lipophilic xenobiotics in living organisms. (*Source:* Walker, C.H. (1994). In *Introduction to Biochemical Toxicology.* Appleton and Lange. With permission.)

simplified model identifies processes that are important from a toxicological view. The interplays between processes determine the toxic effects of pollutants. For any particular chemical, interspecific differences in the operation of these processes will lead to corresponding differences in toxicities between species (selective toxicity).

The model identifies five types of sites as follows—sites of uptake, metabolism, action, storage and excretion; the arrows identify the movement of chemicals between them.

Once a chemical is within an organism, the four types of site that it may then reach are as follows:

1. *Sites of toxic action.* Here the toxic form of a pollutant interacts with an endogenous macromolecule [e.g., protein or DNA] or structure [e.g., membrane], and this molecular interaction leads to the appearance of toxic manifestations in the whole organism. THE CHEMICAL ACTS UPON THE ORGANISM.

2. *Sites of metabolism.* These are enzymes that metabolize xenobiotics. Usually metabolism causes detoxication, yet in a small but highly significant number of cases it causes activation. THE ORGANISM, ACTS UPON THE CHEMICAL.

3. *Sites of storage.* Here the pollutant exists in an inert state. THE CHEMICAL IS NOT ACTING UPON THE ORGANISM NEITHER IS IT BEING ACTED UPON.

4. *Sites of Excretion.* Excretion may be of the original pollutant but often it is of a biotransformation product [metabolite or conjugate]. After terrestrial organisms have been exposed to lipophilic pollutants, excretion is very largely of biotransformation products, not of original compounds.

In this simple model each of these boxes represents one of these types of site. In reality, however, there is often more than one form for any type of site; for example, a detoxifying enzyme may exist in several different forms. Furthermore, each type of site may have more than one location—in different tissues or cellular organelles, for example.

After uptake, pollutants are transported to different compartments of the body by blood and lymph (vertebrates) or hemolymph (insects). Movement into organs and tissues may occur via diffusion across membranous barriers or, for extremely lipophilic compounds, by transport with lipids. Uncharged molecules that have reasonable balances between oil and water solubility tend to move across membranous barriers by passive diffusion. This happens if they are not too large (molecular weight < 800) and an optimal octanol–water partition coefficient (K_{ow}) for doing so exists (for explanation of partition coefficient, see Sections 3.3 and 5.1.2). Some very lipophilic compounds are transported when they are dissolved in lipoproteins. After partial degradation, fragments of lipoprotein are taken into cells such as hepatocytes by endocytosis, carrying the associated lipophilic molecules with them. Most xenobiotics are distributed throughout the compartments of the body after uptake. Quantitative aspects of this are described in Section 5.1.7.

Many of the organic pollutants discussed in this book are highly lipophilic (hydrophobic), i.e., they have high K_{ow} values. If not metabolized, they will be stored in fat depots or at other lipophilic sites such as membranes or lipoproteins. Such storage of potentially toxic lipophilic xenobiotics may be protective in the short term. In the long term, however, release from storage may occur and cause toxic effects. Delayed toxicity may be observed some time after initial exposure to a xenobiotic, as in the case of organochlorine insecticides such as dieldrin.

Because of their marked tendency to move into hydrophobic locations (e.g., membranes, fat depots), xenobiotics with high K_{ow} values are not excreted directly in the feces or urine of terrestrial organisms to an important extent. Their efficient elimination depends on their biotransformation to water-soluble metabolites and conjugates (Section 5.1.5) that are then excreted readily in feces and/or urine. The thick arrow through the middle of Figure 5.1 emphasizes the importance of this process for terrestrial animals. In aquatic organisms, loss by direct diffusion into the ambient water (e.g., across gills of fish) represents an important mechanism of excretion for lipophilic xenobiotics.

The model can be subdivided into two parts. The processes of uptake, distribution, and metabolism constitute the toxicokinetic component. (In the case of drugs, this would be referred to as the pharmacokinetic component.) Molecular interactions at the site of action are part of the toxicodynamic component (pharmacodynamic component in pharmacology). The operation of toxicokinetic processes determines how much of a toxic compound reaches the site of action (this may be the original xenobiotic or an active metabolite of the same). By contrast, the nature and degree of interaction between the toxic compound and the site of action will determine the toxic response that is produced (toxicodynamic component). Sometimes it is convenient to consider these two elements separately when investigating the mechanisms that underlie toxicity. A xenobiotic may be particularly toxic to a defined species for either or both of the following reasons.

1. The toxicokinetics are such that a high proportion of the active form of the xenobiotic reaches the site of action.

2. The toxicodynamics are such that a high proportion of the xenobiotic that reaches the site of action will interact there to produce a toxic response.

Conversely, another species may be insensitive to a xenobiotic because neither of these components operates in a way that favors toxicity.

Toxicokinetic aspects of the model will now be discussed in more detail. Toxicodynamic aspects will be discussed in Section II of the text (Chapter 7), which is concerned with effects of pollutants upon individual organisms.

TABLE 5.1

Major Routes of Uptake for Organic Pollutants

Organisms	Uptake Route	Sources
Terrestrial vertebrates	Alimentary tract	Food and ingested water
	Skin	Contaminated surfaces
	Lungs	Droplets and particles in air, vapor
Terrestrial invertebrates	Alimentary tract	Food and water
	Cuticle (insects)	Contaminated surfaces
	Body wall (slugs, worms)	Contaminated environment, e.g., soil
	Tracheae	Droplets and particles in air, vapor
Fish	Gills	Pollutants dissolved or suspended in ambient water
	Alimentary tract	Food
Aquatic mammals and birds	Alimentary tract	Food
		Small amounts from ambient or ingested water
Aquatic amphibians	Alimentary tract	Food
	Skin	Small amounts from ambient water
		Pollutants in ambient water dissolved or suspended
Aquatic invertebrates	Alimentary tract	Food; some from ambient water
	Respiratory surfaces	Pollutants dissolved or suspended in ambient water
Plants	Leaves	Pollutants in droplets or particles[a]
		Vapors
	Roots	Pollutants dissolved in soil water[a]

[a] Important route of uptake for herbicides, systemic insecticides, and fungicides.

5.1.2 Processes of Uptake

Table 5.1 summarizes the most important routes of uptake. The movement of organic molecules into an organism is usually the consequence of passive diffusion across natural barriers. This is how passage across plant or insect cuticles, vertebrate skins, or membranes lining the gut, lungs, or tracheae usually occurs. Also, very lipophilic molecules may be absorbed from the gut in association with fat (bulk transport).

The movement of organic molecules across natural barriers by passive diffusion requires that the molecules have optimal solubility properties. To move effectively across such barriers, the molecules must first have some affinity for the barrier—which is usually of lipophilic character. Also, they must have some affinity for the water inside the barrier. Thus, the molecules should have a reasonable balance between lipid solubility and water solubility as indicated by their K_{ow} values (Section 3.1).

Octanol is a lipophilic (hydrophobic) solvent that is immiscible with water. For efficient movement across lipophilic barriers, K_{ow} values should be close to 1. Values much below 1 indicate high water solubility and very low lipid solubility. Values much higher than 1 indicate very high lipid solubility (lipophilicity) but very low water solubility. The relationship between K_{ow} and rate of movement through a lipophilic barrier is indicated in Figure 5.2.

Although the K_{ow} value generally indicates the likelihood that a molecule will be taken up efficiently by passive diffusion, it must be emphasized that there are no optimal K_{ow} guarantees that rapid uptake in all situations; other factors must be considered as well. As noted earlier, molecules above a certain size cannot diffuse readily through biological membranes, so this argument does not apply to them. The composition and temperature of lipophilic barriers determine their state of fluidity and therefore the ease with which

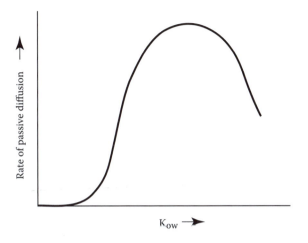

FIGURE 5.2
Passive diffusion of xenobiotics across a biological membrane. They move through the membrane into water on other side.

molecules can diffuse into them. At low temperatures, lipid bilayers can lose their fluidity, making diffusion through them difficult or impossible.

Another factor must be borne in mind with passive diffusion of pollutants that are weak acids or bases. Here, a state of equilibrium exists between charged and uncharged forms, determined by the pH of the ambient medium (Figure 5.3). Usually only the uncharged form will readily cross a lipophilic barrier. Thus the uptake of weak acids is favored by low pH, but the uptake of many weak bases is favored by high pH. Herbicides that are weak acids (e.g., 2,4-D, MCPA) penetrate plant cuticles rapidly if they are in a medium of low pH. Within the alimentary tracts of mammals, weak acids tend to be absorbed in the stomach (pH 1 to 2) and weak bases in the duodenum, where the pH is much higher.

Returning to Table 5.1, the different types of organism will now be considered separately. Terrestrial vertebrates and invertebrates take up lipophilic pollutants from the alimentary tract or across the skin or cuticle. Pesticides represent a very important category of pollutants in agricultural ecosystems where there can be substantial exposure to potentially toxic compounds by either or both of these routes. In general, uptake across the cuticle of insects is likely to be more important than uptake across the skin of vertebrates. This is because insects are much smaller and have much higher ratios of surface area to body volume than do vertebrates (i.e., they have much more absorbing surface per unit volume). The mobility of the organism is an important factor in determining the rate of uptake across cuticle or skin. With invertebrates, mobile predatory species will tend to come into

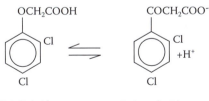

FIGURE 5.3
Equilibrium of weak acid.

contact with more pesticide on soil or plant surfaces than will more sedentary species. In the case of vertebrates, movement between different locations in agricultural areas will determine the extent to which they come into contact with pesticides.

Soil organisms such as earthworms, collembola, and mites may be continuously exposed to persistent pesticides in soil. This raises the issue of the availability of compounds that are bound to clay and organic matter. Although compounds dissolved in soil water are freely available for uptake, the availability of bound organic compounds is not well under-stood. These comments refer particularly to pesticides, but similar considerations apply to other organic pollutants in terrestrial ecosystems.

Organic pollutants may also be absorbed via the respiratory systems of vertebrates and invertebrates. Absorption by this route readily occurs with pollutants in the gaseous state. A more complex situation exists with pollutants associated with droplets or particles that may be deposited in the respiratory tract. Examples include smokes and dusts emitted from factories, residences, and internal combustion engines, pesticidal sprays and dusts applied to agricultural land, and rain contaminated with airborne pollutants. As yet, little is known about the extent to which terrestrial animals take up organic pollutants by this route.

In contrast to their terrestrial counterparts, aquatic vertebrates and invertebrates are exposed directly to many pollutants dissolved or suspended in surface waters. Uptake across respiratory surfaces or skin represents an important route of entry for many dis-solved aquatic pollutants. However, as with soils, we know little about the extent to which pollutants are taken up when bound to sediments or suspended particles.

Uptake from food may also be important for aquatic organisms, particularly predatory birds, mammals, and reptiles at the top of the marine food chain that do not appear to take up pollutants directly from water to an important extent. They do not have gills and their skins are not thought to be very permeable to organic molecules (compare the more permeable moist skins of some amphibians).

Descriptions of the routes of uptake of pollutants would be incomplete without a brief mention of the transfer of pollutants from parent to offspring. Pollutants may be trans-ferred across the placentae of mammals into developing embryos. In birds and reptiles, lipophilic compounds are transported with lipids into eggs and subsequently into devel-oping embryos. Also, some lipophilic pollutants are secreted into mammalian milk and thus passed to offspring during suckling.

Plants can absorb pollutants across their leaf cuticles and through their roots. These processes are well characterized for translocated herbicides and systemic insecticides and fungicides (Hassall, 1990). Gases can be absorbed through stomatal openings. Movement through leaf cuticles is by passive diffusion and depends on the K_{ow} values of pollutants.

5.1.3 Processes of Distribution

In vertebrates, absorbed pollutants may travel in the bloodstream and to a lesser extent in the lymph. Where absorption occurs from the gut, much of the absorbed pollutant will ini-tially be taken to the liver by the hepatic portal system. A high proportion of the circulat-ing pollutant will then be taken into hepatocytes (first-pass effect). Entry into hepatocytes may be by diffusion across the membrane or by co-transport with lipoprotein fragments taken up by endocytosis.

Absorption via lungs or skin may lead to a somewhat different initial pattern of distri-bution, as the blood travels first to tissues other than the liver. Within blood and lymph, organic molecules are distributed among different components according to their solu-bility properties. Highly lipophilic compounds (high values of K_{ow}) will associate with lipoproteins and membranes of blood cells and show little tendency to dissolve in blood water. Conversely, more polar compounds (low values of K_{ow}) tend to dissolve more in water and will associate less with lipoproteins and membranes of blood cells.

Movement of organic molecules into the brain is of particular concern in toxicology as this is the site of action of many highly toxic substances (see Chapter 7). To enter the brain, organic molecules must cross the blood–brain barrier consisting essentially of membranes that lie between blood plasma and the brain. In general, lipophilic compounds can cross this barrier, but polar molecules (very low K_{ow}) cannot.

In invertebrates, movement of organic pollutants is in the hemolymph. Otherwise, distribution follows a course similar to that described for mammals. Plants transport absorbed pollutants in the phloem (symplastic transport) or in the transpiration stream of the xylem (apoplastic transport). Molecules entering via the roots may be transported to the aerial parts of the plant in the xylem. Those entering via leaves may be taken to other parts of the plant via the phloem (Hassall, 1990).

5.1.4 Storage

Xenobiotics may locate where they cannot interact with their sites of action and are not subject to metabolism. Of particular importance are lipophilic (hydrophobic) environments, especially fat depots, but also lipoprotein micelles and cell membranes that lack sites of action or enzymes that can metabolize specific xenobiotics. (It should be emphasized that no general rule applies here—an inert membrane for one xenobiotic may contain a site of action or detoxifying enzyme for another.) Many lipophilic xenobiotics can be stored in depot fat and some are stored by binding to proteins, for example, the binding of the warfarin rodenticide to serum albumin.

The amount of depot fat in vertebrates is subject to considerable variation. When food is plentiful, fat depots may be built up—in some cases to the point where they account for 20% or more of total body weight. For example puffin (*Fratercula arctica*) chicks are balls of fat when they leave the nesting burrows and find their way to the sea. When food is scarce and during illness, egg laying, and migration, an organism's fat depots are utilized for energy. The rapid mobilization of depot fat will bring a rapid release of stored pollutants into the bloodstream, and the pollutants will find their ways to sites of action and metabolism. Thus, in the short term, storage of lipophilic pollutants in fat depots may minimize their toxic effects but release after long-term storage may lead to toxic effects.

The problem of delayed toxicity was well illustrated in studies on the effects of the dieldrin insecticide on eider ducks (*Somateria mollissima*) in the Netherlands in the 1960s (Koeman and Van Genderen, 1972). During the breeding season, females died of dieldrin poisoning, but males did not. At the onset of the breeding season, females build up considerable fat depots that are then utilized during egg laying (large clutches of eggs are produced). In the present case, significant quantities of dieldrin were laid down in fat depots. Because of the size of the fat depots, the concentrations in other tissues were not particularly high. The dieldrin concentrations in fat were 10 to 20 times greater than in tissues such as liver or brain. With the mobilization of fat depots, however, blood dieldrin levels rose rapidly from about 0.02 to 0.5 µg/ml. Birds were poisoned because of the corresponding increase of concentrations of the insecticide in the brain.

Similar effects are to be anticipated in other situations where mobilization of fat depots leads to the relatively rapid release of stored lipophilic compounds of high toxicity. Other organochlorine insecticide residues (e.g., p,p′-DDT, heptachlor epoxide), organomercury compounds, polychlorodibenzodioxins (PCDDs), and polychlorinated biphenyls (PCBs) can be redistributed similarly. Also, such compounds may cause delayed toxicity during illness, starvation, or migration.

The importance of storage of persistent lipophilic compounds in depot fat is evident from many analyses of vertebrate samples from terrestrial, marine, and freshwater ecosystems. A wide range of organochlorine compounds can be identified using techniques such as capillary gas chromatography. Concentrations are particularly high in predators at the top of food pyramids.

5.1.5 Metabolism

The enzymic metabolism of most lipophilic xenobiotics occurs in two phases (see Timbrell, 1999):

$$\text{Xenobiotic} \xrightarrow{\text{Phase}} \text{Metabolite} \xrightarrow[\substack{\text{Endogenous}\\\text{compound}}]{\text{Phase}} \text{Conjugate}$$

$$\xrightarrow{\hspace{2cm}} \text{Increasing polarity}$$

The initial (phase I) biotransformation involves oxidation, hydrolysis, hydration, or reduction in the great majority of cases and normally leads to the production of metabolites containing hydroxyl groups. The hydroxyl group introduced in this initial step is necessary for most of the subsequent conjugation reactions that constitute the second stage (phase II) of biotransformation. These two phases lead to a progressive increase in water solubility—from a lipophilic xenobiotic to a more polar metabolite and then to an even more polar conjugate. Most conjugates are negatively charged (anions), have appreciable water solubility, and are readily excreted in bile and/or urine. After exposure of vertebrates and insects to lipophilic xenobiotics, most of the excreted products are conjugated.

The scheme above represents a simplification of the real situation. Phase I may involve more than one step. Also, some xenobiotics (especially if they already possess hydroxyl groups) may undergo conjugation directly. The relationship between metabolism and excretion will be explained in Section 5.1.6.

In most cases, biotransformation leads to a loss of toxicity (detoxification) and is protective to the organism. However, in a small but highly significant number of cases, metabolism leads to an increase in toxicity (activation). In particular, oxidation in phase I leads to the production of reactive metabolites that can bind to cellular macromolecules. Oxidation of organophosphorous insecticides such as dimethoate, diazinon, malathion, disyston, chlorpyriphos, and others leads to the production of reactive metabolites (oxons) that can phosphorylate and thereby inhibit acetylcholinesterase of the nervous system (Chapter 7). Oxidation of carcinogens such as benzo(a)pyrene, aflatoxin, and vinyl chloride leads to the formation of reactive metabolites that can bind to DNA (Chapter 7). Thus relatively inert molecules that do not cause toxic effects alone are converted to reactive metabolites with very short biological half-lives which can cause cellular damage.

One of the curious features of biochemical toxicology is that many of the most destructive types of molecules are reactive metabolites that are difficult or impossible to detect because of their short biological half-lives. Proof of their existence and the damage that they cause often depends on identification of the modifications they cause to cellular macromolecules, e.g., inhibited acetylcholinesterase or damaged DNA (see Chapter 10).

TABLE 5.2

Enzymes that Metabolize Lipophilic Xenobiotics (Phase I)

Enzymes	Principal Location	Co-Factors	Substrates
Microsomal monooxygenases (mixed function oxidases)	Endoplasmic reticulum of many animal tissues, especially livers of vertebrates; hepatopancreas, fat bodies, and guts of invertebrates	NADPH/NADH, O_2	Most lipophilic xenobiotics of molecular weight < 800
Carboxyl esterases	Endoplasmic reticulum of many animal tissues; cytosol, serum and plasma of vertebrates	None known	Lipophilic carboxyl esters
A esterases	Endoplasmic reticulum of certain cell types of vertebrates; mammalian serum and plasma (associated with high density lipoprotein)	Ca^{2+}	Organophosphate esters
Epoxide hydrolases	Endoplasmic reticulum of animal cells; some in cytosol	None known	Organic epoxides
Reductases[a]	Endoplasmic reticulum and cytosol of several types of animal cells	NADH/NADPH	Organonitro compounds, some organohalogens, e.g., p,p´-DDT

[a] Many different enzymes can express activity at low levels of dissolved oxygen; include certain flavoproteins and hemoproteins.

The remainder of this section will cover the major enzymes concerned with biotransformation, using organic pollutants as examples. Table 5.2 lists the major classes of enzymes that metabolize xenobiotics in vertebrates and invertebrates (Gibson and Skett, 1986). Many of the enzymes responsible for phase I biotransformations are located in the endoplasmic reticulum, notably of the liver (vertebrates), hepatopancreas, fat body, and gut (invertebrates). Lipophilic xenobiotics tend to move into the endoplasmic reticulum, but their more polar biotransformation products tend to partition into the cytosol. Conjugating enzymes such as sulfotransferases and glutathione-S-transferases in the cytosol can then conjugate the metabolites. The conjugates are readily excreted in urine and/or bile. Thus the increase in polarity resulting from the sequential biotransformation shown leads to movement first from membrane to cytosol and then from cytosol to urine and/or bile. Although many of the types of enzymes shown in Table 5.2 are also found in plants, they exert fewer activities in comparison with those found in animals.

Of the enzymes responsible for phase I biotransformations, the microsomal monooxygenases (mixed function oxidases) are the most versatile (Waterman and Johnston, 1991; Lewis, 1996). They are found in all vertebrates and invertebrates in the endoplasmic reticulum of a variety of tissues. Low levels have been found in plants. Unlike other phase I enzymes, microsomal monooxygenases do not restrict their catalysis of biotransformations to any particular functional group. They can metabolize the great majority of lipophilic xenobiotics so long as they are not too large (large molecules would be unable to fit into the available space adjacent to the catalytic site). The main exceptions are the highly halogenated compounds such as certain PCBs and PBBs. Microsomal monooxygenase does not work effectively against C–halogen bonds (e.g., C–Cl, C–Br).

Oxidations by the microsomal monooxygenase system depend on the activation of molecular oxygen (O_2) after it is bound to an associated hemoprotein known as cytochrome P_{450}. Activation is accomplished by the transfer of electrons to bound O_2, which

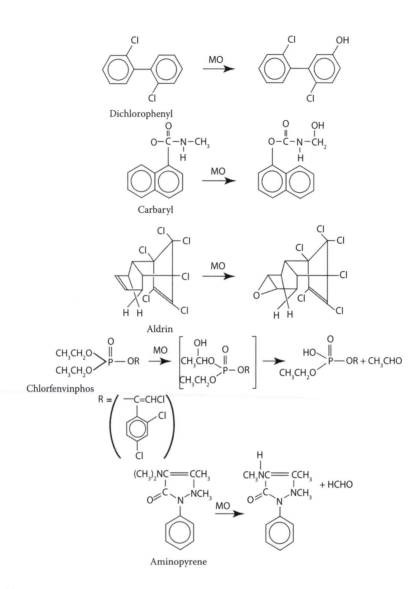

FIGURE 5.4A
Phase I biotransformations. MO = microsomal monooxygenase. (Continued.)

then splits; one atom is used to oxidize the substrate (e.g., organic pollutant) and the other to form water. Electrons for this purpose come from NADPH (sometimes from NADH). The overall reaction can be written as $NADPH_2 + X + O_2 \rightarrow XO + H_2O + NADP$ where X = substrate (Walker, 2009; Section 2.3.2.2).

The hemoprotein cytochrome P_{450} exists in many forms that have contrasting yet overlapping substrate specificities (Lewis, 1996; Livingstone and Stegeman, 1998; Nelson, 1998). Some forms of cytochrome P_{450} are readily inducible. The appearance of a lipophilic xenobiotic inside the body can trigger the synthesis of more cytochrome P_{450}, leading to more rapid biotransformation of xenobiotics. This is normally protective, enabling an organism to rapidly eliminate the xenobiotic that caused induction. However, in some cases (e.g., highly chlorinated PCBs), the chemical causing induction is not metabolized by the induced cytochrome P_{450}.

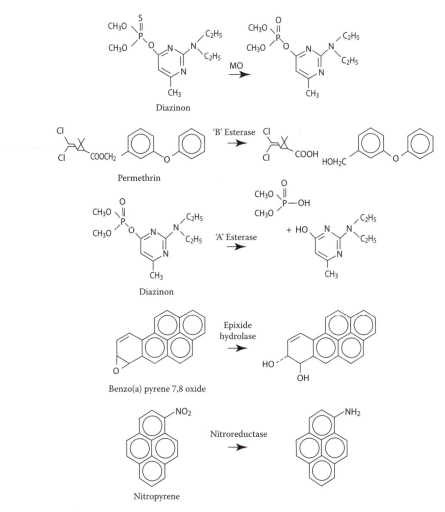

FIGURE 5.4B (CONTINUED.)
Phase I biotransformations. MO = microsomal monooxygenase.

Although oxidation by cytochrome P_{450} causes detoxification in most cases, often introducing hydroxyl groups to facilitate conjugation, there are some important exceptions to this rule. The oxidative desulfuration of organophosphorous insecticides that possess thion groups (diazinon, dimethoate, disyston, and malathion) leads to the formation of oxons which are active anticholinesterases (see Figure 5.4). Some organochlorine insecticides of the cyclodiene group are converted to stable and toxic epoxides (Figure 5.4). Thus aldrin is converted to dieldrin, and heptachlor to heptachlor epoxide. Also, a number of carcinogens are activated by the same system. Polycyclic aromatic hydrocarbons such as benzo(a)pyrene (Figure 5.4) and aflatoxin B are converted into epoxides that are strongly electrophilic and can bind to DNA. Nitrosamines are converted to methyl radicals and other reactive species when oxidized by cytochrome P_{450}. Vinyl chloride is also activated by cytochrome P_{450}. In all these examples, enzymatic attack causes an increase in toxicity rather than detoxification.

Hepatic microsomal monooxygenase (HMO) activity toward xenobiotic substrates varies substantially among groups, species, and strains. In omnivorous and herbivorous

BOX 5.1   ARYL HYDROCARBON (AH) RECEPTOR-MEDIATED TOXICITY

One particular form of cytochrome P_{450}, cytochrome P_{4501A}, metabolizes "flat" molecules such as polycyclic aromatic hydrocarbons (PAHs) and coplanar PCBs and can generate active metabolites such as the epoxides mentioned above and certain hydroxy PCBs that act as thyroxine antagonists (see Figure 7.6). Substrates of this type frequently interact with the so-called arylhydrocarbon (Ah) receptor in the cytosol, causing the induction of cytochrome $P_{450}1A$ (induction is discussed in Section 7.2). In this way, some potentially toxic molecules can promote their own activation! The interaction of coplanar PCBs and dioxins with the Ah receptor leads to a complex and poorly understood pattern of responses culminating in toxic effects and an increase in the quantity and activity of cytochrome $P_{450}1A$. This phenomenon is called Ah receptor-mediated toxicity (see Ahlborg et al., 1994; Safe, 2001; Walker, 2001). Examples of this type of toxicity occurring under field conditions are discussed in Sections 13.6.5 and 16.2.

mammals, activity is inversely related to body weight (Figure 5.5A); small mammals tend to have more activity per unit body weight than large mammals. Fish have relatively low HMO activities and show no clear relationship to body weight (Figure 5.5B). Birds exhibit variable activities. Omnivorous and herbivorous species have HMO activities similar to those of mammals of similar body size (on average a little lower). Activities of fish-eating birds and specialized predators (e.g., sparrowhawk, *Accipiter nisus*) are similar to those of fish (Walker, 1980, 1998a; Ronis and Walker, 1989). However, birds generally and fish-eating birds in particular show the same relationships between HMO activity and body weight observed in mammals (Walker, 1980).

These differences are explicable in terms of the detoxifying function of HMO. Terrestrial vertebrates depend on HMO for the effective elimination of lipophilic xenobiotics, but fish have far less dependence on metabolic detoxification because they excrete uncharged lipophilic molecules into ambient water by passive diffusion. (The same argument can be applied to amphibians that eliminate xenobiotics by diffusion across skin.) Small mammals have much higher surface area/body volume ratios than large mammals. Consequently, they ingest food and associated xenobiotics much more rapidly than large mammals do to obtain sufficient metabolic energy to maintain body temperature. Thus they need higher levels of detoxifying enzymes than large mammals because they take in xenobiotics more rapidly.

This argument does not apply to poikilotherms, so it is not surprising that the small amount of HMO of fish is not obviously related to body size, in contrast to the situation in mammals and birds. With specialized predators such as fish-eating birds and bird-eating raptors such as the sparrowhawk, the need for detoxification is small because their food does not contain many xenobiotics. Indeed, the food is similar in composition to the predators, especially in the case of bird-eating predators such as the sparrowhawk and peregrine falcon. Plants by contrast contain many compounds that act as xenobiotics to animals. Some of these compounds have the function of protecting plants against grazing. Thus the variations in HMO activity shown in Figure 5.5 can be explained in terms of the requirements of different species for detoxification of liposoluble xenobiotics.

Turning now to esterases, Aldridge (1953) distinguished A esterases that hydrolyze organophosphates and B esterases that are inhibited by them. Carboxyl esterases are B types and constitute an important group of detoxifying enzymes that hydrolyze lipophilic carboxyl

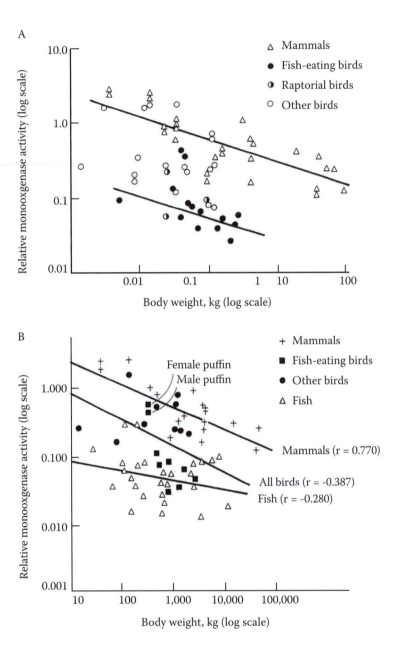

FIGURE 5.5
Relative monooxygenase activities of mammals, birds, and fish. (A) Mammals and birds. (*Source:* Ronis, M.J.J. and Walker, C.H. (1989). *Reviews in Biochemical Toxicology* 10, 310–384. With permission.) (B) Mammals, birds, and fish. Monooxygenase activities toward a range of substrates were measured in hepatic microsomes and expressed in relation to body weight. Relative activities were calculated in relation to those obtained for male rats using the same substrate; the value for a male rate was set at 1. Each point represents one species (males and females of a single species are sometimes entered separately).

esters to form acids and alcohols (Figure 5.4). They are widely distributed in nature, found in membranes (especially endoplasmic reticulum) from a variety of tissues in all animals so far investigated. They are also found in cytosol and in vertebrate plasma and serum. A number of forms have been recognized, showing contrasting but overlapping substrate specificities. Purified carboxyl esterases from mammalian liver endoplasmic reticulum metabolize both xenobiotic and endogenous esters. In mammals, the range of different forms of carboxyl esterases differs considerably among tissues. Liver, for example, contains a wide range of forms; blood and muscle have smaller numbers of forms.

The A esterase enzymes can hydrolyze organophosphorous triesters and diesters that possess oxon (P=O) groups (Figure 5.4B). These esters are lipophilic and do not appear to hydrolyze organophosphates that are ionized. The known substrates are mainly organophosphorous insecticides, although some forms of A esterase can hydrolyze nerve gases (e.g., soman and tabun). In vertebrates, A esterase activity is found in the endoplasmic reticulum of livers and other tissues. In mammals, activity is also found in serum and plasma, some of it in association with high density lipoprotein. The A esterases take a number of forms. Those that hydrolyze organophosphorous insecticides are calcium dependent and lose their activity in the presence of chelating agents that bind calcium.

Species show marked differences with regard to A esterase activities. In contrast to mammals, birds have little or no serum or plasma A esterase (Walker and Thompson, 1991; Walker et al., 1991b). Some species of insects have no measurable A esterase activity.

Epoxide hydrolases are found principally in the endoplasmic reticulum of animals; mammalian liver is a particularly rich source. A different form of the enzyme is found in cytosol. Epoxide hydrolases hydrate a wide range of aromatic and aliphatic epoxides to form trans diols and do not require cofactors (Figure 5.4). Epoxide hydrolases of the endoplasmic reticulum hydrate epoxides generated by microsomal monooxygenases. In some cases, this represents a protective function, because the epoxides are strong electrophiles that can form adducts with cellular macromolecules (see Chapter 7). Epoxide hydrolases can metabolize endogenous and xenobiotic substrates. Epoxides of steroids and of insect juvenile hormone are examples of endogenous substrates. As with monooxygenases and esterases, these enzymes generate metabolites with hydroxyl groups that are then available for conjugation.

Reductases are enzymes that can catalyze the transfer of electrons to organic molecules such as nitroaromatic and organohalogen compounds (Figure 5.4). Whether enzymes operate as reductases often depends on the availability of oxygen. Freely available oxygen can act as an electron acceptor (i.e., electrons flow to oxygen rather than to the organic molecule). Both flavoproteins and hemoproteins (e.g., cytochrome P_{450}) have been shown to act as reductases when oxygen levels are low. The flow of electrons to oxygen leads to the formation of oxygen radicals (e.g., superoxide anion and hydroxyl radical) that can cause cellular damage. This question will be discussed later. When nitroaromatic compounds are reduced, they are converted to amines. Reduction of halogenated compounds leads to dehalogenation (e.g., conversion of p,p'-DDT to p,p'-DDD).

Of the phase II enzymes responsible for conjugation, glucuronyl transferases are largely confined to membranes—especially the endoplasmic reticulum of vertebrate livers. These enzymes catalyze the interactions of lipophilic organic molecules (xenobiotic and endogenous) with labile hydrogen atoms (usually of hydroxyl groups) and glucuronic acid to form glucuronides (Figure 5.6). The glucuronic acid is supplied by a nucleotide cofactor, uridine diphosphate glucuronic acid (UDPGA), synthesized in the cytosol. Thus the glucuronyl transferase enzyme and its associated lipophilic substrate receive the polar co-factor from the hydrophilic environment of the cytosol. Glucuronides exist largely as anions at cellular

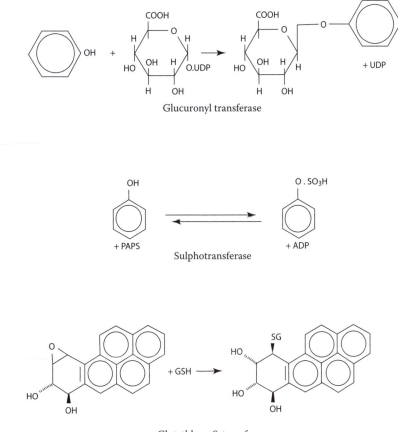

Glucuronyl transferase

Sulphotransferase

Glutathlone-S-transferase

FIGURE 5.6
Phase II biotransformations.

pH (approximately 7.4) and move away from the membrane and into the cytosol when they are formed (Table 5.3).

Glucuronides are subject to hydrolysis by glucuronidases—these enzymes occcur in the alimentary tract—and can reverse the reaction shown in Figure 5.6 with a release of the hydroxyl compound that was originally conjugated. The consequences of this will be covered when we discuss excretion. Glucuronyl transferases, like other enzymes concerned with xenobiotic metabolism, exist in a number of forms.

Sulfotransferases exist in the cytosols of various cell types (especially liver) in a number of forms. They catalyze the transfer of the sulfate group from phosphoadenine phosphosulfate (PAPS) to xenobiotic and endogenous substrates that possess free hydroxyl groups (Figure 5.6). The co-factor PAPS is generated in the cytosol. The resulting sulfate conjugates exist as anions with appreciable water solubility. Like glucuronides, sulfate conjugates are subject to enzymic hydrolysis, in this case catalyzed by sulfatases.

Glutathione-S-transferases are found in the cytosol of many cell types, especially vertebrate livers. A number of forms are known in vertebrates and invertebrates. These enzymes catalyze the conjugation of reduced glutathione to a variety of xenobiotics that are electrophilic (Figure 5.6). This conjugation would proceed naturally but very slowly in

TABLE 5.3

Enzymes that Metabolize Xenobiotics (Phase II)

Enzyme	Principal Location	Co-Factors	Substrates
Glucuronyl transferases	Endoplasmic reticulum of many animal cells	UDP-glucuronic acid	Organic compounds with free OH groups; some organic compounds with free –SH or NH_2
Sulfotransferases	Cytosol of many animal cells	Phosphoadenine phosphosulfate	Organic compounds with free OH groups
Glutathione-S-transferases	Cytosol of many types of animal cells; endoplasmic reticulum	Reduced glutathione	Foreign electrophiles, including some organohalogens and organic epoxides

the absence of the enzyme. The enzyme speeds the reaction by binding the reduced glutathione in close proximity to the xenobiotic.

Conjugations with glutathione do not depend on the presence of hydroxyl groups. Important substrates include certain organohalogen compounds and organophosphorous insecticides (Walker, 2009). The glutathione conjugates so formed often undergo further modification before excretion. In vertebrates, further metabolism of the glutathione moiety leads to the formation of mercapturic acid conjugates that are usually the predominant excreted forms. Both glutathione conjugates and the mercapturic acids derived from them are anionic.

Conjugation in phase II metabolism promotes excretion and, with few exceptions, fulfills a detoxifying function (see Section 5.1.6 for further discussion). The conjugating enzymes discussed here are very important in vertebrates and invertebrates. However, a number of other types of known conjugates are often group specific. This is particularly true of peptide conjugation.

Before leaving the subject of metabolism, we should mention oxyradical formation. As discussed earlier when we described reductases, electrons can be passed on to molecular oxygen, causing the generation of highly reactive oxyradicals. Examples include the superoxide anion ($•O_2^-$), the hydroxyl radical ($•OH$), and hydrogen peroxide (H_2O_2). Descriptions of the routes by which these are formed lie outside the scope of this book. The significant point from an ecotoxicology view is that pollutants may promote the formation of such oxyradicals and thus lead to cellular damage. Some organic pollutants (e.g., paraquat herbicide and nitroaromatic compounds) may pass electrons on to oxygen to form oxyradicals. Additionally, certain refractory compounds may stabilize superoxide (e.g., PCBs which bind strongly to cytochrome P_{450} but are not themselves metabolized; they combine with cytochrome P_{450}). Because such molecules cannot interact with the oxyradicals, the lives of the oxyradicals are prolonged, thus increasing the chance that they will escape the domain of the P_{450} to cause cellular damage elsewhere).

Cells possess enzyme systems that detoxify oxyradicals such as superoxide dismutase, catalase, and peroxidase. However, they are not necessarily able to cope with increased rates of formation of oxyradicals caused by the actions of organic pollutants. This is an important but difficult area of ecotoxicology about which far too little is known.

5.1.6 Sites of Excretion

The excretion of xenobiotics and their metabolites and conjugates has been studied in some detail in vertebrates. Far less is known about these processes in invertebrates. Insects have

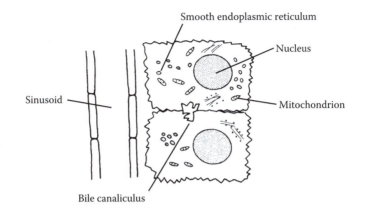

FIGURE 5.7
Excretion in bile. Transverse section of liver cell showing bile canaliculus.

received more attention than other groups. This section mainly covers vertebrates, with some references to invertebrates.

Many aquatic vertebrates can excrete lipophilic xenobiotics by diffusion into ambient water. Fish can do this across gills, and amphibia such as frogs excrete across permeable skin. Aquatic birds have no permeable membranes in contact with water; skin and feathers appear not to be readily permeable to pollutants. The skins of aquatic mammals (whales, porpoises, and seals) seem to be relatively impermeable to such compounds.

The effective elimination of lipophilic xenobiotics by terrestrial vertebrates and invertebrates depends upon converting the compounds into water-soluble metabolites and conjugates that can be excreted (some highly lipophilic compounds are excreted to a limited extent in milk by mammals and in eggs by birds, reptiles, and invertebrates). Vertebrates excrete these biotransformation products in bile and/or urine.

Considering excretion in urine first; soluble metabolites and conjugates are removed from blood in the glomerular filtrate that then passes along the proximal and distal tubules of the kidney. During this movement along the renal tubules, some xenobiotics pass into the tubular lumen from the plasma by passive diffusion. Also, some active transport of weak acids and bases may occur across the walls of the proximal tubules into the lumen and some reabsorption of xenobiotics from the tubular lumen into the plasma may occur. Eventually, urine collects in the bladder and is voided independently of the feces in the case of mammals but is combined with feces in birds, reptiles, and amphibia.

Conjugates and metabolites may also be excreted via bile. They first move across the plasma membranes of the hepatocytes into bile canaliculi (Figure 5.7). Bile then passes into the main bile duct and is eventually released into the duodenum where conjugates may pass completely through the alimentary tract to be voided with feces. As conjugates are usually polar (most are anions), they are not readily reabsorbed by passive diffusion across the wall of the alimentary tract.

If, on the other hand, they are broken down in the gut, e.g., by the action of enzymes such as glucuronidases and sulfatases, the metabolites (or sometimes original xenobiotics) so released may be reabsorbed into the bloodstream by passive diffusion because they are uncharged and can readily cross membranous barriers. The reabsorbed compounds are then returned to the liver and reconjugated and the cycle is repeated. This process is known as enterohepatic circulation. Some xenobiotic metabolites may be recycled many

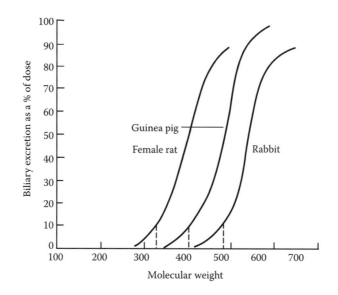

FIGURE 5.8
Excretion routes of anionic conjugates. (*Source:* Moriarty, F., Ed. (1975). *Organochlorine Insecticides: Persistent Organic Pollutants.* Academic Press. With permission.)

times before they finally appear in feces. Enterohepatic circulation may have toxicological consequences. Recycled metabolites may have toxic effects. Also, metabolites that have low toxicities may be transformed into toxic secondary metabolites in the gut (e.g. by microbial action). In mammals, excretion via urine does not raise the problem of enterohepatic circulation or further biotransformation in the gut.

The extent to which excretion occurs in bile or urine depends on the molecular weight of the xenobiotic and the species in question. As noted above, most excreted conjugates are organic anions. When their molecular weights are below 300, excretion is predominantly in urine. Above molecular weight 600, it occurs mainly in bile. Between these limits, the preferred rate of excretion depends on the species and the molecular weight of the anionic conjugate (Figure 5.8). Threshold molecular weights have been proposed for anionic conjugates in different species (Figure 5.8). These are weights above which there is likely to be appreciable excretion into bile (< 10% total excretion in bile). Certain weights have been proposed: rat 325 (± 50), guinea pig 425 (± 50), and rabbit 475 (± 50). Thus rats show a greater tendency than rabbits to excrete via bile and may thus be more susceptible to the toxic actions of certain compounds for the reasons stated above.

The situation in terrestrial invertebrates appears similar to that for vertebrates, except that the hepatopancreas or fat body serves the function of the liver.

5.1.7 Toxicokinetic Models

The processes by which pollutants are taken up, distributed, stored, metabolized, and excreted are summarized in Figure 5.1. From a toxicological view, a living organism contains five kinds (categories) of sites, each of which is ascribed a different function—uptake, storage, toxic action, metabolism, and excretion. This simple model is purely descriptive and not quantitative. It does not define the rates at which transfers and biotransformations occur.

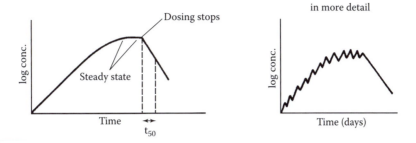

FIGURE 5.9
Kinetics of uptake and loss. When an animal is continually dosed with a chemical, the log concentration increases until a steady state is reached. When dosing stops, the log concentration falls linearly over time if first-order kinetics apply (as in a one-compartment model).

In quantitative toxicokinetics, the central issue centers on the rates at which processes proceed, the determination of which can lead to the development of both descriptive and predictive models. In toxicokinetic terms, an organism can be divided into compartments that usually represent specific organs and tissues. Each compartment contains a discrete quantity of xenobiotic that is subject to a particular rate of transfer and biotransformation. These compartments may or may not correspond to the sites defined in Figure 5.1.

Multicompartmental models represent ideals that are seldom realized in practice. Although it is desirable to have maximum information about the kinetics of specific xenobiotics in compartments of the body that are of toxicological interest, a large amount of work is involved in determining kinetic constants even for a single pollutant or species. In ecotoxicology, such an approach is clearly of little value because the discipline covers a large range of organisms and organic pollutants. Interest is largely restricted to simple models that treat a whole organism as a simple compartment. The following account will be concerned principally with these models.

If an organism is continuously exposed to a constant level of an organic pollutant, the concentration of the pollutant in the whole organism will increase over time (Figure 5.9). An aquatic organism may be exposed to a pollutant dissolved in ambient water; a terrestrial organism may ingest a pollutant in its food. Initially, the rate of increase in the tissue concentration of a pollutant will be rapid, but it will then begin to tail off and will eventually reach a plateau, so long as a lethal concentration is not reached before that. When the pollutant concentration reaches a plateau, the system is said to be in a steady state. The rate at which a pollutant is taken up is balanced by the rate at which it is lost via metabolism and/or direct excretion. If exposure to the pollutant ceases, its tissue concentration will begin to fall. In the simplest situation in which the kinetics of loss can be described by a single rate constant (one-compartment model), the rate of loss is proportional to tissue concentration (Figure 5.10A). First-order kinetics apply and we can use the equation:

$$\frac{-dC}{dt} = KC \text{ or } C_t = C_o e^{-Kt}$$

where C is the concentration in the whole organism, C_0 is the concentration when exposure ceases, C_t is the concentration at time t, K is the rate constant for loss, and $-dC/dt$ is the rate of loss of pollutant from the organism. This negative exponential decline in concentration is illustrated in Figure 5.10B. It follows that the log concentration of the xenobiotic (log C)

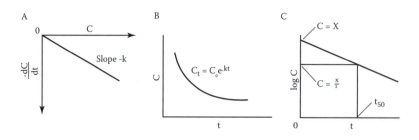

FIGURE 5.10
Kinetics of loss. (A) Rate of loss of chemical is proportional to tissue concentration. (B) Tissue concentration of chemical falls exponentially over time. (C) Log of tissue concentration of pollutant falls linearly over time. C = tissue concentration. t = time (minutes or hours). t_{50} = half-life. x = initial tissue concentration.

is linear with respect to time (Figure 5.10C). The biological half-life (t_{50}) is the time it will take for the concentration to fall by one-half, and this can be readily determined from the plot of log C versus time.

Sometimes a whole organism does not behave as a single compartment. In such cases, more complex equations are needed to describe the rate of loss after cessation of exposure. An example of a two-compartment model of PCB toxicokinetics in wild birds can be found in Drouillard et al. (2001). The model is calibrated for male American kestrels but can also be used where seasonal fluctuations in fat content are known. Sometimes the rate of loss is biphasic and more complex modelling is required to describe the loss. An example from inorganic metabolism can be found in the nickel toxicokinetics of carabid beetles and earthworms (Laskowski et al., 2009).

Figure 5.9 represents a simplified one-compartment model that assumes that the rate of uptake of xenobiotic is constant and that the state of the organism remains the same. The organism is continuously exposed to a xenobiotic until a steady rate is reached, after which dosing is discontinued. In practice, the rate of uptake is not absolutely constant. For example, if the source of the xenobiotic is food, diurnal variations in the rate of ingestion will occur. The activities of enzymes may change because of induction or inhibition with consequent change in the rate at which a xenobiotic is lost—and ultimately in the steady-state concentration for any defined rate of uptake of xenobiotic (Figure 5.10). In reality, a change of xenobiotic concentration over time is usually more complex than the simple situation shown in Figure 5.9.

5.1.8 Toxicokinetic Models for Bioconcentration and Bioaccumulation

Toxicokinetic models have been developed to describe and predict the degrees of bioconcentration or bioaccumulation of organic pollutants in animals continuously exposed to xenobiotics in the steady state (Moriarty, 1975, 1999; Moriarty and Walker, 1987; Walker, 1990a). It is important to emphasize the value of steady-state models. The tissue concentrations in the steady state represent the maximum levels expected based on a certain level of exposure. Further, they are not time dependent. Bioconcentration or bioaccumulation factors determined before the steady state is reached are of little value; they apply only to particular periods of exposure and do not indicate the maximum concentration that may be reached. The following definitions will be used for bioconcentration factor (BCF) and bioaccumulation factor (BAF):

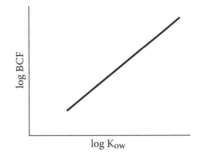

FIGURE 5.11
Bioconcentration factors and K_{ow} values.

$$BCF = \frac{\text{conc. in organism}}{\text{conc. in ambient medium}}$$

$$BAF = \frac{\text{conc. in organism}}{\text{conc. in food (or ingested water)}}$$

For the reasons stated, these factors should be determined when a system is in the steady state. Bioconcentration factors are important in aquatic ecotoxicology, where ambient media represent major sources of organic pollutants. Bioaccumulation factors are critical in terrestrial ecotoxicology where food and ingested water are major sources of organic pollutants.

For aquatic organisms (e.g., the edible mussel, *Mytilus edulis*), a close relationship has often been demonstrated between bioconcentration factors for lipophilic organic pollutants and their K_{ow} values (Figure 5.11). A simple model applies. Most of the uptake and loss of xenobiotic are accomplished by passive diffusion, with metabolism playing only a minor role. At the steady state, the BCF is determined by the partition between water and the hydrophobic components of the organism. With nonpolar xenobiotics of high lipophilicity, the partition lies very much in the direction of the organism. In other words, high BCF values are associated with high values for K_{ow}. (Sometimes, K_{ow} values are transformed to pK_{ow} values by converting them to $\log_{10}$ values.)

In more complex situations in which other sources of uptake and loss become more important, this simple model will break down. Thus with fish, BCF values for certain pollutants fall well below the values predicted by a plot such as Figure 5.11 because they are rapidly metabolized, i.e., they are less than would be expected from K_{ow} values. The rate of loss is then greater than would be expected from passive diffusion alone.

For terrestrial organisms, a different approach is required to predict BAFs. In the simplest situation, both uptake and loss may be ascribed to single processes, each of which will be governed by one rate constant (Figure 5.9). With strongly lipophilic compounds, food may represent the main source of pollution, whereas metabolism may represent the main source of loss (Table 5.4; Walker 1987, 1990a). It should be possible to develop predictive models for bioaccumulation in the steady state using rate constants for these two processes so long as metabolism is simple. As yet, however, no such models have been validated.

There is much interest in predictive models that will enable reasonable assessments of the extents with which various organic pollutants may be bioconcentrated or bioaccumulated by different organisms. Such models could be of great assistance in assessing risks

TABLE 5.4

Model System for Bioaccumulation of Lipophilic Pollutants

Organism Type	Uptake Routes			Loss Routes	
	Diffusion	Food	Water	Diffusion	Metabolism
Aquatic					
Mollusks	+++++	(+)		+++++	
Fish with substantial enzyme activity	++++	+ → +++		++++	+ → +++
Terrestrial					
Various predators		++++	(+)		+++++

Note: The importance of routes of uptake and loss is indicated by a scoring system on a scale of + to +++++.
Source: Walker, C.H. (1975). *Environmental Pollution by Chemicals*, 2nd ed. Hutchinson. With permission.
More may be learned about one compartment models in ecotoxicology by consulting the web site www.es.ucsb.edu/faculty/muller/toxicokineticses120

of environmental chemicals (Chapter 6). They would need to be simple and economical to operate. The major routes of uptake and loss of organic pollutants by organisms from terrestrial and aquatic habitats are summarized in Table 5.4. The routes must be considered in developing predictive models. To date, the only model that meets these criteria is the one that uses K_{ow} values to predict bioconcentration by certain aquatic organisms.

More may be learned about one-compartment models in ecotoxicology by consulting www.es.ucsb.edu/faculty/muller/toxicokineticses120.

5.2 Organic Pollutants in Terrestrial Ecosystems

The routes by which the pollutants reach the land surface and terrestrial animals and plants were described in Chapter 2. Pollutants may remain on a land surface, enter terrestrial food chains, or be transported to air or water. Contamination of the land surface and soil occurs from dumping at landfill sites and in the areas surrounding industrial operations. Agricultural soils are of particular concern because they are treated with pesticides that are designed to be highly toxic to certain types of organisms. After pesticides enter soils, questions arise about residues in crops, contamination of drinking water, and possible effects on soil organisms and soil fertility. The fate of organic pollutants in soil will be discussed before we deal with their movement through terrestrial food chains.

5.2.1 Fate in Soils

Pesticides are the most important pollutants in agricultural soils, and these may reach the soil directly or by transfer of residues from plants to which they have been applied. Pesticides are applied as sprays, granules, or dust. Some pesticides are sufficiently soluble to be fully dissolved in spray water, but in most cases they are formulated as emulsifiable concentrates (dissolved in an oily liquid) or wettable powders (fine particles mixed with an inert diluent) because of their low water solubility. In the latter case, droplets (emulsifiable concentrate) or particles (wettable powder) are suspended in water. The availability of the pesticide is dependent on formulation and on rates of release from particles, droplets, or granular formulations.

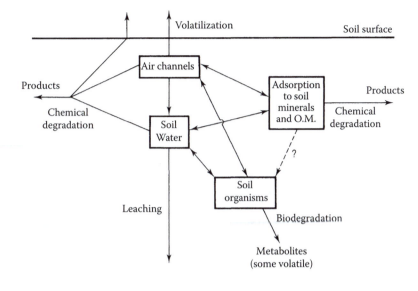

FIGURE 5.12
Fates of pollutants in soil.

Apart from pesticides, soils are sometimes contaminated by hydrocarbons, PCBs, and other industrial chemicals. Such compounds are usually far less toxic than pesticides, and pollution problems associated with them are more localized. The fate of pesticides and other organic pollutants in soils will now be discussed before we consider their transfer along food chains.

Soils are complex associations containing living organisms, mineral particles, and organic matter. The clay fractions of the minerals and the so-called humus of the organic matter are colloids (diameter < 0.002 mm) which, because of their small size, have large surface area: volume ratios. When organic compounds enter soils, they become distributed through soil water, soil air, and the available surfaces of soil minerals and organic matter. Where they are in the form of or associated with particles or droplets, some time will elapse before the individual molecules are distributed among these compartments, for example, when pesticides formulated as wettable powders are applied as sprays.

The distribution of organic compounds in soil depends on their physical properties, especially solubility (e.g., K_{ow} values), vapor pressure, and chemical stability (Figure 5.12). Regarding chemical stability, organic pollutants are broken down by hydrolysis, oxidation, isomerization, and by the action of light (photochemical breakdown) if they are on the surface. Usually, breakdown leads to loss of toxicity, but occasionally the products are already highly toxic (e.g., the isomerization of organophosphorous malathion to isomalathion). Polar compounds (hydrophilic compounds of low K_{ow}) tend to dissolve in water and are adsorbed to soil colloids only to a limited degree. At times organic compounds that exist as ions serve as exceptions to this rule because they become strongly associated with sites on soil colloids that bear opposite charges. Thus the paraquat herbicide exists as a cation and binds strongly to negatively charged sites of clay minerals. Conversely, compounds of low water solubility (high K_{ow}) tend to become strongly adsorbed to the surfaces of clay and soil organic matter and exist only at very low concentrations in soil water. Compounds of high vapor pressure tend to volatilize into the soil air or directly from the soil surface into the atmosphere. If they enter soil air, they may be retained in the soil for some time but will eventually pass into the atmosphere.

The binding of organic molecules to soil colloids restricts their movement in the soil and their availability to soil organisms. Thus compounds of high K_{ow} (i.e., lipophilic compounds) applied to soil show little tendency to be leached down through the soil profile by water. Some soil-acting herbicides (e.g., simazine) exhibit depth selection in their herbicidal actions because of this phenomenon. They are only toxic to surface-rooting weeds, because they are not carried far enough down the profile to be taken up by deeper rooting plants. Limitation of availability to soil organisms restricts the rates of biotransformation and the toxicities of lipophilic compounds. Thus chemically stable lipophilic compounds often have long half-lives in soil because they are tightly bound to clay and/or organic matter and are metabolized very slowly.

By contrast, hydrophilic compounds (e.g., soluble herbicides such as 2,4-D and MCPA) that are not strongly bound to soil clay or organic matter move more freely in soil and are readily available to soil organisms. They tend to be carried down the soil profile by water. Also, they tend not to be very persistent because soil organisms metabolize them relatively rapidly.

Active forms of some organochlorine insecticides such as *p,p´*-DDT and dieldrin are lipophilic compounds that are only slowly metabolized. Even when freely available, they are metabolized slowly because they are poor substrates for detoxifying enzymes. The loss of compounds such as these from soils is biphasic (Figure 5.13). Application of a pesticide is immediately followed by a period of relatively rapid loss if the pesticide is present as particles or is dissolved in an oily solvent. It is volatilized or simply blown away as dust (Edwards, 1976). During this period, molecules of the insecticide become adsorbed to soil colloids and the initial rapid loss is succeeded by a period of slow exponential loss because the adsorbed insecticide only slowly becomes amenable to loss by evaporation or metabolism. During the period of slow loss, the half-lives for these compounds can run into years, even tens of years (see Table 5.5). The half-lives depend on the compound, soil type, and climate. Compounds with higher vapor pressures tend to be lost more rapidly than compounds of lower vapor pressures. Rapidly biotransformed compounds tend to disappear more quickly than compounds that resist biotransformation. Persistence tends to be greatest in heavy soils that are high in clay and/or organic matter. Persistence is also favored by low temperatures. Under tropical conditions, rates of loss due to volatilization, chemical breakdown, and biotransformation tend to be faster than in cooler, more temperate conditions.

Hydrophilic compounds such as the 2,4-D and MCPA herbicides and the carbofuran insecticide show a fundamentally different pattern of loss from that just described (Figure 5.13). Immediately after application to soil, the rate of loss is relatively slow. However, after a lag phase, the rate accelerates and the compound is quickly lost. Typically, it disappears within days or weeks. If more of the compound is immediately added to the soil, it also disappears quickly. The soil has become enriched. Strains of microorganisms that can metabolize the organic compound have developed in such soils. If no more compound is added to the soil, the enrichment will be lost and the soil will revert to its original state. The enrichment phenomenon may be a problem; certain pesticides lose their effectiveness if they are used too often (carbofuran is a case in point). The fates of organic compounds in soil—movements and decay curves—have been predicted with some success using models that incorporate parameters such as K_{ow} and vapor pressure.

5.2.2 Transfer along Terrestrial Food Chains

Organisms in terrestrial ecosystems may take up pollutants from their food and ingested water or directly from air, water, or solid surfaces with which they come into contact.

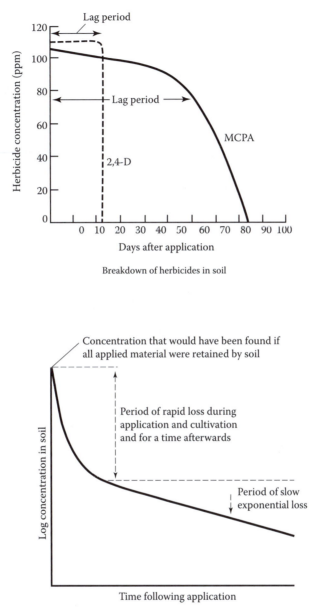

FIGURE 5.13
Loss of pesticides from soil. (*Source:* Walker, C.H. (1975). *Environmental Pollution by Chemicals,* 2nd ed. Hutchinson. With permission.)

Uptake from ambient water is not as important a route as it is for aquatic organisms, although certain soil organisms (e.g., earthworms) may acquire pollutants this way. Uptake from ingested food or water is a very important route of uptake—often the major route for terrestrial vertebrates. The passage of organic pollutants and their stable biotransformation products along food chains are matters of great importance.

TABLE 5.5

Persistence of Organochlorine Insecticides in Soils

Compound	50% Loss (Years)	95% Loss (Years)
Dieldrin	0.5 to 4	4 to 30
p,p′-DDT	2.5 to 5	5 to 25
Lindane (γ-BHC)	Approximately 1.5	3 to 10

Note: As the rate of loss from soils falls into different phases, true half-lives cannot be determined. Estimates have been made of the time required for a specific percentage of an applied dose to disappear.

Source: Walker, C.H. (1975). *Environmental Pollution by Chemicals,* 2nd ed. Hutchinson. With permission.

The passage of compounds that have long biological half-lives along a food chain may lead to biomagnification at some or all stages. Typically, the highest concentrations of pollutants are found in predators of the higher trophic levels of the food pyramid. Also, because of the mobility of some animals (especially birds), the pollutants may be transported to areas far removed from the points where they were originally released, e.g., between Africa and Europe. For a discussion of the problem, see Balk and Koeman (1984).

Relatively high levels of persistent lipophilic pollutants in terrestrial predators have often been reported. Examples include organochlorine insecticides (Figure 5.14), PCBs, and methyl mercury. Reliable data about concentrations of persistent pollutants in organisms of different trophic levels of the same terrestrial ecosystem at the same time are insufficient.

Levels of persistent organochlorine insecticides were determined in organisms from different trophic levels of terrestrial ecosystems in Britain during the 1960s. Figure 5.14 shows the example of birds from different trophic levels. Like data from the marine environment (Figure 5.15; Section 5.3), these point to a general upward trend in residues of compounds such as p,p′-DDE (a metabolite of p,p′-DDT) and dieldrin with movement along the food chain. However, there is a danger of oversimplifying the complex situation that existed when very large temporal and spatial variations in exposure to these compounds were found in the field. For example, grain-eating birds acquired lethal doses of dieldrin in some areas (e.g., agricultural areas of eastern England), where they could feed on grain that had been dressed with it (Turtle et al., 1963; Moore and Walker, 1964); in nearby areas where the chemical dieldrin was not used, the exposure of the same species was almost zero.

Also, biomagnification did not necessarily occur at each step in the food chain, even for highly persistent compounds. Dressed seed might carry a dieldrin residue of 100 μg/g but a grain-eating bird or mammal could not normally build up such a concentration in its tissues because lethal dieldrin poisoning would normally occur at levels well below this. Further, species such as the sparrowhawk and peregrine tend to feed selectively. In the case under consideration, they may have tended to select prey with the highest residue concentrations in their tissues, i.e., individuals showing sublethal symptoms of poisoning. They may have been exposed not to average concentrations of pollutant in the tissue of the prey species but to the highest concentrations in surviving members of that species.

One important issue is the tendency of predators to bioaccumulate persistent pollutants with long biological half-lives present in prey when exposed to the pollutants over long periods. Data suggest that predatory birds exposed to compounds such as dieldrin

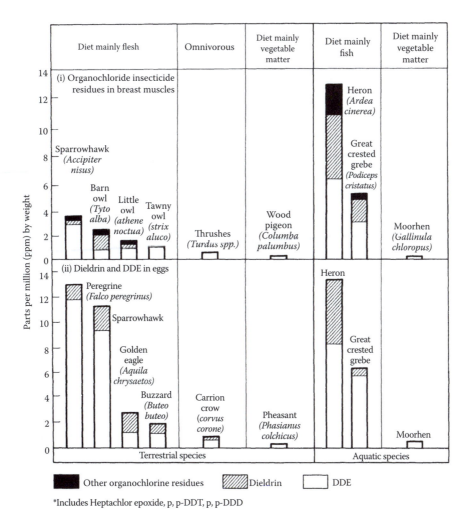

Parts per million (ppm) by weight

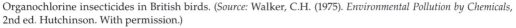

FIGURE 5.14
Organochlorine insecticides in British birds. (*Source:* Walker, C.H. (1975). *Environmental Pollution by Chemicals,* 2nd ed. Hutchinson. With permission.)

or *p,p′*-DDE can achieve bioaccumulation factors of 5- to 15-fold in relation to their prey if a steady state is reached. As noted earlier, predatory birds are thought to be particularly efficient bioaccumulators of lipophilic xenobiotics because they have poorly developed oxidative detoxification systems.

In summary, certain persistent lipophilic pollutants are transferred along terrestrial food chains, sometimes reaching their highest concentrations in predators of the highest trophic levels. This constitutes a major problem with certain pesticides used as seed dressings (aldrin, dieldrin, methyl mercury). The concentration at the first step of the food chain is already high and grain-eating species can quickly acquire lethal doses if they feed on dressed grain. The persistence of most of these compounds is due to a combination of high lipophilicity (high K_{ow}) and resistance to metabolic detoxification (especially in species such as specialized predators that are already deficient in detoxifying enzymes). Lipophilic compounds that are readily metabolized tend not to be persistent in terrestrial

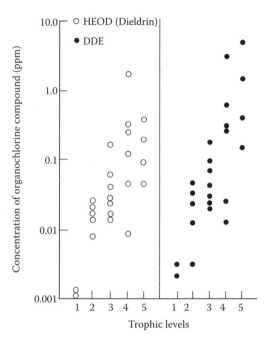

FIGURE 5.15

Organochlorine insecticides in the Farne Island ecosystem. (*Source:* Walker, C.H. (1975). *Environmental Pollution by Chemicals,* 2nd ed. Hutchinson. With permission.)

animals. Any toxic effects appear to be acute—of short duration and limited to the species immediately exposed and the area where the compounds are released.

5.3 Organic Pollutants in Aquatic Ecosystems

As in the case of soils, the fate of organic pollutants entering aquatic ecosystems depends on their physical properties, especially lipophilicity, vapor pressure, and chemical stability. Compounds that lack stability, for example, those that tend to be hydrolyzed, present few problems as aquatic pollutants unless their transformation products are toxic. Also, volatile compounds tend not to persist in aquatic ecosystems. Polarity is again important in determining distribution and persistence. Hydrophilic compounds tend to be dissolved in and distributed throughout surface water. Lipophilic compounds tend to become associated with particulate matter, notably of sediments. They may also exist on water surfaces, e.g., dissolved in surface oil films.

5.3.1 Pollutants in Sediments

Like soils, the sediments of rivers, lakes, and seas are associations of organic and inorganic particles and living organisms. Organic pollutants may be adsorbed to the particles of sediments, and this limits their mobility and availability to bottom-dwelling organisms. It is often uncertain to what extent pollutants can be taken up directly by animals from the adsorbed state or whether the pollutants must first move into aqueous media before they become available. The extent to which an adsorbed pollutant can be taken up directly depends on the nature of the chemical, the nature of the surfaces to which it is bound, the strength of the binding, the species taking it up and, in some cases the temperature, pH, and oxygen content of the ambient water.

The oxygen content of water can be important in determining the natures and rates of chemical and biochemical transformations. As the oxygen content declines, the tendency is for oxidative transformations to be replaced by reductive ones (Section 5.1.5). Sediments can differ widely in oxygen content. At the bottoms of deep seas, conditions are anaerobic; in shallow fast-moving streams oxygen levels are relatively high. Sediments of intertidal zones along the seashore experience fluctuating oxygen levels based on tidal movements.

5.3.2 Transfer along Aquatic Food Chains

The same general considerations apply to both aquatic and terrestrial food chains (Section 5.2.2) except that exchange diffusion between organisms and ambient water is a complicating factor for aquatic food chains. This makes the interpretation of data such as those in Figure 5.15 more difficult. The residue levels shown for the persistent organochlorine compounds dieldrin and p,p'-DDE (a metabolite of p,p-DDT) were measured in organisms sampled from different trophic levels of the marine ecosystem around the Farne Islands, Northumberland, UK from 1962 to 1964 (Robinson et al., 1967). This shows a strong relationship between the log concentrations of the residues and the trophic levels of the organisms in which they were determined. The lowest levels are in the plants (brown algae) in trophic level 1; the highest are in vertebrate predators of trophic levels 4 and 5.

It is interesting that p,p'-DDE is subject to a steeper gradient than dieldrin. p,p'-DDE is metabolized more slowly and has a longer biological half-life than dieldrin in the animals studied. It appears as if biomagnification occurred at every stage of the food chain. However, aquatic invertebrates and fish at trophic levels 2 to 4 obtained unknown proportions of their residue burdens directly from water and/or sediment and not from food. Also, as noted earlier, selective predation may occur. Eaten prey may contain more organochlorine residues than the average value shown in Figure 5.15. These points aside, the predators appear to be achieving a substantial biomagnification (BAFs considerably greater than 1, assuming that most of the pollutant burden comes from food). The apparent BAF calculated from the data in Figure 5.15 lies between 50 and 60 for the fish-eating shag (*Phalacrocorax aristotelis*; Walker, 1990b). The variation in residue levels in the principal prey species, the sand eel (*Ammodytes lanceolatus*), was not large. Even if the animals fed on the most contaminated individuals, a several-fold BAF is still suggested.

The fish-eating birds shown to bioaccumulate substantial levels of persistent organochlorine compounds in this and other studies revealed low activities of their monooxygenase systems that are mainly responsible for their detoxification (Ronis and Walker, 1989; Walker, 1990b; Figure 5.5). The species in question include the shag, cormorant (*Phalacrocorax carbo*), guillemot (*Uria aalge*), and razorbill (*Alca torda*). In general, a deficiency in a detoxification system favors bioaccumulation.

There is a general concern about the build-ups of relatively high levels of persistent pollutants in predators at the top of aquatic food chains (for a review, see Walker and Livingstone, 1992). In addition to the organochlorine insecticides, persistent polychlorinated biphenyls (PBBs) and polychlorinated dibenzodioxins (PCDDs) have also created concern. In addition to sea birds, marine mammals such as seals and cetaceans (porpoises, dolphins, and whales) have been shown to bioaccumulate these compounds and to pass them to their offspring via milk. Like sea birds, predatory marine mammals have poorly developed monooxygenase detoxification systems.

This trend has also been observed in freshwater ecosystems, where relatively high levels of persistent organochlorine compounds have been found in predatory birds such as herons (*Ardea cinerea*) and mammals such as otters (*Lutra lutra*) in Western Europe and in predatory birds such as the double-crested cormorant (*Phalacrocorax auritus*) in the Great Lakes of Canada and the US.

Lipophilic pollutants that have relatively short half-lives (e.g., PAHs) do not show the same tendency to pass along food chains or be biomagnified. Some invertebrates of the lower trophic levels (e.g., *Mytilus edulis*, mussels) bioconcentrate and/or bioaccumulate PAHs because they have little ability to metabolize them. On the other hand, fish, birds, and mammals metabolize them rapidly by monooxygenase attack, so they have little tendency to reach higher trophic levels and thus bioaccumulate there. Although PAHs do not raise the problem of biomagnification, some PAHs and other readily degradable compounds are subject to metabolic activation (Section 5.1.5).

5.4 Summary

Organic pollutants are examples of xenobiotics — compounds that are foreign to organisms. A model is presented to describe the fates of xenobiotics in living organisms, as seen from the toxicological view. In this model, five types of sites with which xenobiotics can interact are identified: sites of uptake, metabolism, storage, action, and excretion. In animals, xenobiotics are circulated in blood and lymph (vertebrates) or in hemolymph (invertebrates); in plants, they move in the phloem and/or xylem; and are distributed throughout different organs and tissues.

Lipophilic xenobiotics move across membranous barriers by simple diffusion or by transport with certain macromolecules (e.g., lipoprotein fragments) that can traverse membranes. Metabolism of lipophilic xenobiotics by animals occurs in two phases. Phase I (oxidation, hydrolysis, hydration, or reduction) yields metabolites of greater polarity than the parent compound. Phase II involves conjugation to produce polar conjugates that are usually readily excreted anions. A variety of monooxygenases and hydrolases are responsible for phase I biotransformations. In phase II biotransformations, conjugation is usually with cellular anions such as glucuronide, sulfate, or glutathione. The excretion of water-soluble metabolites and conjugates completes the process of metabolic detoxification. Sometimes, phase I metabolism causes activation rather than detoxification, as in the case of certain oxidations of benzo(a)pyrene and organophosphorous insecticides.

Persistent lipophilic pollutants with long biological half-lives can undergo biomagnification as they pass along terrestrial or aquatic food chains. They can reach especially high concentrations in vertebrate predators at the tops of food chains. Examples include organochlorine insecticides such as dieldrin and DDT, many PCBs, dioxin (TCDD), methyl mercury, and tributyl tin. Specialist predators, for example, fish-eating birds and bird-eating birds, have poor oxidative detoxification systems and efficiently bioaccumulate organochlorine compounds. These compounds are often highly persistent in soil.

Water-soluble and readily biodegradable compounds do not usually cause problems of persistence and bioaccumulation.

Further Reading

Crosby, D.G. (1998). *Environmental Toxicology and Chemistry*. Good account of environmental chemistry.

Environmental Health Criteria. A range of monographs on pollutants published by the World Health Organization (over 160 titles). Some detail the fates of specific pollutants. Selected titles are in the references at the end of this book.

Hodgson, E. and Levi, P. (1994). *Introduction to Biochemical Toxicology*, 2nd ed. Multiauthor text covering main aspects of biochemical toxicology including toxicokinetics of xenobiotics and modes of action of poisons.

Moriarty, F. (1975). *Organochlorine Insecticides: Persistent Organic Pollutants*. Information about the fates and environmental behaviors of the most researched persistent organic pollutants.

Schürmann, G. and Markert, B., Eds. (1998). *Ecotoxicology*. Text that deals in depth with many topics in the first section of this book.

Timbrell, J.A. (1999). *Principles of Biochemical Toxicology*, 3rd ed. Clear and readable account of the basic principles.

Walker, C.H. and Livingstone, D.R. (1992). *Persistent Pollutants in Marine Ecosystems*. Detailed account of the fates and levels of persistent organic compounds in marine food chains.

www.es.ucsb.edu/faculty/muller/toxkineticses120. Useful case study of one-compartment toxicokinetic models in the aquatic environment. Explains why small organisms accumulate toxicants faster than larger ones and seem to be more sensitive); Why high K_{ow} compounds accumulate slowly; and why environmental variability affects smaller animals more than larger ones.

Section II

Effects of Pollutants on Individual Organisms

6

Testing for Ecotoxicity

The first five chapters discussed the fates of chemicals in the environment and within living organisms. This chapter begins the second section of this book in which the emphasis shifts to questions about the effects of chemicals upon individual organisms. Ecotoxicology is concerned with the harmful effects of chemicals in the environment, especially where the chemicals may be related to adverse changes at the population, community, or ecosystem level. We use the term *ecotoxicity* in this chapter to refer here to toxic effects that are relevant in ecotoxicology.

This chapter will first address harm and explain how toxicity is measured before dealing with more specific issues such as what constitutes ecotoxicity, how it can be measured or predicted through testing procedures, and that test data may be used in preparation of an environmental risk assessment. It is important to understand toxic effects before moving on to the more complex issues in Section 3 concerning the relationship between toxic effects on individuals and consequent adverse effects at the levels of population, community, and ecosystem.

6.1 General Principles

Of central importance in both toxicology and ecotoxicology is the relationship between the quantity of chemical to which an organism is exposed and the nature and degree of consequent harmful (toxic) effects. Dose–response relationships provide the basis for assessment of hazards and risks presented by environmental chemicals. This simple basic concept immediately raises questions about defining poisons because effects depend on doses.

Paracelsus (1493–1541) recognized the dilemma and stated, "All substances are poisons; there is none that is not a poison. The right dose differentiates a poison and a remedy" or "the dose maketh the poison." Thus no chemical is poisonous if a dose is low enough. Conversely, all chemicals are poisonous if doses are high enough (even apparently harmless substances such as sugar and salt can be toxic at high doses). It is advisable to remember this principle when sensational articles in the media cite "poisonous" or "toxic" substances in the environment without pointing out that the levels are far too small to exert toxic effects.

Toxicity can be measured in a number of ways. Most commonly, the measure (end point) is death, although interest in more sophisticated indices is growing. The desire to minimize lethal toxicity testing with vertebrate animals is a significant driving force. Biochemical, physiological, reproductive, and behavioral effects can provide measures of toxicity. Until now, much ecotoxicity testing conducted to meet statutory requirements (see Section 6.5 concerning risk assessment) has utilized lethality as the end point.

It has become increasingly clear that population declines may be the consequences of sublethal rather than lethal effects of pollutants. Examples in this text include the effects of tributyl tin on reproduction of dog whelks (Section 16.3), p,p'-DDE on eggshell thinning in birds of prey (Section 16.1), and reproductive failures of fish caused by EE2, an endocrine disruptor (Kidd et al., 2007; Walker, 2009). Evidence also indicates that neuro-toxic pesticides such as pyrethroids and neonicotinoids can affect the communications of bees via disruption of foraging (Walker, 2009). It has been suggested that this effect may relate to recent declines in bee populations.

Not surprisingly, suggestions have been made to direct ecotoxicity testing toward more ecologically relevant sublethal end points such as effects on reproduction or behavior rather than lethality. Such a proposal is in accord with the aims of animal welfare orga-nizations such as FRAME that seek to reduce the suffering of laboratory animals during toxicity testing. Better science could be more cost effective and more animal friendly.

Many toxicity tests provide estimates of doses (or concentrations in food, air, or water) that will cause a toxic response at the 50% level, e.g., the median lethal dose that kills 50% of a population. It is also possible—and this approach is gaining in popularity—to estab-lish the highest concentration or dose that will not cause an effect.

Several terms used in relation to toxicity testing require definition. First, in lethal tox-icity testing, LD_{50} represents the median lethal dose. LC_{50} represents the median lethal concentration. In toxicity tests that determine these values, it is also possible to determine the highest doses or concentrations that cause no toxicity: the no-observed-effect dose (NOED) and no-observed-effect concentration (NOEC), respectively. These values can be determined only in situations where a higher dose or concentration produced an effect in the same toxicity test. Figure 6.1 illustrates these points; it shows the determination of an LC_{50} after 96 hours of exposure.

If a test end point is an adverse response other than death, an EC_{50} or ED_{50} is determined. The concentration or dose producing the effect in 50% of the population is determined. As with lethal toxicity testing, it is possible to determine NOEC or NOED via this approach. However, values for NOEC or NOED are meaningful only in a test in which a higher dose has been shown to produce an effect.

In addition to toxicity tests on live animals covered in this chapter, other methods can evaluate the toxic properties of chemicals. They stem from understanding the modes of action of chemicals. For example, bacterial mutagenicity assays (e.g., the Ames test) help

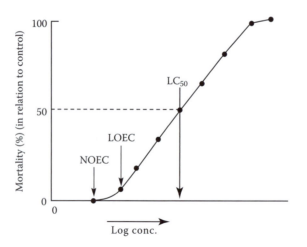

FIGURE 6.1

Toxicity after 96-hour exposure in an aquatic toxicity test. Note that NOEC can be determined only where LOEC is known; otherwise there would be no indications of toxic concentrations. NOEC = no-observed-effect concen-tration. LOEC = lowest-observed-effect concentration. LC_{50} = median lethal concentration at 96 hours.

identify substances that can act as carcinogens or mutagens in mammals. Also, the study of the relationship between structure and toxicity [quantitative structure–activity relationships (QSARs)] can aid the identification of toxic molecules. Further examples are given in Section 6.7.2. These approaches become more viable as molecular mechanisms of toxicity become better known, and they can reveal the *molecular characteristics* that cause chemicals to interact adversely with cellular macromolecules.

An environmental chemical can enter a living organism by one or more routes of uptake. Depending on the chemical, species, and environmental conditions, one route of uptake may be dominant or more than one may be significant. The efficiency of uptake and the degree of toxic effect differ by route. For example, for most lipophilic compounds, absorption from the gut of a vertebrate is faster than absorption from the skin. Thus, toxicity is usually higher with oral administration than with topical application.

When tests are performed on terrestrial animals, it is common to apply single measured doses, e.g., orally, topically (i.e., applied to skin or cuticle), or by injection into tissues or body fluids. There may also be continuous exposure to a constant concentration of chemical in food, in air, or in soil (soil organisms). With aquatic organisms, direct uptake from water is predominantly from water or food, and toxicity testing usually involves continuous exposure to defined concentrations rather than the administration of single doses. In selecting the route of administration of chemicals in toxicity testing, account needs to be taken of the major route(s) of uptake in nature.

There can be very large differences between groups of organisms and between species in their susceptibility to the toxic action of chemicals; there can also be large differences between different strains of the same species—as in the case of strains of insects that have become resistant to insecticides (see Section 13.6). The selective toxicity ratio (SER) is expressed in terms of the median lethal dose (MLD) in this way:

$$\text{SER} = \frac{\text{MLD (or concentration) for species A}}{\text{MLD (or concentration) for species B}}$$

Selectivity ratios of beneficial organisms and pests must be considered when evaluating the environmental safety of pesticides (Table 6.1). Some examples of selectivity are discussed in Hodgson and Levi (1994) and Walker (1983).

The ability of chemicals to be so selective makes it difficult to extrapolate toxicity data from one species to another. For practical reasons, it is possible to perform toxicity tests only on very limited numbers of species. Thus toxicity tests are very rarely carried out on species thought to be at risk in the field. When regulatory authorities make decisions about releases of chemicals into the environment, they base their decisions on toxicity data from surrogate species whose susceptibilities to certain chemicals may be very different from those of field species. For example, estimation of the toxicity of a chemical to wild birds in a risk assessment exercise will be based on data from two or three laboratory species at most, although more than 9,000 very diverse species exist globally. In another example, tests on the pH preferences of eight species of springtails (Insecta: Collembola) by Van Straalen and Verhoef (1997) showed that some species had much greater tendencies to settle in strongly acidified soils than did the standard test *collembolan Folsomia candida* that is clearly not representative of all springtail species.

This problem of extrapolating toxicity data obtained from one species to another is also a feature of human toxicology. For humans, the surrogate species are rats and mice. Generally, the difficult practice of extrapolating species becomes easier with a better

TABLE 6.1

Selective Toxicities of Some Pesticides

Compound	Acute Oral LD$_{50}$ (mg/kg)			Topical/Dermal LD$_{50}$ (mg/kg)		
	Rat	Bird	SER	Rat	Insect	SER
Organophosphorous insecticides				*Housefly*		
Dimethoate	250	26	(4) 9.9	925	0.20	4.6×10^3
Fenitrothion	462	332	(4) 1.4	> 3,000	5.7	526
Dichloros	27	9.6	(2) 2.8	488	0.80	610
Diazinon	450	4.5	(4) 100	850	1.9	447
Malathion	1,650	685	(3) 2.4	> 4,000	17.4	> 230
Pirimiphos-methyl	1,400	162	(3) 8.6			
Pirimiphos-ethyl	138	6.5	(2) 21			
Organochlorine insecticides				*Housefly*		
DDT	400	923	(3) 0.43	2,500	14.0	179
Dieldrin	40	91	(7) 75	75	1.0	75
γ-HCH	200	118[a]	1.7	750	3	250
Carbamate insecticides				*Housefly*		
Carbaryl	500	990	(5) 0.5	> 4,000	> 500	
Baygon (propoxur)	135	26	(6) 5.2	> 2,400	25	96
Carbofuran	6	2.4	(4) 2.5		7	
Aldicarb	6	3.6	(4) 1.6	1	6	0.16
Zectran	39	12	(6) 3.2	3.2		
Pyrethroids				*Bee*		
Permethrin	500	> 13,000	(4) < 0.04	> 2,500	0.017	$> 1.47 \times 10^5$
Cypermethrin	250	> 10,000[b]	< 0.25	> 4,800	0.11	$> 4.4 \times 10^4$
Fenvalerate	451	> 4,000	(3) < 0.11	4,000	0.21	2.4×10^4
Deltamethryn	129	4,000[b]	< 0.03	> 800	0.035	2.3×10^4

Source: Walker, C.H. (1994). In *Introduction to Biochemical Toxicology.* Appleton and Lange. With permission.

Note: For birds, an average LD$_{50}$ is usually given. Figures in parentheses indicate numbers of species used to calculate averages. Where ranges of LD$_{50}$ values appear in the original source, an average has been calculated to simplify presentation. Selectivity ratio (SER) = (LD$_{50}$ rat)/(LD$_{50}$ other species).

[a] Value for pheasant.

[b] Value for duck.

understanding of the mechanisms responsible for toxicity. Such understanding facilitates interspecies comparisons and can aid the development of models to predict toxicity to individual species. Data from in vitro studies can contribute to these models, for example, metabolic rate constants (Chapter 5) may be used in models that estimate tissue concentrations and half-lives of organic chemicals.

Multiple species tests represent a more sophisticated approach to ecotoxicity testing than single species tests, but they are at an early stage of development. They will probably not become routine for some years because of the complexities of the interactions between species. Tests using a simple predator–prey system, e.g., mites preying on Collembola (Axelsen et al., 1997; Hamers and Krogh, 1997), provided useful data, but results from tests with three or more species are difficult to relate to field conditions.

We will not describe toxicity tests that have been widely used to aid risk assessments (Section 6.4). The values obtained by toxicity testing (e.g., LD$_{50}$ and LC$_{50}$) are dependent on

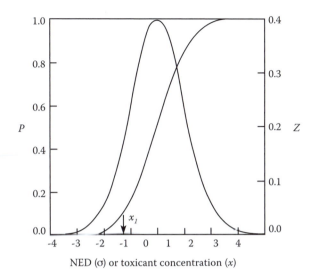

NED (σ) or toxicant concentration (*x*)

FIGURE 6.2
Plot of standardized normal frequency distribution (*P*) with its integral, the cumulative normal frequency function (*z*).

the conditions under which tests are performed. Factors such as formulation of chemical, route of dosing, feeding regimen, temperature, humidity, and animal health influence the values obtained. The interpretation of toxicity data requires caution because data (i) are determined under a well-defined set of conditions that may differ considerably from those in the natural environment; and (ii) are not very reproducible because it is difficult to control all the variables mentioned. At best, standard toxicity data give a measure of toxicity under closely defined operating conditions. This may lead to giving toxicity data more significance than they actually have.

The transformation of percentage kills by statistical procedures requires brief discussion. The underlying assumption is that the data can be fitted to a statistical distribution that will give a straight line when plotted against the log of the dose (or concentration) of the chemical. This is illustrated in Figure 6.2, which shows a normal Gaussian distribution in a toxicity test in relation to the dose of chemical given. If the median concentration is ascribed a 0 value, the values to both sides of 0 can be measured as normal equivalent deviates (NEDs). Comparisons can then be made between measured doses or concentrations producing a particular response and those that would be expected if the data followed the normal distribution. In practice, data from toxicity tests usually depart from the normal distribution when approaching the extremes of 0 and 100% (as in Figure 6.1). It is possible to correct for these deviations by converting percentage kills into NEDs or the related probit units. When these values are plotted against dose or concentration, a straight line is obtained, as in Figure 6.8. Probit values for percentage responses can be found in tables.

6.2 Determination of Toxicities of Mixtures

Much toxicity testing is performed on single compounds. This testing is a necessary component of environmental risk assessments of pesticides and industrial chemicals.

Sometimes testing is carried out on relatively pure samples, but materials tested often contain appreciable quantities of other compounds. In the case of pesticides, the technical product used in formulations may contain appreciable quantities of by-products of manufacturing. Furthermore, the marketed formulation may contain additives such as carrier solvents, emulsifiers, and stabilizers, all of which may affect the formulation's toxicity. In environmental risk assessment, tests must be performed on the products actually released into the environment if a realistic estimation of toxic impact is to be achieved.

Matters become more complicated when considering the pollution that actually exists in the environment. Sewage, factory, and pulp mill effluents released into surface waters often contain complex mixtures of pollutants. So, too, do contaminated sediments and soils that further complicate testing because of the presence of highly persistent lipophilic compounds with long biological half-lives (Chapter 5). Although toxicity is usually additive, the possible of potentiation arises when animals or plants are exposed to mixtures; the toxicity of a mixture can greatly exceed the total toxicities of its component chemicals (Chapter 9).

Tests of environmental samples such as water, sediment, soil, and air can measure the toxicity of mixtures. If these tests are accompanied by chemical analyses, the measured toxicity may be compared with toxicity predicted from the detected residues of chemicals (toxicities of individual chemicals determined by analysis are calculated and are then combined to yield predicted toxicity for the mixture). The measured toxicity may differ markedly from the predicted toxicity.

Such discrepancies arise from several possible causes. In addition to the questions of potentiation and antagonism, a chemical analysis may be incomplete because of failure to detect the presence of certain toxic molecules. This can be a particularly difficult problem for complex mixtures of organic pollutants at low concentrations. Another issue in soils and sediments is availability. To what extent is a strongly adsorbed chemical available to aquatic organisms in contact with sediments or to terrestrial organisms dwelling in soil? In general, relating chemical analysis to toxicity is relatively easy when only a small number of chemicals (up to three) express toxicity.

The assays described in the next section may be used to evaluate the toxicities of mixtures in industrial products and environmental samples. We must emphasize that it can be very difficult to attribute measured toxicity to particular compounds or combinations of compounds present in environmental samples. During regular monitoring of effluents containing complex mixtures of chemicals, the composition of which may change radically over time, it is desirable to analyze large numbers of samples. Rapid, inexpensive tests including bioassays (Section 6.7.2), have an important role here.

6.3 Toxicity Testing with Terrestrial Organisms

6.3.1 Introduction

It has been stated that toxicity testing with terrestrial animals is simpler than with aquatic animals since only one route of exposure (the gut) must be considered. Although this is the case with common tests on rats and mice, it is certainly not the case with newer tests with invertebrates in which exposure via an external medium (air, soil) is important also. In this section, ecotoxicological tests with invertebrates, plants, and birds will be described.

6.3.2 Invertebrate Testing

Two ecotoxicological tests using earthworms and bees are in widespread use and have been developed to test the effects of chemicals on "representative" terrestrial invertebrates. Several other tests are at various stages of development. The methods of testing on three types of organism will be covered here: widely used standard tests with earthworms (*Eisenia* species), a test using *Folsomia candida* (Collembola) springtails, and a standard test using bees. The earthworm tests mainly measure effects of chemicals that pass across the body surface. The springtail test measures effects on reproduction, mainly through contact poisoning of adults, eggs, or juveniles that hatch from the eggs. The bee test is designed to assess the effects of chemicals on "beneficial insects." For further details of these and other tests using enchytraeid worms, mites, beetle, centipedes, millipedes, and nematodes, see Løkke and Van Gestel (1998).

6.3.2.1 Testing with Earthworms

The earthworm test (Box 6.1) developed by the Organisation for Economic Cooperation and Development (OECD) is the most established method and is a legislative requirement in some countries before new chemicals can be released into the environment. Several approaches have been adopted (Greig-Smith et al., 1992a; Sheppard et al., 1998). The most widely used species is *Eisenia fetida* (Figure 6.3A) because it is easy to culture in a laboratory and has a reproductive cycle of about 6 weeks at 20°C. However, this species is native to the southern Mediterranean and survives only in compost or manure heaps or in heated glass houses during the winter in northern latitudes. Thus care should be taken in extrapolating the results of tests with *Eisenia* to other species of earthworm.

One problem with the standard OECD method is that the worms are not fed during the experiment and tend to lose weight. In more recent studies, a small pellet of cow or horse manure placed on an artificial soil surface allows the worms to gain weight. Fed worms also exhibit higher reproductive rates than starved specimens. The bioconcentration factor (concentrations of chemical in worms versus concentration in soil) can also be determined

---

BOX 6.1   OECD EARTHWORM TOXICITY TEST

The basic test uses an artificial soil made of 70% sand, 20% kaolin clay, and 10% sphagnum peat (by weight). Water is added to constitute 35% (by weight) of the final soil, and the pH is adjusted to 6.3 with calcium carbonate. Ten adult worms are added to each of the four replicates of a control and an ascending series of concentrations of the chemicals under test. The worms are left for 14 days, after which the number of survivors is counted. An LC_{50} for survival can then be calculated.

Some workers run their experiments for a longer period to allow the worms to reproduce (Spurgeon et al., 1994). The reproductive rate is easy to assess by counting the number of cocoons produced (Figure 6.3B). Effects on reproduction are detected at lower concentrations of metal in soils than effects on growth and survival, and it is thought that the former approach is more ecologically relevant for analyzing potential effects in the field (Figure 6.4). Interestingly, evidence also indicates that the artificial soil may be deficient in copper (Figure 6.4C), although further work is required to substantiate this suggestion (Hopkin, 1993c).

(A)

(B)

FIGURE 6.3

(A) Group of earthworms (*Eisenia fetida*) in standard OECD soil. Adults are approximately 8 cm in length. (B) Juvenile *Eisenia fetida* emerging from a cocoon approximately 2 mm in diameter. (Photographs courtesy of Steve Hopkin.)

from such experiments and can help predict the exposure of earthworm predators to potential toxins in the field.

When the results of toxicity tests using laboratory soils are compared with those in natural soils from the field, it is clear that chemicals are invariably much more bioavailable to worms in artificial media. Indeed, for metals such as zinc, worms are affected at concentrations 5 to 10 times lower in artificial soils than in field soils (Spurgeon et al., 1994). Uptake of metals by worms is more closely related to the concentrations in soil pore water than

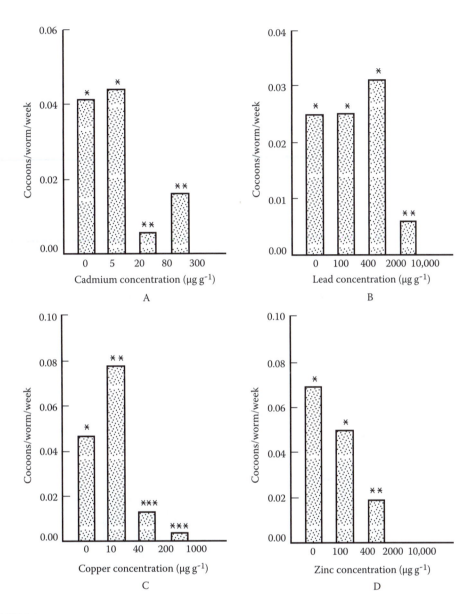

FIGURE 6.4
Rate of cocoon production by *Eisenia fetida* exposed to (A) cadmium, (B) lead, (C) copper, and (D) zinc (µg/g dry weight). Bars with the same number of asterisks (*) were not significantly different at $P < 0.05$. (*Source:* Spurgeon, D.J. et al. (1994). *Environmental Pollution* 84, 123–130. With permission.)

to total content (Janssen et al., 1997). Accumulation rates tend to be relatively high in soils with low pH and low organic matter (Spurgeon and Hopkin, 1996). Artificial soil has a lower organic matter content than many field soils and the clay used has a relatively low cation exchange capacity. To ensure environmental protection, the test includes an in-built safety factor that should be considered when results of laboratory tests are extrapolated to field conditions.

A biochemical test assesses the integrity of membranes of lysosomes in earthworm coelomocytes (Weeks and Svendsen, 1996). When worms are stressed, the lysosomal membranes become leaky. The extent of membrane disruption is measured by the time taken for a neutral red dye to diffuse out of the lysosomes into the cytoplasm of coelomic fluid removed from the coelom and spread onto a microscopic slide. The neutral red retention time was used successfully as a biomarker assay to measure the impact of pollution on earthworms in the aftermath of a plastics fire in England (Svendsen et al., 1996).

6.3.2.2 Tests with Springtails

Springtails (Collembola) are among the most abundant soil arthropods (Hopkin, 1997). Although the springtail test (Box 6.2) with *Folsomia candida* (Figures 6.5 and 6.6) has not become a legislative requirement as has the earthworm test, many laboratories worldwide use this species to assess the effects of chemicals on nontarget soil arthropods (Wiles and Frampton, 1996; Crouau et al., 1999; Fountain and Hopkin, 2001, 2004a, 2004b). Comparison of a range of test organisms by Ronday and Houx (1996) concluded that *Folsomia candida* (Figure 6.5) was the most suitable animal for soil quality assessment. The species can also be used for assessing interactions of chemical effects and natural stressors such as drought (Sorensen and Holmstrup, 2005) and interactions between toxicity and pH (Van Gestel and Koolhaas, 2004). For a comprehensive review of the use of *Folsomia candida* as a standard arthropod, see Fountain and Hopkin (2005).

6.3.2.3 Tests with Beneficial Arthropods

Toxicity tests on insects have been used widely to establish the effectiveness of insecticides against pest species, many of which belong to the Lepidoptera, Diptera, Hemiptera, and Coleoptera orders. Pesticide manufacturers often use these tests for screening new compounds. Such tests generally use the lethal end point and are valuable for identifying resistance to pesticides in the field. Similar testing procedures have also been used to

FIGURE 6.5
Adult and juvenile *Folsomia candida*, a parthenogenetic springtail. The largest specimen is 2 mm in length. (Photograph courtesy of Steve Hopkin.)

FIGURE 6.6
Folsomia candida with a batch of eggs. (Photograph courtesy of Steve Hopkin.)

BOX 6.2 TEST ON SPRINGTAILS

The most widely used species is a parthenogenetic strain of *Folsomia candida* (Figure 6.5) that has a reproductive cycle of 3 to 4 weeks at 20°C. A similar experimental design to that used in the earthworm test is constructed with four replicates each of a control and an ascending series of concentrations of the test chemical. The chemical is mixed in the same formulation of artificial soil as described for the earthworm test in Box 6.1.

Ten adult springtails of the same age are placed in a small container about the size of a plastic vending machine cup (about 10 cm in height, 6 cm in diameter) along with a small amount of dried yeast for food. They are left for at least 3 weeks and then the soil is flooded. All the springtails including any offspring produced during the experiment float to the surface of the water. Each container is photographed on film or captured digitally and the images are projected onto a large screen on which the number of progeny are counted. Developments in image analysis software make it possible to count and measure the lengths of all animals automatically (Martikainen and Krogh, 1999). The concentration that causes a 50% (or other percentage) reduction in reproduction compared with reproduction in controls can then be calculated.

If a larger number of replicates is prepared, a proportion of the containers can be flooded at intervals to produce time responses (Figure 6.7). The test is currently being developed to include more sophisticated measures of effects at the population level (Herbert et al., 2004; Smit et al., 2004).

For earthworms, chemicals tend to be more toxic in artificial soil than are their toxic equivalents in the field. Better laboratory–field comparability can be obtained by aging soils after application of contaminants and percolating them with water before conducting the test (Bongers et al., 2004; Smit and Van Gestel, 1998).

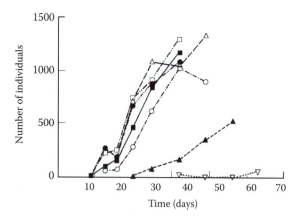

FIGURE 6.7
Total numbers of *Folsomia candida* individuals at different levels of cadmium in artificial soil. Blank (□), 34.8
(●), 71.3 (Δ), 148 (■), 326 (○), 707 (▲), and 1491 (▽) μg Cd/g dry weight. (*Source:* Crommentuijn, T. et al. (1993).
Ecotoxicology and Environmental Safety 26, 216–227. With permission.)

establish toxicity to beneficial insects including natural parasites and predators of insect
pests. Toxicity testing procedures for insects will be reviewed in this section.

A common procedure is to use a microsyringe to apply a solution of a test chemical
directly to the outside of an insect. This allows administration of a known dose to each
insect and a topical LD_{50} can be estimated using the same procedure described later for
vertebrates (Section 6.3.3). Commonly, the chemical is dissolved in an organic solvent such
as acetone. Tests of this kind are useful for ranking the toxicities of chemicals and determining resistance. However, the exposure levels are unrealistic and do not reflect events
in the field. With insecticides and other pesticides, insects are exposed to formulations—
sprays, granules, and dusts—under field conditions. Thus in more realistic testing protocols, insects are exposed to insecticidal deposits on the surfaces of the dishes in which
they are contained or on their food. After such exposures, LC_{50} (median lethal concentrations) can be calculated; these measurements will be encountered again in the context of
aquatic toxicity testing. The German regulatory authority for pesticides published details
of formal laboratory tests for beneficial insects other than bees, and these are described
by Jepson in Calow (1993). Because of concern about the effects of pesticides on nontarget
insects, a number of laboratories test new chemicals on honey bees (Gough et al., 1994;
Box 6.3).

6.3.2.4 Automated Videotracking

A promising development is the ability to quantify aspects of the behaviors of terrestrial
invertebrates in response to pollution by automated video tracking. Arenas are established in
which the substrate is contaminated to different degrees with chemicals. Animals are released
into these arenas and their movements are monitored by video camera. Sophisticated analysis of the data collected can then be performed to reveal effects on parameters such as speed,
turning rate, and time spent motionless. For example, woodlice (Isopoda) collected from leaf
litter contaminated with residues of a plastics fire spent significantly less time walking and
showed less tendency to turn than did controls (Sørensen et al., 1999). Such behavioral effects
often occur at concentrations well below those at which effects on growth and reproduction
can be detected (e.g., in wolf spiders exposed to cypermethrin; Baatrup and Bayley, 1993).

BOX 6.3   TOXICITY TESTS WITH BEES

The standard method is to collect workers from a hive, anesthetize them with carbon dioxide, and maintain them in cylinders of wire mesh (ten per container) with three replicates of each test concentration plus controls. Each container has a feeding bottle supplied with a sugar solution containing a known concentration of a test chemical. The containers are kept at constant temperature (typically 25°C) and are checked at 1, 2, 4, 24, and 48 h for mortality. In an additional test, contact toxicity is measured by applying the chemical directly to the thorax. However, it is not possible to apply results of such tests uncritically to all species of bees because of their differing lifestyles (Thompson and Hunt, 1999).

A field-based method can also be used but requires considerably more space. A beehive is maintained in a large enclosed tunnel of polythene or netting and a pollen- and/or nectar-bearing crop is sprayed with the test chemical. Bees that return to the hive and die inside during the night are ejected in the morning by healthy workers. The level of mortality can thus be simply determined by counting the number of dead bees in a tray placed under the hive entrance.

6.3.3 Vertebrates

The toxicity of chemicals to mammals, birds, and other vertebrates has commonly been measured as a median lethal dose (LD_{50}). In routine toxicity testing, a single oral dose provides an estimate of acute oral LD_{50}. The procedures are described in many standard texts (for example, Calow, 1993) and will only be outlined here. Groups of animals are given doses of a test chemical over a range of values that centers on a rough estimate of LD_{50} obtained via preliminary tests. The percentage of animals that die in each group over a fixed period following dosing is then plotted against the log of the dose in milligrams per kilogram. To obtain a straight-line relationship between dose and mortality, it is necessary to transform percentage kills into normal equivalent deviates (NEDs) or probit values (probit analysis); see Figure 6.8.

Values of LD_{50} are sometimes obtained using other dosing methods. Chemicals may be injected into blood, muscle, or the peritoneal cavity or they may be applied directly to the skin (dermal LD_{50}). Sometimes they are continuously administered in the food or water over a fixed period, in which case toxicity is expressed as a median lethal concentration in food or water over a stipulated period such as a 5-day oral LC_{50}.

Birds are sometimes used for reproductive toxicity tests. The end points may be clutch size, shell thickness, hatchability of eggs, embryotoxicity, and viability of chicks. For further details on avian toxicity tests, see Walker in Calow (1993).

Two fundamental problems surround statutory toxicity testing on vertebrates for purposes of risk assessment. The first is that for cost reasons only a very small number of species can be tested. Thus one or two bird species may act as surrogates for unrelated species, about 9,000 of which exist worldwide. It is rarely possible to perform tests on the species thought to be most at risk. This is in marked contrast to toxicity testing related to human health hazards in which several species are used as surrogates for a single other species—humans. A similar situation exists with ecotoxicity testing using mammals or fish. Birds, mammals, and fish exhibit large species differences in toxicity so toxicity data obtained from surrogates constitute only rough guides to toxicity in species that are not

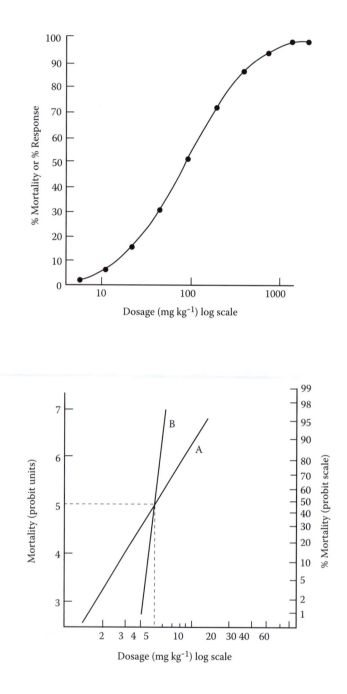

FIGURE 6.8
Determination of LD$_{50}$. For details, see text.

closely related. The second problem is ethical. Western countries face growing opposition to the use of vertebrate animals in toxicity testing, as will be discussed in Section 6.7

In toxicity testing of pesticides, it is common to chose two species of birds, typically the mallard duck (*Anas platyrhynchus*) and either the Japanese quail (*Coturnix coturnix japonica*) or northern bobwhite quail (*Colinus virginianus*). The Japanese quail is favored in Europe and the northern bobwhite quail in the US. Some testing may also be done on feral pigeons (*Columba livia*). Mammals fare rather better than birds because toxicity data relevant to human hazards is available including data on rats, mice, rabbits, guinea pigs, and primates (Calow, 1993, 1994).

During the course of this type of toxicity testing, it is possible to establish values for NOED or NOEC, i.e., the highest dose or concentration that produces no lethal effect. At present, the use of the LD_{50} tests is decreasing partly in response to ethical considerations. The British Toxicology Society proposed an alternative procedure that requires far fewer animals and would merely classify compounds as harmful, toxic, or very toxic (Timbrell, 1995).

6.3.4 Plants

A wide variety of methods can assess the toxicity of chemicals to plants. In this section, tests to assess the effects of metals will be covered in detail since most of the work has been on metals. However, the basic principles of these tests are applicable to other materials and to assessments of the effects of acid rain and gaseous pollutants.

Plants that naturally contain very high concentrations of specific metals are known as metallophytes (Baker and Proctor, 1990). Some metallophytes can grow naturally on metal-contaminated soils. Others have evolved genetically distinct strains that are resistant to metals and can grow far better in contaminated situations than nonresistant plants of the same species (Schat and Bookum, 1992). Resistant plants may also grow more successfully in nutrient-poor conditions in acidic mine waste (Ernst et al., 1992; Section 13.6). Interest is growing in using tolerant plants to revegetate old mine sites (Plenderleith and Bell, 1990).

A number of techniques have been developed to measure metal tolerance of plants (Baker and Walker, 1989), for example, comparing the responses of tolerant plants from contaminated sites and expressing the result as a percentage of the performance of nontolerant plants from clean areas. The principal parameters measured are seedling survival, biomass, shoot growth, and root growth.

Seedling survival is the number of plants surviving from seed after a specified time. It is a better measure than straightforward germination. Many nonresistant plants successfully germinate in metal-contaminated soil but subsequently fail to grow and remain stunted.

Rates of dry matter production and the biomass yield of resistant plants are generally found to be lower than rates in their nonresistant counterparts grown in uncontaminated soil. This reduction is believed to be due to the energy expenditure in metal tolerance mechanisms such as compartmentalization of metals in intracellular compartments (Vasquez et al., 1994; Section 13.6.2). Resistant plants cultivated in contaminated soil or nutrient solutions exhibit growth rates exceeding those of nonresistant plants. Other ways of measuring this difference are comparing shoot and root lengths. Shoot lengths can be compared in plants grown in soil, but measuring differences in root length is easier if the plants are reared in nutrient solutions in clear containers. Regular monitoring of root lengths can be conducted without disturbing the plants. This approach has been used to demonstrate resistance, as it allows the composition of the liquid medium to be closely controlled (Table 6.2).

TABLE 6.2

Tolerance of Three Grass Species from Hallen Wood
Contaminated by Aerial Deposition from Smelting and
from Uncontaminated Midger Wood

Species	Hallen Plants (%)	Midger Plants (%)
Dactylus glomerata	105.9	57.4
Holcus lanatus	113.8	28.1
Deschampsia	82.2	37.5

Source: Martin, M.H. and Bullock, R.J. (1994). In *Toxic Metals in Soil–Plant Systems.* John Wiley. With permission.

Note: Tolerance calculated from mean lengths of roots in 2 ppm Cd per mean lengths of roots with no Cd. Measurements made after 14 days in full-strength Hoaglands culture solution.

Deleterious effects on plants of gaseous pollutants such as ozone can be detected by methods in addition to those cited above. Yellowing of leaves (chlorosis) is characteristic of stress. The critical air concentrations can be determined by exposing replicates to different levels of the gas under test in separate chambers, although such experiments are expensive. The degree of chlorosis can be measured by counting the number of leaves affected or by determining the leaf area exhibiting the yellow coloration (Mehlhorn et al., 1991). Different varieties of a species may exhibit different degrees of chlorosis at the same chemical concentration and may be useful for biological monitoring of pollution (e.g., tobacco; Chapter 11).

6.4 Toxicity Testing with Aquatic Organisms

The basic principles of aquatic toxicity testing are similar to those described for terrestrial organisms. However, the main routes of uptake that influence some aspects of test designs raise specific questions.

Direct uptake from water is a major route for aquatic organisms—across the gills of fish or the permeable skins of amphibia. Uptake can also occur as food passes through the alimentary system. Bottom-dwelling organisms are exposed to residues in sediments. The relative importance of these routes of uptake differs among organisms and chemicals and depends on environmental conditions. In some cases, all these routes may operate in one organism at one time.

6.4.1 Tests for Direct Absorption from Water

Much of the toxicity testing of aquatic organisms (*Daphnia* and *Gammarus pulex* fish) concerns direct absorption of chemicals from water. The chemicals may be in solution, in suspension, or both. Organisms are exposed to different concentrations of the chemical in water to determine values for median lethal concentration. Although absorption is primarily directly from water, it is not possible to exclude completely the contamination of food and potential uptake from this source.

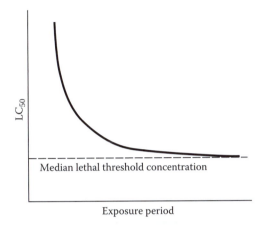

FIGURE 6.9
Relationship of LC$_{50}$ to exposure period.

One difficulty in aquatic toxicity testing is maintaining a constant concentration of chemical in water. Chemicals are lost from water due to (i) absorption and metabolism by the test organism and (ii) volatilization, degradation, and adsorption from the water. Where the rate of loss is relatively low, tests may be performed using static or semistatic systems. With static systems, the water is not changed for the duration of the test. With semistatic systems, the water is replaced at regular intervals (usually every 24 hours). A better but more complex and expensive method for renewing test solutions is a continuous flow (flow-through) system that constantly renews the test solution, thus ensuring a constant concentration of the test chemical and preventing the build-up of contamination from feces, algae, and mucus. If organisms are exposed to a chemical for a sufficiently long time, steady-state concentrations will be reached in the tissues (Chapter 5).

The toxic effect of a chemical depends on the concentration in the tissues at the site of action (Chapters 5 and 7). This in turn depends on the concentration of the chemical in water and the period of exposure. Thus the median lethal concentration (LC$_{50}$) is related to the exposure period (Figure 6.9). With increasing exposure time, the LC$_{50}$ decreases until the median lethal threshold concentration (threshold LC$_{50}$) is reached. At this point, further increases in exposure period cause no change in mortality. It may reasonably be supposed that when the threshold LC$_{50}$ is reached, the system is in a steady state, i.e., tissue concentration no longer increases over time.

In testing aquatic toxicity, preliminary ranging tests are necessary to obtain a rough estimate. Small groups of test organisms (typically two or three individuals per group) are exposed to a wide range of concentrations of the test chemical on a log scale. The results of the ranging test can be used to plan a full toxicity test that exposes larger numbers of test animals to a narrower range of concentrations centering on the LC$_{50}$ estimated from the ranging test. The mortality percentages for all test groups are recorded over various time intervals for the duration of the test. The results can be used to determine LC$_{50}$ values at different exposure times. Data can be initially plotted in two ways:

1. At one fixed exposure concentration, a series of chosen survival times can be plotted against the percentage of the sample surviving at each time (e.g., 25% of a sample survive 4 days). This can then be repeated for each of the remaining concentrations to determine median survival periods (median periods of response).

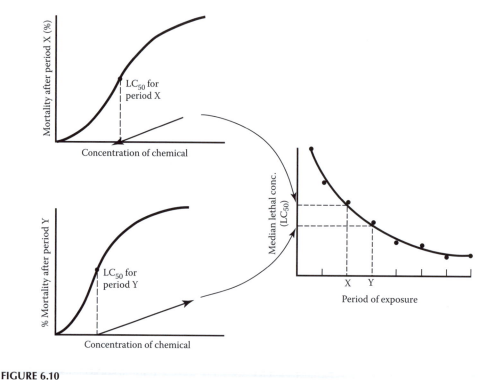

FIGURE 6.10
Determination of LC_{50}. For details, see text.

2. At one fixed exposure period, the percentage mortality can be related to exposure concentration. From this, the median lethal concentration can be calculated for each individual exposure period (Figure 6.10).

From these data, two further plots can be produced. Using data from method 1, median survival time can be plotted against concentration of chemical. Using data from method 2, the median lethal concentration can be related to exposure period, and from this LC_{50} values can be estimated for particular exposure periods, e.g., LC_{50} at 96 hours.

Our discussion to this point has been restricted to lethal toxicity testing. It should be emphasized, however, that end points other than lethality can be used. More generally, median effective concentrations (EC_{50}) can be determined for a variety of nonlethal end points.

As the toxic effect of a chemical depends greatly on the period of exposure, two important issues are (i) the period over which a toxicity test should be conducted and (ii) the exposure periods for the estimation of EC_{50} or LC_{50}. The answers depend on the organism tested and the purpose of the test. Tests with *Daphnia* are commonly of only 24 to 48 hours' duration in static systems. By contrast, fish toxicity tests are usually longer. Simple screening tests commonly require relatively short exposures. Typical measures of toxicity are LC_{50} at 48 hours for *Daphnia* and LC_{50} at 96 hours for fish. Such data indicate the toxicity of one chemical in relation to others (i.e., ranking of toxicity). Longer exposure periods may be required to yield toxicity data required to evaluate water quality; exposure may be continued until a median threshold concentration has been established (Figure 6.9). Some aquatic toxicity tests work to nonlethal end points; examples appear in Boxes 6.4 and 6.5.

BOX 6.4 *DAPHNIA* REPRODUCTION TEST

Neonates of *Daphnia magna* less than 24 hours old are used (Figure 6.11A). One neonate is placed into 60 to 80 ml of test solution in a 100-ml beaker. Usually ten replicates are used for five test concentrations plus a control. Thus 60 neonates are used per test. The test is performed at 20°C with a 16 hours of light to 8 hours of darkness photoperiod. Every 2 or 3 days, the numbers of surviving organisms and young are determined (static renewal system). The adults are then moved to fresh test solution and the young are discarded.

Daphnia are fed on microalgae (typically *Chlorella* or *Scenedesmus*). There is no general standard concerning food supply and variations in quantity and quality can affect the outcomes of tests. For example, deficiencies of certain trace elements in food may not become apparent until several generations have been cultured (Caffrey and Keating, 1997). Usually, the first brood is observed after 8 to 10 days, with subsequent broods appearing at approximately 2-day intervals. The duration of the test is 21 days (five broods). The reproduction data obtained are used to calculate lowest observed effect concentration (LOEC) and NOEC data. Comparison is made of the numbers of offspring per surviving female in the treated groups and the reproductive outputs of controls.

Care should be taken to standardize clones used for the test. Baird et al. (1990) found a more than 100-fold difference in the LC_{50} values for cadmium in two clones of *D. magna*.

BOX 6.5 ALGAL TOXICITY TEST

The essential feature of an algal toxicity test is to determine the effects of a chemical on a green algal population growing exponentially in a nutrient-enriched medium for 72 or 96 hours (Lewis, 1990). Cell density is determined using either a direct measurement (microscopic counting) or one of several indirect techniques (spectrophotometric method or electronic particle counter). Controlled experimental factors include the initial cell density, test temperature, pH, light quality, and intensity. Typically, an EC_{50} value and the NOEC are calculated based on growth inhibition. The EC_{50} value is the concentration of the test substance that causes a 50% reduction of the effect parameter.

The most common test species is the freshwater green coccoid *Selenastrum capricornutum*—the only species recommended by the US Environmental Protection Agency for monitoring toxicity effects of effluents. At least ten other species have been used. However, Lewis (1990) concluded that data were insufficient to support the environmental relevance of results using the standard method in the present format. The main problem is that interspecific variation in response to chemicals is significant, and the field validation of most laboratory-derived results is lacking (a problem common to most tests). The way forward is probably to conduct simultaneous multiple species tests to assess the relative toxicities of different chemicals to multiple species.

BOX 6.6   SEDIMENT TOXICITY TEXT USING *RHEPHOXYNIUS ABRONIUS*

Sediment samples are taken in the field to a depth of 2 to 5 cm below the sediment–water interface. After mixing sediments, subsamples are removed and filtered through 0.5-mm sieves to remove biota before adding to eleven flasks containing aerated seawater to a depth of 2 cm. Samples of contaminated sediments may be diluted with varying proportions of clean sediments to assess their toxicity. Controls are run with samples of clean sediment. Sediment samples may also be spiked with different concentrations of chemicals whose sediment toxicities are to be assessed.

Samples of *R. abronius*, a marine amphipod, are obtained from uncontaminated sites and acclimated at the test density for 4 days before twenty test organisms are added to each flask. The duration of this static test is usually 10 days. Survival is the most commonly used end point. Where fivefold replication was used, there was a 75% certainty of detecting a mean survival change of 15%. The same general principle was followed with chironomid larvae (Figure 6.11B).

A number of factors influence the values obtained in aquatic toxicity testing along with the inherent properties of the chemical and the test organism. A critical question is the availability of the chemical to the test organism. Availability may be limited because of adsorption to particulate matter or to the surface of the tank.

6.4.2 Sediment Toxicity Tests

Sediments represent important sources of pollutants to aquatic organisms. Lipophilic organic pollutants and some metals are strongly retained by sediments in which they have long half-lives. An example of a sediment toxicity test is given in Box 6.6.

The toxicity of chemicals associated with sediments is difficult to assess because of the uncertainty about the total quantity available to the test organism. Strongly adsorbed molecules that exist at low concentrations in water may nevertheless be directly absorbed by aquatic animals via the passage of sediment particles through the alimentary tract or across respiratory surfaces. The development of improved sediment toxicity tests is being actively pursued.

6.5 Risk Assessment

The toxicity data obtained by the testing procedures described here may be used to assess hazards and risks (Walker et al., 1991a; Calow, 1993, 1994; Suter, 1993). For the purposes of this discussion, two definitions will be used: (i) hazard = potential to cause harm; (ii) risk = probability that harm will be caused. Risk assessment depends on a comparison of two factors:

1. The toxicity of a compound expressed as a concentration (EC_{50}, LC_{50}, or NOEC)
2. The anticipated exposure of an organism to the same chemical, expressed in the same units (concentration in water, food, or soil to which the organism is exposed)

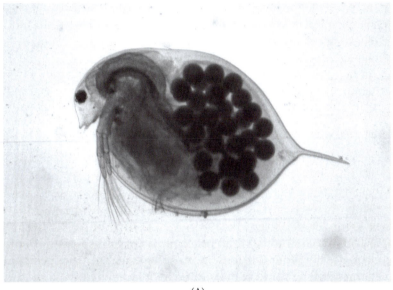

(A)

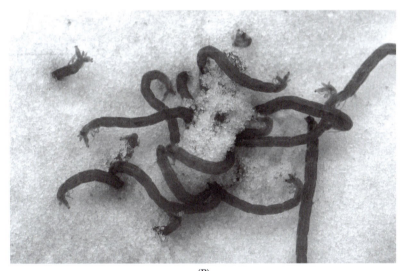

(B)

FIGURE 6.11

Organisms used in aquatic ecotoxicity testing. (A) *Daphnia magna* 2 mm in length. (B) Chironomid larvae 1 cm in length. (Photographs courtesy of Steve Hopkin.)

In hazard assessment, a toxicity test can yield a plot of the frequency of a toxic effect (e.g., mortality) to the dose given (Figure 6.1). From this, NOEC and an EC_{50} can be estimated. The result can be compared with a putative "high" environmental concentration to decide whether a hazard exists. A ranking of compounds according to their toxicities is important at this stage. If toxicity is very low, a compound is not considered hazardous.

In risk assessment, further calculations are carried out to obtain values for the predicted environmental concentration (PEC) and the predicted environmental no-effect concentration (PNEC). The details of these calculations lie beyond the scope of this book. PEC calculations are based on known rates of release and dilution factors in the environment. If, for example, a chemical is used in an industrial process, the level of the industrial effluent is measured or calculated. This figure is then divided by the dilution that occurs in receiving waters (river or lake) to obtain a value for PEC. The PNEC can be estimated by dividing LC_{50} or EC_{50} for the most sensitive species tested in the laboratory by an arbitrary safety factor (often 1000) to compensate for the great uncertainty in extrapolating from laboratory toxicity data for one species to expected field toxicity for other species.

$$\frac{PEC}{PNEC} = \text{risk quotient}$$

If this value is well below 1, the risk is low. If it is 1 or more, the risk is substantial.

We should note the inconsistencies in the use of the terms *hazard assessment* and *risk assessment* in the literature (Calow, 1994). According to some researchers, PEC and PNEC comparisons should be regarded as examples of hazard assessment, not of risk assessment, because they do not represent acceptable measures of probability. In this account, no attempt will be made to make this distinction, and the term *risk assessment* will be applied to all cases.

Clearly, calculations such as these provide only rough estimates of risk. It is necessary to include a large safety factor to allow for uncertainties about environmental toxicity. In addition to uncertainty about toxicities to organisms in the field, another issue is estimating environmental exposure. In terrestrial ecosystems, it is often very difficult to obtain a realistic PEC value. If a mobile species (bird, mammal, or insect) ingests a chemical from its food—and the chemical is distributed very unevenly throughout the ecosystem—it is not possible to estimate exposure with any degree of accuracy. A grain-eating bird may not consume grain treated with a pesticide; a mammal may or may not be in a field when it is sprayed with a pesticide.

Considerations such as these strengthen the case for the development of new strategies using biomarkers for risk assessments. Biomarker assays can provide measures of exposure and sometimes of toxic effects under actual field conditions, e.g., in field trials with pesticides. This issue will be discussed further in Chapter 15.

6.6 Field Testing for Toxicity

If the normal procedures of risk assessment raise uncertainties about the safety of a chemical, further testing may be necessary before a decision can be made about permitted release into the environment. Field trials are sometimes required for pesticides In a full-scale field trial, a pesticide is likely to be applied at a dose and under conditions that are most likely to produce toxic effects (worst case scenario; Somerville and Walker, 1990). A variety of measurements may be made, depending on the chemical, method of application, habitat, climate, and agricultural system. Measurements may include identifying dead animals and birds (corpse counting), estimating population numbers and breeding success, measuring residues in soil, crops, and animals, and, occasionally biomarker assays (cholinesterase inhibition, eggshell thinning). Such trials are expensive and not undertaken lightly. A field study concerned with the toxic effects of a radionuclide is described in Box 6.7.

BOX 6.7   FIELD STUDY TO ASSESS ENVIRONMENTAL EFFECTS OF ^{137}CS ON BIRDS

Lowe (1991) demonstrated the importance of laboratory testing for field extrapolation for proving or disproving a causative relationship between levels of an environmental pollutant and an ecological effect in the area of the Sellafield nuclear waste reprocessing complex in northwest England.

Since 1983, concern had been expressed about the apparent decline in numbers of birds in the Ravenglass estuary in West Cumbria, particularly in the black-headed gull colony on the Drigg dunes. One suggestion was that the decline arose from excessive radiation in the birds' food and their general environment. Lowe studied twelve species of marine invertebrate from Ravenglass. Most of them were known to be important foods for birds, but none showed excessive contamination with radionuclides.

Analysis of a sample of carcasses from the area showed that oystercatchers (*Haematopus ostralegus*) and shelducks (*Tadorna tadorna*) had some of the highest tissue concentrations of ^{137}Cs of all birds but their breeding success and population numbers were unaffected. Black-headed gulls, on the other hand, were found to be feeding mainly inland and were the least contaminated with radionuclides of all the birds at Ravenglass, but their numbers and breeding success were in decline.

Calculations of the total dose equivalent rate to the whole body of the most contaminated black-headed gull amounted to 9.8×10^{-4} mSv/h (about 8.4×10^{-4} mGy/h, whole body absorbed dose rate), and the background exposure dose was of the order of 8.3×10^{-4} mGy/h. As a minimum chronic dose of 1000 mGy day/h was found necessary to retard growth of nestling birds and 9600 mGy over 20 days of incubation was required to cause the deaths of 50% of embryos in black-headed gulls' eggs in laboratory experiments, the concentrations of radionuclides in the food, body tissues, and general environment were at least three orders of magnitude too low to have any effect.

The most likely cause of the desertion of the gullery was the combination of an uncontrolled fox population, the severest outbreak of myxomatosis among rabbits since 1954, and the driest May–July period on record—all in 1984.

Of growing interest are mesocosms—small, carefully controlled systems that simulate conditions in the natural environment (Section 14.3.3). Experimental ponds are examples of mesocosms. Ponds of standard size can be established and become colonized by plants, insects, and vertebrates. The effects of chemicals on the pond communities can then be tested. Such mesocosms serve as "half-way houses" between closely controlled laboratory tests and extensive, loosely controlled full-scale field trials. One advantage is that they allow replication, which is not usually possible in full-scale field trials (Ramade, 1992; Crossland, 1994).

6.7 Alternative Methods in Ecotoxicity Testing

In recent years, opposition to the use of animals in testing procedures that cause suffering has been mounting. In particular, toxicity tests of vertebrates that use lethality as the end point have been criticized (Balls et al., 1991; Van Zutphen and Balls, 1997; Walker, 1998b; Walker et al., 1998; Calabrese and Baldwin, 2003). These objections have been raised to test both human risk assessment and ecotoxicity.

Along with such ethical considerations, existing practices in ecotoxicity testing have been criticized strongly on scientific grounds. Organizations actively involved in the quest for alternative methods include the Fund for the Replacement of Animals in Medical Experiments (FRAME) and the European Center for the Validation of Alternative Methods (ECVAM).

The limited value of estimates of environmental risk based on data from standard toxicity tests was explained in the previous section.

The ultimate concern in ecotoxicology is about the effects of pollutants at the levels of population, community, and ecosystem. Estimates of the probability that certain individuals may experience toxic effects tell us virtually nothing about this. Indirect effects, such as the decline of the grey partridge on agricultural land as a consequence of the use of herbicides (Chapter 12), are not predictable from conventional ecotoxicity tests. Indeed, the herbicides normally used on agricultural land have very low toxicity to partridges and other farmland birds and would appear very safe in a risk assessment exercise. The development of alternative methods for ecotoxicity testing could serve two purposes—improving the quality of the science and reducing suffering caused to animals. Alternative testing procedures fall into two major categories:

1. Alternative methods for estimating toxicity to vertebrates
2. Alternative methods and strategies that lead to more ecologically relevant end points

The first approach most directly addresses the immediate question of replacing tests that cause suffering to vertebrate animals; the second deals with the longer term issue of developing a more ecological approach to environmental risk assessment. If strategies and tests that lead to more ecological end points are successfully developed, it may be possible to redirect limited resources away from current testing practices toward others that provide better assessments of environmental risk. These two approaches will now be briefly described.

6.7.1 Alternative Methods for Estimating Toxicity to Vertebrates

6.7.1.1 Toxicity Testing on Live Vertebrates

It is frequently asserted that it is important to continue with some toxicity tests on live vertebrates because of the uncertainties of estimating vertebrate toxicity by other means. Even in this situation, however, steps can be taken to reduce suffering. The three major aims of FRAME, ECVAM, and related organizations are the replacement, reduction, and refinement of toxicity tests on animals (the three Rs). The last two can still be followed where vertebrate toxicity testing continues. The proposals include testing procedures that work to lethal end points but reduce the number of animals used. The so-called fixed dose procedure provides an example of this approach (Section 6.3.3). Of greater long-term interest is the use of other end points including biomarker responses in tissue samples such as blood that can be obtained by nondestructive sampling.

Biomarker responses can give early warning of the operation of a toxic mechanism or process before overt symptoms of intoxication are shown. Thus, from the point of view of animal welfare, the use of such tests rather than lethal toxicity tests is attractive. Other arguments favor this type of testing rather than lethal toxicity tests on other grounds:

the scientific value of the data generated by tests and reduction of the costs of testing in the longer term.

As noted elsewhere in this text, sublethal effects can be important in causing population declines and may be measured using mechanistic biomarker assays. Also—and of fundamental importance—mechanistic biomarker assays can directly measure the degree of operation of a toxic mechanism at varying levels of exposure, from early sublethal effects to effects serious enough to cause death. Furthermore, biomarker assays can be employed in field studies.

An advantage of using mechanistic biomarker assays in ecotoxicity testing is that only a relatively small number of mechanisms of toxicity are involved and assays can, in principle, be based on the operations of these mechanisms As explained in Chapter 1, the members of a few major classes of pesticides share a common mode of action. Of these, the insecticides have caused the greatest concern about environmental side effects. Most commercial insecticides are neurotoxic; the main groups now used are the pyrethroids, neonicotinoids, organophosphates, and carbamates. The pyrethroids act on sodium channels; the neonicotinoids on cholinergic receptors; and the organophosphorous and carbamate insecticides on acetylcholinesterase (Chapters 7 and 8). The inhibition of acetylcholinesterase by organophosphorous insecticides has already been widely covered in ecotoxicological studies. In principle, it follows that only a few—perhaps four or five—mechanistic biomarker assays would have been needed for the standard risk assessment of most insecticides currently marketed if appropriate biomarker assays had been available. At present, this approach is feasible for only a limited number of pollutants for which mechanistic biomarker assays are available. In time, however, the approach should become more widely applicable as new mechanistic biomarker assays become available.

6.7.1.2 Toxicity Testing on Nonvertebrates

Some correlations have been shown between the toxicities of a group of related chemicals to vertebrates and their toxicities to nonvertebrate species. For example, toxicity to fish sometimes correlates well with toxicity to *Daphnia*. Such correlations are likely where a compound exerts a similar mode of action in vertebrates and nonvertebrates.

However, toxicity is dependent on toxicokinetic factors (e.g., metabolism) that differ greatly among vertebrates and other groups, and toxicity expressed as LD_{50} or LC_{50} can vary greatly between vertebrates and invertebrates even for compounds with similar modes of action. Over a wide range of environmental chemicals of differing modes of action, tests with nonvertebrate species are not reliable guides to vertebrate toxicity.

6.7.1.3 Toxicity Testing on Cellular Systems

Vertebrate cells, for example, hepatocytes of mammals, fish, and birds, are sometimes used to measure toxicity. They may contain reporter genes that mediate a characteristic response upon exposure to chemicals that operate toxic mechanisms.

One example is the use of genetically modified mouse hepatocytes that emit light when dioxins or planar PCBs interact with the Ah receptor (CALUX system; Murk et al., 1997). Such systems can be useful in screening for the presence of chemicals operating a particular toxic mechanism in a vertebrate species. However, they do not provide reliable estimates of median lethal doses found by common toxicity tests.

6.7.1.4 Predictive Models

The best known examples of predictive models are quantitative structure–activity relationships (QSARs) used to predict toxicity from physicochemical properties of environmental pollutants. At the present stage of development, QSARs cannot be regarded as
workable alternatives to toxicity tests, although they can give valuable information and are
likely to be of greater predictive value as the technology improves (Box 6.8).

A more promising approach over the long term may be the incorporation of in vitro data
obtained from cellular systems (see Section 6.7.1.3) into more sophisticated models that utilize toxicokinetic parameters. This approach is now being considered to obtain better predictions of human toxicities of drugs and environmental chemicals utilizing data obtained
from human in vitro systems. In theory, such testing should help overcome the serious
problem of making interspecies comparisons when evaluating toxicity—a more difficult
problem in ecotoxicology than in human toxicology! Many species that are not available
for present toxicity testing can be studied in vitro. Practical, ethical, and economic factors
all limit the numbers and varieties of species that can be used for ecotoxicity testing.

6.7.2 Alternative Approaches toward More Ecological End Points

The ultimate concerns in ecotoxicology are chemical effects at and above population levels
in the field. The whole question of effects at these higher levels is the subject of Section III
of this text. In the context of toxicity testing, however, the possibility of using data relating
to such effects in the field instead of or along with toxicity data obtained in a laboratory for
individual species will now be briefly considered.

6.7.2.1 Field Studies

Field studies encompass a wide range of activities from various types of monitoring to
field trials with pesticides. In theory, studies of the effects of chemicals on individuals,
populations, and communities in the field directly address the basic issues of ecotoxicology. The overriding problem with field studies is the very limited ability to control variables such as temperature, rainfall, air and water movements, and animal migrations, all
of which can influence the effects of environmental chemicals. Such variables can be controlled only to a limited extent in field trials, where pesticides are applied in a controlled
way and effects are monitored, followed by comparisons with control areas or groups.
The use of biomarkers to measure responses of individual organisms to a chemical can
link exposure to a chemical with a change at the population level (e.g., decline of numbers,
decline of reproductive success, and increased mortality, as will be explained in Chapter
15). It is also possible to bring contaminated material, for example, soil from a field, into a
laboratory and then to expose animals under more controlled conditions.

Many field studies analyze an ongoing pollution problem rather than controlled releases
of chemicals (see examples in Chapters 12 and 15). Some degree of experimental control is
still possible, for example, the deployment of "clean" organisms along pollution gradients.
Residues of chemicals detected by analysis can be related to expected biomarker responses
and these responses in turn can be related to population changes. Changes in populations and communities caused by chemicals may be identified by biotic indices such as
river invertebrate prediction and classification (RIVPACS; Wright et al., 1993) or structural
analysis of communities (see Chapter 14). The development of resistance to chemicals may
involve genetic changes that can be measured and related to their original cause.

BOX 6.8   QUANTITATIVE STRUCTURE–ACTIVITY RELATIONSHIPS (QSARS)

Interest in predicting the toxicities of molecules based on their chemical characteristics and their structures has been ongoing. As understanding of the molecular basis of toxicity has grown, it has become possible to design new pesticides and biocides based on chemical structures (Hodgson and Kuhr, 1990, Chap. 3; Calow, 1993, Chap. 14). When the molecular mode of action is known, the designs of novel pesticides can be guided by considering their structures and the properties of their sites of action.

Molecules can be designed, for example to: (i) bind to active sites of enzymes, and act as inhibitors for the reactions the enzymes normally catalyze or (ii) interact with receptor sites on neurotransmitters. Thus, EBI fungicides have been designed to bind to certain forms of cytochrome P_{450} and inhibit ergosterol biosynthesis. Neonicotinoid insecticides have been discovered that will bind to the nicotinic receptor for acetylcholine, and thereby disrupt synaptic transmission at cholinergic synapses (Section 8.4.1). Similarly, new drugs have been discovered based on this principle.

In the field of toxicity testing, interest is growing in models based on quantitative structure–activity relationships that predict toxicity. This approach has the potential to reduce the toxicity testing required—clearly a desirable objective from the point of view of costs and reduced numbers of laboratory animals needed for testing. However, the way forward is not easy. Species exhibit large differences in toxicity as consequences of various sites of action and metabolism routes of xenobiotics. Thus, a model developed for laboratory rats may not be of much value for predicting toxicity in predatory birds.

Perhaps the most successful models of this kind were developed to analyze narcotics (sometimes called physical poisons). The models utilize measures of lipophilicity such as K_{ow} and predict the distribution of these chemicals in aquatic organisms (Section 5.1.8). Narcotics exhibit relatively low toxicity; they exert their effects by reaching high concentrations inside certain biological membranes. Successful models predict the toxicity of narcotics to fish (Konemann, 1981). For example the following equation relates the hydrophobicities of aliphatic, aromatic, and alicyclic narcotics to their toxicities to fish:

$$\text{Log } 1/\text{LC}_{50} = 0.871 \log K_{ow} - 4.87$$

Highly toxic compounds such as pesticides usually act at specific sites within organisms at very low concentrations. Toxicity depends then on the characteristics of sites of metabolism and action within specific organisms—and these vary greatly based on species, strain, sex, and age. Thus, simple models of this type are only of limited value here. More complex models that consider species differences are needed to obtain better models to predict toxicity.

6.7.2.2 Microcosms and Mesocosms

Model populations or communities established in s laboratory or in the field (Section 6.6) may simulate but not exactly reproduce real world conditions. They can be used to run controlled experiments with adequate replication and demonstrate the effects of chemicals on ecological processes such as the carbon and nitrogen cycles (Chapter 14). The major problem is interpretation of results and relating them to the real world.

6.7.2.3 Theoretical Models

Data from biomarker studies, field studies, and microcosms and mesocosms can be incorporated into mathematical models that attempt to predict the effects of chemicals at or above population levels. This approach is at an early stage of development. It will be covered briefly in Section III of this text.

6.8 Summary

Concern about possible environmental impacts of new and existing chemicals has led to the development of a range of methods for testing their biological effects. End points of such tests include mortality, reproduction, and even behavior. The most widely used organisms are algae, earthworms, springtails, honeybees, *Daphnia*, fish, and birds. Several other animals have been proposed in recent years, and alternative testing methods are under active development. There are, however, problems in extrapolating the results of such tests to field conditions. Tests serve as useful safety nets for screening the most toxic chemicals but are unlikely to predict ecological effects on all species.

Further Reading

Balls, M. et al., Eds. (1991). *Animals and Alternatives in Toxicology*. Broad view of alternative test methods, including ecotoxicity tests.

Calow, P., Ed. (1993, 1994). *Handbook of Ecotoxicology*, Vols. 1 and 2. Comprehensive text, describing most common types of ecotoxicity testing and their scientific bases. Useful description of risk assessment.

Hudson, R.H., Tucker, R.K., and Háegele, M.A. (1984). *Handbook of Toxicity of Pesticides to Wildlife*.

Lewis, M.A. (1990). *Review of algal toxicity testing*.

Løkke, H. and Van Gestel. K., Eds. (1998). *Handbook of Soil Invertebrate Toxicity Tests*. Comprehensive manual.

Newman, M.C. and Unger, M.A. (2003). *Fundamentals of Ecotoxicology*, 2nd ed. Contains a useful chapter comparing ecological and human risk assessments.

Suter, G.W. (1993). *Ecological Risk Assessment*.

Timbrell, J. (1995). *Introduction to Toxicology*, 2nd ed. Straightforward account of toxicity and toxicity testing.

Walker, C.H. (1998b) and Walker, C.H. et al. (1998). Reviews of alternative methods for ecotoxicity testing.

7

Biochemical Effects of Pollutants

7.1 Introduction

When pollutants enter living organisms they cause a variety of changes (for general accounts, see Guthrie and Perry, 1980; Hodgson and Levi, 1994; Timbrell, 1995). These changes (bioeffects) are generally classified as those that serve to protect organisms against the harmful effects of the chemical and those that do not. Some examples of both types are given in Table 7.1.

We first consider protective responses. Some protective mechanisms work by reducing the concentrations of free pollutants in cells, thereby preventing or limiting interactions with cellular components that may be harmful to the organism. Organic pollutants often cause the induction of enzymes that can metabolize them (see Chapter 5). One of the most important of these enzymes is the monooxygenase system, whose function is to increase the rate of production of water-soluble metabolites and conjugates of low toxicity that can be rapidly excreted. In this case, metabolism causes detoxification. However, in a small but important number of cases, metabolism leads to the production of active metabolites (e.g., of carcinogens or organophosphorous insecticides) that can cause more damage to cells than the original compounds. Figure 7.1 depicts these reactions.

Another mechanism that reduces the bioavailability of pollutants is binding to another molecule. This can lead to excretion or storage. The metallothioneins are examples of proteins that can bind metal ions and become induced upon exposure to high metal concentrations. Inducible proteins can bind organic pollutants; this mechanism provides the basis of resistance to some drugs by removing them from cells.

In addition to protective mechanisms that control the levels of free pollutants, other responses are concerned with repairing damage caused by pollutants. The release of stress proteins falls into this category. When organisms are exposed to chemical insult or heat shock, stress proteins are released to repair the cellular damage. Similarly, if pollutants cause damage to DNA, repair mechanisms can come into play (Figure 7.2). There are many examples of homeostatic mechanisms such as these that restore cellular systems to their normal functional states after chemicals cause toxic damage.

Pollutants also cause changes unrelated to any protective function (Table 7.1). These include alterations in the levels of enzymes, receptors, and reactive intermediates and changes in DNA. In many cases, such changes cause no obvious harm to the organism and appear unrelated to toxicity, although they can seldom be regarded as beneficial. Indeed, as with protective mechanisms, the changes carry energy costs (Chapters 8 and 13).

In others cases, molecular changes have serious consequences at the level of cells or entire organisms. They may adversely affect the health of an organism so seriously that death will ensue. Here the biochemical changes are initiated at the site of action and constitute the molecular basis of toxicity. The question of the relationship between biochemical changes of this kind and toxic manifestations at higher levels of organization will be returned to later (Chapters 10 and 15). The changes caused when pollutants enter living

TABLE 7.1

Protective and Nonprotective Responses to Chemicals

Type of Effect	Example	Consequences
Protective	Induction of monooxygenases	Usually increase rate of metabolism of pollutant to more water-soluble metabolite and thus increase rate of excretion; occasional activation
	Induction of metallothionein	Increases rate of binding with metals to decrease bioavailability
Nonprotective (may or may not lead to toxic manifestations)	Inhibition of AChE	Toxic effects seen above 50% inhibition
	Formation of DNA adducts	May cause harmful effects if leading to mutation

organisms may be specific for certain types of chemicals or they may be nonspecific. Some examples, in decreasing order of specificity, are given in Table 7.2. The use of these changes to assess the impacts of chemicals is considered in Chapter 10 on biomarkers (Walker, 1992).

Similarly, responses to any pollutant may be specific to certain groups of organisms or may occur in nearly all organisms. The degree of induction of different cytochrome P_{450} forms in response to particular pollutants shows considerable phylogenetic variation among species. Thus some inducing agents such as barbiturates produce marked responses in mammals but no responses in fish or some birds. In the next section, we will discuss two categories of biochemical effects (i) those that exert protective functions and (ii) those that represent the molecular modes of action of pollutants. We will consider the consequences of these effects at the levels of the cell and the whole organism.

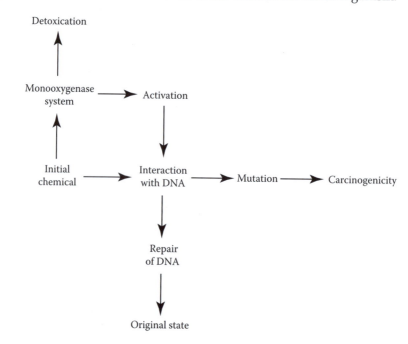

FIGURE 7.1

Pathways of activation and detoxification of chemicals.

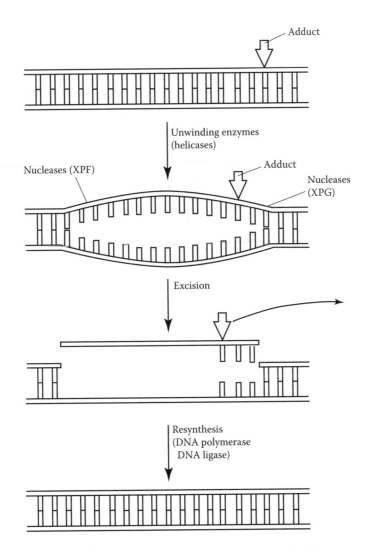

FIGURE 7.2
Mechanism of DNA repair after adduct formation to remove covalently bound adduct.

TABLE 7.2

Specificities of Responses

Biological Response (Biomarker)	Chemicals	Comment
Inhibition of ALAD	Lead	Specific for lead
Inhibition of vitamin K cycle	Anticoagulant rodenticides, e.g., warfarin	Antagonists for vitamin K
Inhibition of AChE	Organophosphorous compounds and carbamates	Specific to these two classes; OPs can be separated from carbamates by strength of binding to AChE
Induction of monooxygenases	Organochlorines, polynuclear aromatics	Many man-made and natural chemicals cause induction; particular P_{450} isoforms may be induced by certain types of pollutants

7.2 Protective Biochemical Responses

When the concentration of a xenobiotic or inorganic ion exceeds a certain level in cells, it may trigger responses designed to protect the organism against potential toxic effects. Very commonly, this response is an increase in the quantity of a protein that can facilitate the removal of the molecule or ion. In the case of lipophilic xenobiotics, a number of enzymes are induced to increase the rate of biotransformation of the molecule to water-soluble and readily excretable metabolites and conjugates. (Induction involves an increase in the activity of an enzyme as a consequence of an increase in its cellular concentration produced in response to a chemical.) Prominent among these enzymes are the monooxygenases of the endoplasmic reticulum of vertebrates and invertebrates (Chapter 5) that contain cytochrome P_{450} as their catalytic centers. A number of inducible forms exist in the livers of vertebrate animals. Induction involves an increase in both cytochrome P_{450} and the associated enzyme activities. In mammals, a number of inducible forms of cytochrome P_{450} belong to a single gene family designated family 2 (Nebert and Gonzalez, 1987).

Many lipophilic xenobiotics are both inducers and substrates for this type of cytochrome P_{450}. Another group of enzymes designated cytochrome P_{450} family 1 has a much more restricted range of inducers and substrates. These enzymes interact particularly with flat (coplanar) molecules such as polycyclic aromatic hydrocarbons (PAHs), coplanar PCBs, and dioxins. Although metabolism by a cytochrome P_{450} usually causes detoxification, exceptions occur. In particular, induction of a cytochrome P_{450} of family 1 can cause increased activation of carcinogens (e.g., certain PAHs) and coplanar PCBs that can act as thyroxine antagonists. The induction of different types of cytochrome P_{450} in diverse groups of animals is reviewed in Livingstone and Stegeman (1998).

When certain metal ions exceed a critical cellular level, another type of protein is induced. Metallothioneins can increase in concentration after exposure to various metals. These are binding proteins, rich in SH groups, that can lower cellular concentrations of metal ions such as Cu^{2+}, Cd^{2+}, and Hg^{2+} by sequestration (Hamer, 1986).

These responses are concerned with prevention of toxic damage by the simple strategy of removing potentially harmful xenobiotics and ions before they interact to a significant degree with their sites of action. Other responses repair damage after it occurs. Important examples are the production of stress proteins (Korsloot et al., 2004) and the operation of DNA repair mechanisms. The effects of chemicals and heat shock can cause the release of stress proteins into cells. Their functions are repairing cellular damage and protecting cellular proteins against denaturation.

Evidence indicates that the induction of stress proteins in response to chemicals and other stressors is a component of an integrated cellular stress defense system that includes metallothioneins, enzymes responding to oxidative stress such as superoxide dismutase, and detoxifying enzymes such as microsomal monooxygenases. A recent text by Korsloot et al. (2004) presents evidence for the existence of an integrated defense system of this kind in arthropods. The induction of stress proteins has been proposed as the basis for biomarker assays, including microarray analyses (see Box 15.1). When chemicals damage DNA by forming adducts, the activity of DNA repair mechanisms increases (Figure 7.2). "Unscheduled" DNA synthesis in response to the action of genotoxic compounds provides the basis for biomarker assays.

7.3 Molecular Mechanisms of Toxicity

An understanding of mechanisms of toxicity at the molecular level is important for several reasons. First, it can pave the way for developing antidotes to the adverse effects of pollutants. Second, the understanding of the mechanism can lead to the development of assays that can demonstrate and measure the deleterious effects of chemicals. Both benefits are important in human medicine. Unsurprisingly much of the research in this area has been in the field of medical toxicology.

Mechanisms of toxicity have often been investigated in rats, mice and other mammals that serve as surrogates for humans. Less work has been done with birds, fish, and reptiles—animals of interest in ecotoxicology. Some work has been done with insects related to insecticides and pest control, including the investigation of mechanisms of resistance. An example is the inhibition of acetylcholinesterase by the active forms of an organophosphorous insecticide (OP) shown in Figure 7.3. The OP interacts with a hydroxyl group that is a functional part of the enzyme. The phosphorylated enzyme produced has no activity, i.e., it cannot hydrolyze the natural substrate acetylcholine. Another area of interest is

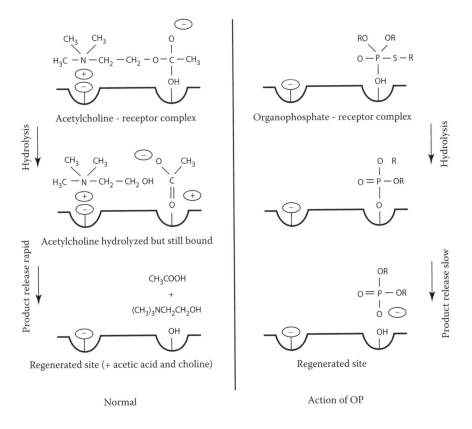

FIGURE 7.3

Mechanism of action of AChE. Under normal conditions, acetylcholine binds to acetylcholinesterase and is broken down (hydrolyzed) to yield acetic acid and choline that break away from the enzyme. Organophosphates bind to hydroxyl groups belonging to the serine amino acid, which is part of the binding site shown on the right side of the enzyme surface. When this happens, the enzyme is inhibited and can no longer hydrolyze acetylcholine.

developing new biocides. If the site of action of a pesticide is known, a molecular model can be constructed. Other molecules that will fit the same site and interact there can then be identified. Molecules that fit the model can then be synthesized and tested as novel biocides (Box 6.8).

Some of the most toxic compounds known are highly reactive and readily form covalent bonds with their sites of action. Frequently, reactive molecules of this type are generated by enzymatic attacks on compounds that alone are unreactive and considered nontoxic. This is the case for many chemical carcinogens such as benzo(a)pyrene (BaP), dibenzo(Ah), anthracene, aflatoxin, acetylaminofluorene, and others. The enzymes involved (often in the monooxygenase system that usually protects organisms) can sometimes generate the reactive compounds that cause cellular damage. This phenomenon is considered in more detail in Section 7.4.1 on genotoxic compounds.

Toxic effects may arise from molecular interactions that do not lead to the formation of covalent bonds. The reversible binding of a molecule to a site on a cellular macromolecule can lead to toxicity. The target may be a receptor site for a chemical messenger, such as a receptor for acetylcholine or GABA (gamma aminobutyric acid), or a pore through a membrane that normally allows the passage of ions (e.g., the blockage of Na^+ channels by tetrodotoxin).

To gain a full understanding of the toxic effects of chemicals, it is necessary to link initial molecular interactions to consequent effects at higher levels of organization. The extent to which such a molecular interaction occurs is generally related to the dose of the chemical received, although the relationship is rarely a simple one. Dose–response relationships are complex for several reasons. Low-level exposure, up to a certain threshold, may produce no measurable interactions. In some cases, protective mechanisms remove the chemical before it can reach its site of action. For this reason, the results of in vitro and in vivo experiments often exhibit major differences. In other cases, when the dose exceeds a certain threshold level, protective mechanisms may come into play that reduce the amount of chemical reaching its active site. An example of this is the induction of metallothionein that decreases the bioavailability of toxic metals such as cadmium.

Another reason for a lack of effect at low levels of exposure is that many systems have reserve capacities. The carbonic anhydrase enzyme that catalyzes the conversion of carbon dioxide to carbonic acid must be inhibited by more than 50% before physiological effects are seen. Similarly, the inhibition of brain acetylcholinesterase must usually exceed 50% before overt physiological disturbances are seen (Table 7.1). However, subtle disturbances of behavior have been observed in fish at lower levels of inhibition (see Section 8.4.2). In other cases, especially in carcinogenicity, it has been argued that no minimum safe level exists. In theory, a single molecular interaction can initiate an entire process that leads to the development of cancer.

Because of the complexity of the interactions involved in toxicity, dose–response relations may not be straightforward. One example is DDE-induced eggshell thinning in predatory birds (Section 16.1). Extensive shell breakage starts to occur when shell thinning occurs to the extent of 17% to 20%. Clearly, other complications can apply to relating doses to toxic manifestations at higher levels of organization (e.g., toxic effects seen at cell function or whole organism level). The question is more than the relationship between dose and degree of molecular interaction (e.g., the percentage inhibition of an enzyme). A complex relation may involve the degree of molecular interaction and consequent higher level effects, possibly because of the interventions of protective mechanisms such as the induction of stress proteins. In studying responses to toxic molecules, it is very important to construct appropriate dose–response curves over a whole range of exposures that are likely to occur. It should not be assumed that a straight line relationship found at low levels of exposure can be extrapolated to higher levels of exposure.

7.4 Examples of Molecular Mechanisms of Toxicity

Molecular interactions between xenobiotics and sites of action that lead to toxic manifestations may be highly specific for certain types of organisms or nonspecific. In the simplest case, molecules are highly selective between two species because one species exhibits a certain site of action that does not occur in the other. For example, the pesticide dimilin acts on the site of formation of chitin and thus affects only arthropods that utilize chitin to form their exoskeletons. The organophosphorous insecticides that act on the nervous system are toxic to all animals but have little or no toxicity for plants.

Animals have sites of action in their nervous systems (in the example given, a form of acetylcholinesterase, discussed in more detail in Section 7.4.2), whereas plants have no nervous systems and no similar sites of action. Some compounds show little selective toxicity and may be regarded as general biocides. An example is dinitroorthocresol (DNOC) and related compounds that act upon mitochondrial membranes and cause the uncoupling of oxidative phosphorylation. This system is found in all eukaryotes, and uncouplers of this type can run down the gradient of protons across mitochondrial membranes, thus inhibiting or preventing the synthesis of ATP.

Some of most subtle examples of mechanisms of selectivity have been found in resistant strains of insects, a subject dealt with in Chapter 13. Two strains of the same species can have different forms of the same site of action—one susceptible to a toxic molecule, the other nonsusceptible. Some strains of insects resistant to organophosphorous insecticides have forms of acetylcholinesterase that differ from those of susceptible strains of the same species. The difference between the susceptible and the resistant forms of acetylcholinesterase may be simply a change of one amino acid in an entire molecule. Thus a very small difference in the structure of a site of action can cause a large difference in toxicity. An organophosphorous insecticide may show little tendency to interact with the resistant form of an enzyme. We will now discuss effects in animals and plants.

7.4.1 Genotoxic Compounds

Many compounds that act as carcinogens are known to cause damage to DNA, i.e., they are genotoxic, and it is strongly suspected that the relationship is causal. When cells with damaged DNA divide, mutant cells can result. Some mutant cells are tumor cells that will follow uncoordinated growth patterns and may migrate within the organism to produce secondary growths (metastases) in other locations.

The relationship between DNA changes and harm to organisms is complex. Although adduct formation (covalent binding of a pollutant to DNA) is a good index of exposure, the relationship of adduct formation to harm is less well defined. For the most part, these DNA adducts are short lived. DNA repair mechanisms quickly excise the adducted structures and replace them with the original moiety (Figure 7.2). Sometimes, however, the adducts are relatively stable. When a cell divides, the adducted element is mistranslated and a mutant cell is produced. Thus, whereas data clearly relate the number of DNA–BaP adducts to the amount of cigarette smoking, the relationship between DNA–BaP adducts and lung cancer is less well established.

Certain PAHs such as BaP and dibenzo(Ah) anthracene, acetylaminofluorene, aflatoxin, and vinyl chloride are all examples of genotoxic pollutants. In all cases, the original compound does not interact with DNA. Indeed, the original compounds are relatively stable and unreactive. Enzymatic metabolism (mainly oxidative metabolism by one or more forms of monooxygenase) produces highly reactive and short-lived metabolites that can bind to DNA (Figure 7.1).

This type of molecular interaction is common. A significant number of pollutants are known to have mutagenic properties. Certain PAHs serve as a case in point. PAHs are present in smoke, soot, and crude oil. They are therefore present in urban areas due to traffic, smoke, and soot and in marine locations where oil spillages have occurred.

7.4.2 Neurotoxic Compounds

The nervous systems of both vertebrates and invertebrates are sensitive to the toxic effects of chemicals. Neurotoxins are both naturally occurring and man-made. Among the natural neurotoxins are tetrodotoxin (from puffer fish), botulinum toxin (from the anaerobic bacterium *Clostridium botulinum*), atropine (from deadly nightshade *Atropa belladonna*), and the natural insecticides nicotine (from wild tobacco *Nicotiana tabacum*) and pyrethrin (from the flowering heads of *Chrysanthemum* species) and many more. Among man-made compounds, it is interesting that the five major groups of insecticides—organochlorine, organophosphorous, carbamate, and pyrethroid—all act as nerve poisons.

All these examples disturb the normal transmission of impulses along nerves and/or across synapses (junctions between nerves or between nerve endings and muscle or gland cells) in some way. However, a distinction can be made between compounds that act directly on receptors or pores situated in nerve membranes and those that inhibit the acetylcholinesterase (AChE) of synapses. These two groups will now be considered separately.

The passage of an action potential along a nerve depends on the flow of Na^+ and K^+ across the nerve membrane. During the normal passage of an action potential, Na^+ channels (Figure 7.4) are open for a brief instant, allowing the inward flow of Na^+ ions. The channels then close to terminate the Na^+ flow. Pyrethroid insecticides, natural pyrethrins, and DDT all interact with Na^+ channels to disturb this function. The usual consequence of their interaction is retarded closure of the channels. This can prolong the flow of Na^+ ions across the membrane, leading to disturbance of the normal passage of the action potential. This type of poisoning causes uncontrolled repetitive spontaneous discharges along nerves. Several action potentials instead of only one are generated in response to a single

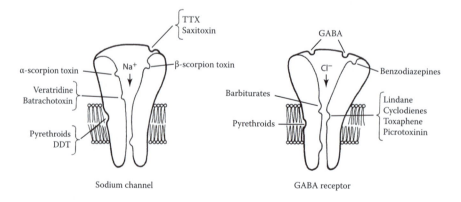

FIGURE 7.4
Sodium channels and GABA receptors. (*Source:* Eldefrawi, M.E. and Eldefrawi, A.T. (1990). In *Safer Insecticides: Development and Use.* Marcel Dekker. With permission.)

stimulus. Uncoordinated muscular tremors and twitches are characteristic symptoms of this type of poisoning.

The interaction of these hydrophobic water-insoluble compounds with Na$^+$ channels is reversible and does not appear to involve the formation of covalent bonds. It is likely that these compounds first dissolve in the lipids of the nerve membranes before interacting with a site on the Na$^+$ channel spanning the membrane.

Another site of action for insecticides and other neurotoxins is the GABA receptor that functions as a Cl$^-$ channel through nerve membranes (Figure 7.4). Chlorinated cyclodiene insecticides and their active metabolites (dieldrin, endrin, heptachlor epoxide), act as GABA antagonists. By binding to the receptors, they reduce the flow of Cl$^-$ ions. So, too, does the insecticide γ-HCH (lindane) and naturally occurring picrotoxin. In vertebrates, these compounds act on GABA receptors of the brain. Convulsions are typical of this type of poisoning.

Receptors for acetylcholine that are located on postsynaptic membranes (i.e., on the other side of the synapses from nerve endings) represent the sites of action for a number of chemicals. Thus nicotine can act as a partial mimic (agonist) of acetylcholine at what are termed nicotinic receptors for acetylcholine. Atropine can act as a competitive inhibitor on muscarinic receptors for acetylcholine. Nicotinic receptors and muscarinic receptors differ in structure, their responses to toxic chemicals, and their locations in the nervous system. In vertebrates, nicotinic receptors are found especially at neuromuscular junctions and on many synapses of autonomic ganglia. By contrast, muscarinic receptors are found on synapses at the endings of parasympathetic nerves—typically on the membranes of smooth muscles or glands.

Nicotine, a natural product, has been used as an insecticide. Neonicotinoid insecticides such as imidacloprid act similarly to nicotine but are more lipophilic than nicotine and more effective as insecticides.

Compounds that inhibit AChE represent one of the most toxic groups of compounds to vertebrates and invertebrates. As noted earlier (Chapter 1), OPs are particularly important. Some were developed for chemical warfare; others were designed as insecticides. Another group of anticholinesterases are the carbamates that exert insecticidal action. (These should not be confused with other carbamates used as herbicides or fungicides that do not act as anticholinesterases.)

The toxic actions of these compounds are indirect in that their primary effect is inhibiting the action of AChEs that have the function of breaking down acetylcholine released into synapses.

Acetylcholine release occurs from the endings of cholinergic nerves, and the acetylcholine functions as a chemical messenger. When an impulse reaches a nerve ending, acetylcholine is released and carries the signal across the synaptic cleft to a receptor on the postsynaptic membrane of a nerve cell, muscle cell, or gland cell (see Section 8.4.1). When acetylcholine interacts with its receptor, a signal is generated on the postsynaptic membrane, so that the impulse (message) is carried on.

For effective neuronal control, it is essential that this signal be rapidly terminated. To achieve this, acetylcholine must be rapidly broken down by AChE in the vicinity of the receptor (Figure 7.3). Anticholinesterases have the effect of reducing or preventing altogether the breakdown of acetylcholine. As a consequence, acetylcholine builds up in synapses, leading to overstimulation of the receptor and the continued production of a signal after it should have stopped. If this situation continues, the signal system will eventually run down, resulting in synaptic block and acetylcholine will no longer be able to relay signals

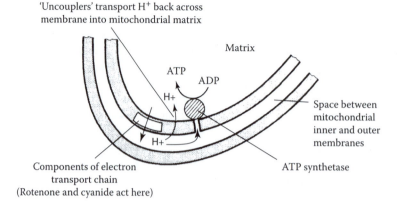

'Uncouplers' transport H⁺ back across
membrane into mitochondrial matrix

Matrix

ATP

ADP

H+

Space between
mitochondrial
inner and outer
membranes

H+

Components of electron
transport chain
(Rotenone and cyanide act here)

ATP synthetase

FIGURE 7.5

Cross section of a mitochondrion showing mitochondrial poison reaction. Protons (H⁺) are actively transported from the matrix across the inner membrane via electron flow along the electron transport chain. These protons can then flow back to the matrix via the ATP synthetase enzyme; the associated energy is used to synthesize ATP in the matrix. Uncouplers such as 2,4-dinitrophenol can carry these protons to the matrix before ATP synthesis occurs. Other poisons, for example, rotenone and CN^-, can inhibit the flow of electrons along the transport chain.

across the synapse. In the case of neuromuscular junctions so affected, tetanus will result; muscles in a fixed state will be unable to contract or relax in response to nerve stimulation.

7.4.3 Mitochondrial Poisons

Mitochondria play a vital role in energy transformation and are found in all eukaryotes. It is not, therefore, surprising that some of the most dangerous nonselective biocides act upon mitochondrial systems.

Uncouplers of oxidative phosphorylation such as 2,4-dinitrophenol fall into this category. When a mitochondrion functions normally, it has an electrochemical gradient of protons across its inner membrane (Figure 7.5). In fact, the inner membrane exhibits an excess of protons on the outside and a deficiency on the inside.

The maintenance of this gradient depends on the impermeability of the mitochondrial membrane to protons. The production of ATP by mitochondria is driven by energy stored in this proton gradient. If the proton gradient is lost, ATP production will cease. The compounds in question can eliminate this gradient, dissipating in the form of heat the associated energy and thus preventing the energy stored in a proton gradient from being used to biosynthesize ATP. Other mitochondrial poisons such as the naturally occurring insecticide rotenone and cyanide ions can inhibit the operation of the electron transport chain of the inner membranes of mitochondria, thus preventing the production of the proton gradient by interacting with components of the electron carrier system.

7.4.4 Vitamin K Antagonists

Warfarin and certain related rodenticides are toxic to vertebrates because they act as antagonists of vitamin K. Vitamin K plays an essential role in the synthesis of clotting proteins in the liver. It undergoes a series of cyclical changes (vitamin K cycle) during the course of which the clotting proteins become carboxylated. After carboxylation has occurred, the clotting proteins are released into the blood, where they play a vital role in the process of clotting, which occurs when there is damage to blood vessels. Warfarin and related compounds have a structural resemblance to vitamin K and strongly compete with it for binding sites, even at low concentrations. This leads to inhibition of the vitamin K cycle and consequently to the incomplete synthesis of clotting proteins. Under these conditions, the levels of clotting proteins fall in blood, and the blood loses its ability to clot. Failure of clotting results in hemorrhage.

Flocoumafen and related compounds, known as second generation anticoagulant rodenticides, were developed to circumvent resistance to warfarin. These rodenticides become lethal when the receptor sites in the liver are blocked. Flocoumafen exhibits dramatic differences in toxicities between rats and quail (LD_{50} 0.25 mg/kg and 300 mg/kg, respectively.). An important factor in this species difference is that the rodenticide is metabolized more rapidly by quail than by rats (Huckle et al., 1989).

7.4.5 Thyroxine Antagonists

The thyroid gland produces two hormones that markedly affect metabolic processes in many tissues. One of these, thyroxine (1–3,3',5,5'-tetraiodothyronine or T_4), binds to the transthyretin (TTR) protein that is part of a transport protein complex found in blood. The other part of the complex consists of a second protein to which is bound retinal (vitamin A; Figure 7.6). Thus transthyretin–thyroxine is attached to the retinol-binding protein (RBP). Certain hydroxy metabolites of the PCB congener 3,3'-4,4'-tetrachlorobiphenyl (3,3'-4,4'-TCB) compete with thyroxine for its binding site on TTR. In particular, the metabolite 4'-hydroxy-3,3',4–5' tetrachlorobiphenyl binds very strongly and effectively competes with thyroxine (Brouwer et al., 1990). These metabolites are formed via metabolism of the original PCB congener by a particular cytochrome P_{450} form of the monooxygenase system (P_{450} 1A1).

Binding produces two consequences. First, thyroxine is displaced and lost from the blood. Second, the retinol complex breaks away, and retinol is lost from the blood. Thus the levels of thyroxine and retinol will fall because of the production of certain PCB metabolites.

7.4.6 Inhibition of ATPases

The ATPases (adenosine triphosphatases) are a family of enzymes (Na^+- and K^+-ATPases, Ca^{2+}-ATPase, etc.) involved in the transport of ions. The osmoregulation of a variety of organisms and the effects of a number of organochlorines on this process have been investigated. The avian salt gland that enables pelagic seabirds to maintain their salt balances in a marine environment is dependent on ATPases. Another ATPase-dependent organ is the avian oviduct. The inhibition of Ca^{2+}-ATPase involved in the transport of calcium by DDE (the persistent DDT metabolite) is considered the basis of DDE-induced eggshell thinning. This phenomenon is considered in detail in Chapter 15.

7.4.7 Environmental Estrogens and Androgens

Although endocrine disruptors have been known for many years, they recently attracted more interest and now represent one of the most controversial environmental issues. Wide ranging reviews of the problem are found in Tyler et al. (1998), Vos et al. (2000), and Janssen et al. (1998). An estrogenic chemical can imitate an estrogen by binding to the estrogen receptor. One mechanism is for the pseudo-estrogen to bind and thus stimulate transcription activity, resulting in the induction of estrogenic processes. Another mechanism is for

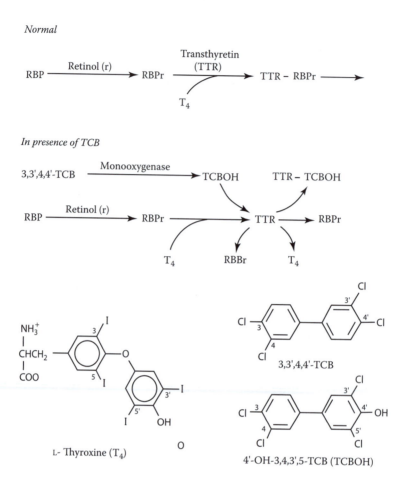

FIGURE 7.6

Mechanism of toxicity of a polychlorinated biphenyl. Retinol binds to retinol-binding protein (RBP) that is then attached to transthyretin (TTR). Thyroxine (T_4) binds to TTR and is transported via the blood in this form. The coplanar PCB 3,3′,4,4′-tetrachlorobiphenyl (3,3′,4,4′-TCB) is converted into hydroxymetabolites by the inducible cytochrome P_{450} called P_{450} 1A1. The metabolite 4′-OH-3,3′,4,5′-tetrachlorobiphenyl (TCBOH) is structurally similar to thyroxine and strongly competes for thyroxine binding sites. The consequences are the loss of thyroxine from TTR, the fragmentation of the TTR–RBPr complex, and the loss of both thyroxine and retinol from blood. (*Source:* Brouwer, A. et al. (1990). *Aquatic Toxicology* 15, 99–105. With permission.)

the compound to act as an antiestrogen, binding strongly to the estrogen receptor and thus blocking the effects of endogenous estrogens. This can result in masculinization of an organism.

Many man-made chemicals are estrogenic including some of the common organochlorine insecticides, tributyl tin, phthalates, and nonylphenols. The estrogenic activity of these compounds varies widely. For example, the *o,p′* isomers of DDT and DDE are much more active than the *p,p′* isomers. In addition, there are the many natural estrogenic compounds. Over 300 plants belonging to more than sixteen families have been shown to contain substances with estrogenic activity. The best known are isoflavonids and the coumestans. The resorcyclic acid lactones are found in fungi. The interest in endocrine-disrupting chemicals led to the development of rapid screening techniques such as transcription of an estrogen-responsive reporter gene in transfected cells, production of vitellogenin in a hepatocyte culture of rainbow trout cells, and recombination of yeast cell cultures containing the human estrogen receptor gene.

The best documented cases of endocrine disruption caused by pollutants are the induction of vitellogenin synthesis in male fish and masculinization of female gastropods. Vitellogenin is a protein synthesized in the livers of female fish and transported to oocytes to form the yolks of the eggs. It is well established that the induction of vitellogenin is under estrogenic control. Elevated levels of vitellogenin were found in fish exposed to sewage treatment effluent (Purdom et al., 1994) after hermaphrodite fish had been reported in treatment lagoons. This phenomenon is considered in more detail in Chapters 10 and 15.

Pollutants with androgenic activities can cause masculinization of female gastropods. Imposex (the development by a female of male characteristics such as a penis and a vas deferens) has been widely reported in marine gastropods near marinas. Detailed studies have been conducted at the Plymouth Marine Laboratory since the first finding of imposex in dog whelks (*Nucella lapillus*) in Plymouth Sound in 1969. Laboratory experiments and in situ transfer experiments show that imposex is initiated in dog whelks by tributyl tin at very low concentrations. This phenomenon is considered in more detail in Chapter 16.

The estrogenic or androgenic activity of pollutants may be due to their direct or indirect actions. On the one hand, they may act as agonists against a receptor as, for example, in the case of the action of ethinylestradiol (EE2) on estrogenic receptors (Chapter 15). On the other, they may disturb the metabolism of the natural transmitter and this may lead to an increased concentration of natural transmitter in the vicinity of the receptor. Tributyl tin, for example, appears to act indirectly by inhibiting the cytochrome P_{450} isozyme aromatase, an enzyme that is involved in the synthesis of testosterone, with consequent elevation of the tissue levels of testosterone (Bettin et al., 1996; Matthiessen and Gibbs, 1998).

By contrast, pollutants may act as antagonists when they bind to estrogenic or androgenic receptors; in other words, they may act as antiestrogens or antiandrogens. In this case, the normal action of the hormone is inhibited because binding to the receptor is retarded. The antiandrogenic action of *p,p'*-DDE in rats has been attributed to this mechanism (Kelce et al., 1995).

7.4.8 Reactions with Protein Sulfhydryl (SH) Groups

The Hg^{2+} and Cd^{2+} ions are toxic to many animals. The main reason appears to be their ability to combine with sulfhydryl (thiol) groups, thereby preventing normal function. Sulfhydryl groups on enzymes and other proteins play important functional roles, e.g., the formation of disulfide bridges and consequent conformational changes in the proteins. With this kind of toxic interaction, it is difficult to establish which sulfhydryl groups on which proteins represent the sites of toxic action.

Organomercury is more lipophilic than inorganic mercury and is distributed differently. It tends to move into fatty tissues and cross membranes, including those of the blood–brain barrier. As a consequence, the toxicity of organomercurial compounds tends to be expressed in the brain, whereas inorganic mercury toxicity is expressed in peripheral tissues. Other organometallic compounds, for example, tetraalkyl lead, also exert their toxic actions in the brain.

7.4.9 Photosystems of Plants

A number of herbicides that show little toxicity to vertebrates and insects act as poisons of the photosynthetic systems of plants. Substituted ureas and triazines are examples. By mechanisms that are not yet clear, they interrupt the flow of electrons through the photosystems responsible for the light-dependent reactions of photosynthesis, i.e., the splitting of water to release molecular oxygen.

7.4.10 Plant Growth Regulator Herbicides

A group of herbicides sometimes called the phenoxyalkanoic herbicides have growth-regulating properties. MCPA, 2,4-D, CMPP, and 2,4-DB are well known examples. The molecular mechanism by which they affect growth and the sites where they act have never been clearly established. However, it is interesting that exposing plants to them causes the production of ethylene—a growth-regulating compound.

These herbicides show very low toxicity in vertebrates and insects. It is presumed that they do not have receptors for growth regulation similar to those of plants. However, significant amounts of TCDD have been found as by-products of synthesis in samples of 2,4-D and 2,4,5-T (Section 1.2.3).

7.5 Summary

Pollutants cause a wide variety of biochemical effects in organisms, broadly classified as protective and nonprotective responses. Protective responses include the induction of enzymes that have detoxifying functions and the induction of proteins that can bind metals. Microsomal monooxygenases are examples of inducible enzymes that usually exert detoxifying functions. However, the situation is complicated by the relatively small (but critically important) number of cases in which monooxygenases cause activation rather than detoxification. Such protective responses bear energy costs that may have detrimental effects on organisms (Chapter 8).

Many responses such as the inhibition of cholinesterase, electrophysiological changes caused by organochlorine insecticides, antagonism of vitamin K, endocrine disruption, and formation of DNA adducts, are nonprotective and in many instances harmful to organisms. It is important to elucidate the mechanisms by which chemicals express toxicity. Understanding the mechanisms of toxicity can provide the basis for biomarker assays and lead to the development of antidotes to poisoning.

Further Reading

Hassall, K.A. (1990). *The Biochemistry and Uses of Pesticides*, 2nd ed. Describes the modes of action of many pesticides.

Hodgson, E. and Kuhr, R.J. (1990). *Safer Insecticides: Development and Use*. Detailed account of the modes of action of certain insecticides.

Hodgson, E. and Levi, P. (1994). *Introduction to Biochemical Toxicology*, 2nd ed. Cites many examples of mechanisms of action of toxicants, including most of those mentioned in this chapter.

Korsloot, A., Van Gestel, C.A.M., and Van Straalen, N.M. (2004). *Environmental Stress and Cellular Response in Arthropods*. Describes the induction of stress proteins and other cellular responses to chemicals in arthropods, presenting evidence for the operation of an integrated system dealing with stressors generally.

Shore, R.F. and Rattner, B.A., Eds. (2001). *Ecotoxicology of Wild Mammals*. Reference work detailing the toxic actions of major pollutants against mammals of different groups. Describes biologies of different groups of mammals before dealing with their responses to toxic effects of chemicals.

Timbrell, J.A. (1999). *Principles of Biochemical Toxicology*, 3rd ed. Long chapter devoted to important biochemical mechanisms in medical toxicology.

Walker, C.H. (2009). *Organic Pollutants: An Ecotoxicological Perspective*. Details biochemical mechanisms of toxicity including chapters on endocrine disruptors and neurotoxicity.

8

Physiological Effects of Pollutants

8.1 Introduction

Pollutants may damage organisms with lethal consequences (as described in Chapters 6 and 7). The effect on the population is then an increase in the mortality rate of at least one age class. Alternatively, there may be damage to, or effects on, the machinery of reproduction or resource acquisition and uptake. These effects are described in detail in this chapter. Where resource uptake is reduced, there are consequent reductions of birth rate and/or somatic growth rate (here referred to jointly as production rate), and these depress the population growth rate.

Detoxication mechanisms generally use resources, including energy, and these resources are consequently not available for production. Thus production is likely to be reduced when detoxication occurs. The overall effects of detoxication mechanisms on production and mortality are considered at the end of this chapter. Note at this stage, however, that either damage or detoxication may result in reduced production in a polluted environment.

With these points in mind, the effects of pollutants will be considered at several different levels of organization, from organelles to whole organisms (Figure 8.1). In this chapter, we are concerned mainly with physiological effects above the biochemical level (covered in Chapter 7). Physiology is defined as the branch of science concerned with the functioning of organisms and the processes and functions of all or part of an organism. Thus, in ecotoxicology, we are concerned with describing the disruptive effects of pollutants on normal physiology; normal referring to the state of the organism when not exposed to pollution although subject to other forms of stress. Abnormal stresses may be completely novel and occur in response to man-made chemicals that have appeared in the environment recently (on the geological time scale), or they may simply be an increase in a response to a substance to which the organism has evolved natural protective mechanisms (e.g., metals).

8.2 Effects of Pollutants at Cellular Level

When a pollutant enters a cell, it may trigger certain biochemical responses that have evolved either to break the chemical down (Chapter 7) or store it in such a way that it is hidden within a compartment, preventing interference with essential biochemical reactions within the cell. For example, invertebrates have clearly defined pathways for metal detoxification (Beeby, 1991; Hopkin, 1989, 1990; Dallinger, 1993; Vijver et al., 2004). Perhaps the simplest case is the epithelium of the digestive system in terrestrial invertebrates. It is usually only one cell in thickness and acts as a barrier between the internal environment of the animal (i.e., blood bathing the organs) and the food in the lumen. Therefore, storage mechanisms and/or methods of exclusion have to be extremely efficient, because

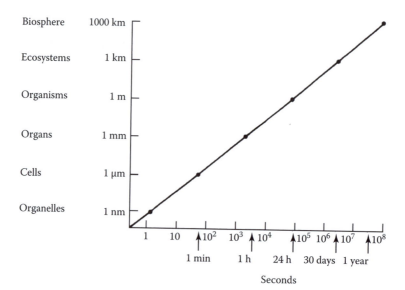

FIGURE 8.1
Relationship between complexities and sizes of natural systems and compartments (black circles) and typical response times to pollutant insults.

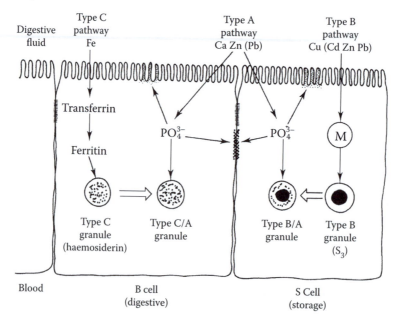

FIGURE 8.2
Three pathways of detoxification of metals from digestive fluids by the B and S cells of the hepatopancreas of the woodlouse *Porcellio scaber*. M = metallothionein. S_3 = sulfydryl group. (*Source:* Hopkin, S.P. (1990). *Functional Ecology* 4, 321–327. With permission.)

terrestrial invertebrates (unlike aquatic organisms) cannot excrete xenobiotics from the blood into the external medium across the respiratory surfaces if they are taken up to excess. In land mammals, lipophilic organics are converted to water-soluble metabolites and conjugates that are then excreted in the bile and urine (see Chapter 5).

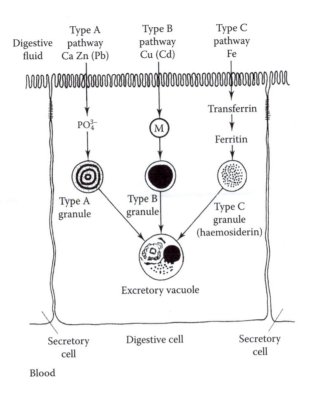

FIGURE 8.3
Three pathways of detoxification of metals from digestive fluids by the digestive cells in the hepatopancreas of the woodlouse-eating spider *Dysdera crocata*. M = metallothionein. (*Source:* Hopkin, S.P. (1990). *Functional Ecology* 4, 321–327. With permission.)

Three main detoxification pathways have evolved for the binding of metals that enter the epithelial cells (Figures 8.2 and 8.3). Although the chemistry of binding appears similar in all terrestrial invertebrates, the subsequent fate of the waste material is controlled by the digestive processes of the animals in question. An example of these differences is provided by the contrasting ways in which the metals are bound in the hepatopancreas of both the terrestrial isopod *Porcellio scaber* and a major predator of isopods, the woodlouse-eating spider *Dysdera crocata* (Figure 8.4; Box 8.1).

Osmoregulation is affected in a wide variety of organisms ranging from crabs to birds and by a wide variety of pollutants (heavy metals, organochlorines, and organophosphorous compounds). The basic mechanisms for water and salt transport in saltwater fish are given in Figure 8.6. Kinter et al. (1972) found that the sodium space increased and the activity of ATPase significantly reduced with DDT and PCBs and the sodium concentration of the serum increased.

The avian salt gland is ATPase-dependent. This enables seabirds to maintain their water balances in a marine environment by ingesting sea water and excreting a highly concentrated salt solution. However, experiments on seabirds did not indicate that organochlorines had a significant effect on osmoregulation. Another ATPase-dependent gland is the avian oviduct. The inhibition of Ca-ATPase is considered to be involved in DDE-induced eggshell thinning, which caused marked declines in many raptorial species. This phenomenon is covered in Chapter 15.

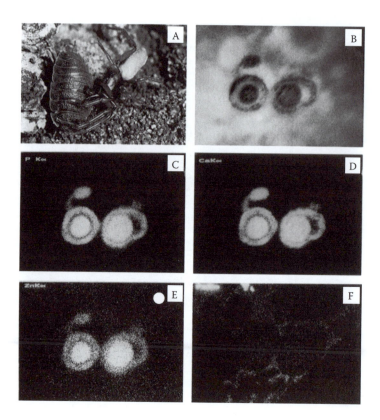

FIGURE 8.4

(A) *Dysdera crocata* attacking a specimen of *Porcellio scaber* 1 cm in length. (B) Scanning transmission electron micrograph (bright-field image) of unstained resin-embedded section (0.5 μm thick) through two type A calcium phosphate granules in a digestive cell of the hepatopancreas of *D. crocata*. Diameter of each granule section = 1 μm. Electron-generated x-ray maps for phosphorus (C), calcium (D), zinc (E), and iron (F) of the type A granules shown in (B). Philips CM12 STEM, EDAX 9900 x-ray analyzer, screen resolution 256–200 pixels, dwell time on each pixel 50 ms, spot diameter 20 nm. (*Source:* Hopkin, S.P. (1990). *Functional Ecology* 4, 321–327. With permission.)

8.3 Effects at Organ Level in Animals

When pollutants are ingested by organisms or pass into the blood across respiratory epithelia or the external surfaces of the body, they may be compartmentalized in organs within the body—accidentally or deliberately. For example, radioactive isotopes of essential elements travel to the same locations in tissues as their nonradioactive counterparts. Radioactive iodine accumulates in the thyroid glands of mammals and may cause thyroid cancer. Cadmium is accumulated in the livers and kidneys of mammals. The symptoms of cadmium poisoning are proteinuria (excretion of proteins in urine) resulting from breakdown of the kidney cells when cadmium levels exceed a critical concentration.

This concept of a critical target organ concentration is important in ecotoxicology. With small organisms, analysis of concentrations of pollutants in whole individuals may be misleading if a contaminant is localized strongly in a specific organ. For example, over 90% of the cadmium, copper, lead, and zinc in woodlice from metal-contaminated sites is contained in the hepatopancreas—an organ that constitutes less than 10% of the weight

BOX 8.1   DETOXIFICATION PATHWAYS OF METALS

The type A pathway is the route of intracellular precipitation of calcium and magnesium as phosphates. In the hepatopancreas of the isopod *Porcellio scaber* (Figure 8.2), zinc and lead may be present in this type A phosphate-rich material that is deposited on the cytoplasmic sides of the cell membranes and around existing metal-containing granules. However, in the spider *Dysdera crocata* (Figure 8.3), the type A material forms discrete granules with characteristic concentric arrangements of layers in thin sections (Figures 8.4B and 8.5). Zinc (Figure 8.4E) and lead are also associated with these granules.

The type B pathway is followed by metals such as copper and cadmium that have affinities for sulfur-bearing ligands. The sulfur-rich type B granules probably contain breakdown products of metallothionein, a cysteine-rich protein involved in the intracellular binding of zinc, copper, cadmium, and mercury, and have their origin in the lysosomal system (Dallinger, 1993; Hopkin, 1993b). Lead may also occur in type B granules. In isopods (but not in spiders), some type B granules may be surrounded by a layer of type A material (type B/A granules, following the convention that the type of material first accumulated in the granule is given precedence), but granules with a type A cores surrounded by type B material have not yet been discovered.

The type C pathway is exclusively for the accumulation of waste iron in isopods and spiders. Type C granules are probably composed of hemosiderin, a break-down product of ferritin. In isopods, type A material may be mixed with the iron-rich type C material to form type C/A granules (Figure 8.2). In *Dysdera crocata*, iron is not found in the type A granules (Figure 8.4F).

No evidence indicates that types A, B, and C materials are remobilized after they are precipitated. Indeed, the only route by which the granules can be excreted is by voiding of the contents of the cell into the lumen of the digestive system for subsequent excretion in the feces. The granules therefore represent a storage detoxification system.

In isopods, type A material occurs in both cell types of the hepatopancreas, but type B and C materials are restricted to the S and B cells, respectively. Large numbers of B cells break down during each 24-hour digestive cycle; S cells are permanent and never void their contents into the lumen of the hepatopancreas. Thus material deposited in the S cells remains there until the isopod dies. In contrast, a spider stores all three types of material in a single digestive cell (Figure 8.3). Large numbers of these cells break down at the end of each digestive cycle and void their contents. The waste contains large numbers of type A, B, and C granules that are excreted subsequently in the feces (Hopkin, 1989).

Thus terrestrial isopods accumulate metals in S cells throughout their lives. In contaminated sites, concentrations in the hepatopancreas reach very high levels. In contrast, a spider regulates the concentrations in the hepatopancreas by excreting metals assimilated in digestive cells. Consequently, concentrations of zinc, cadmium, lead, and copper do not deviate from normal over the long term in the hepatopancreas of *Dysdera crocata*, even when the spiders feed on heavily contaminated woodlice for an extended period.

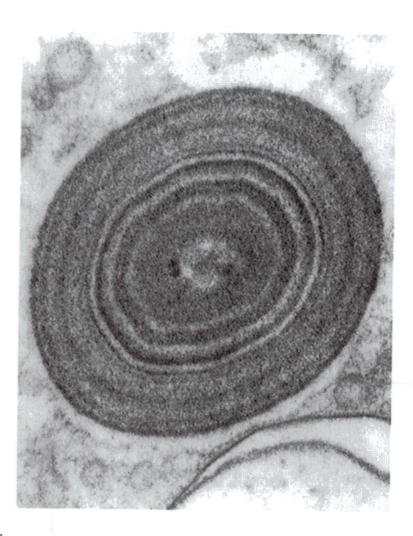

FIGURE 8.5
Concentrically structured type A granule from the hepatopancreas of the woodlouse-eating spider *Dysdera crocata* (diameter 2 μm). (*Source:* Hopkin, S.P. (1989). *Ecophysiology of Metals in Terrestrial Vertebrates.* Elsevier Applied Science. With permission.)

of the whole animal (Hopkin et al., 1989). In regions contaminated heavily with metals, concentrations of zinc can exceed 2% of the dry weight of the hepatopancreas of woodlice. In these cases, the detoxification capacity of the organ is overtaxed, the cells begin to break down, and the woodlouse becomes moribund from zinc poisoning (Figure 8.7).

Cardiovascular and respiratory effects can be viewed as organ effects or, because of their wide-ranging influence on body function, effects at the whole organism level. For convenience, we will first consider organ effects and touch on their role in the expression of whole organism effects in the next section. Specific methods to measure cardiovascular and respiratory responses have been developed and enable cardiovascular monitoring to be carried out noninvasively. Depledge and Andersen (1990) developed a method by which recorded heart rate in reflected light associated with heartbeat can be detected by a transducer. The only requirement is that the heart of the test organism is near enough to

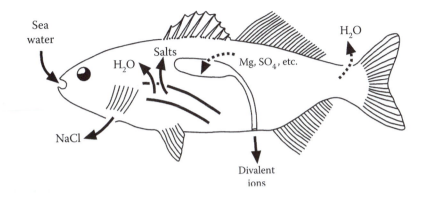

FIGURE 8.6
Salt and water transport in a saltwater fish. (*Source:* Kinter, W.B. et al. (1972). *Environmental Health Perspectives* 1, 169–173. With permission.)

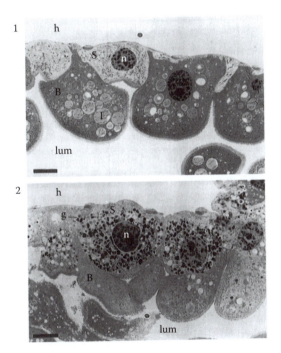

FIGURE 8.7
Light micrographs of B cells (B) and S cells (S) within the hepatopancreas of two specimens of woodlouse *Oniscus asellus* 6 weeks after release from the same brood pouch of a female from an uncontaminated woodland. The S cells of the juvenile reared on leaf litter contaminated with metal pollutants from a woodland near to a smelting works (2) contain far more metal-containing granules (g) than the S cells of the juvenile reared on uncontaminated litter (1). h = haemocoel; lip = lipid; lum = lumen of hepatopancreas tubule. Scale bars: 20 μm. (*Source:* Hopkin, S.P. and Martin, M.H. (1984). *Symposia of the Zoological Society of London.* With permission.)

the body surface to allow detection of the reflected optical signal. This method has been used successfully with bivalves, crabs, and crayfish (Bloxham et al., 1999).

Respiratory responses have been measured in fish and in invertebrates. Respiratory responses are rapid and thus useful for identifying short pollution events. Several parameters may be measured including ventilatory frequency and volume and the exchange of respiratory gases. Oxygen consumption rates show clear dose–response relationships in many organisms exposed to chemicals and can be used as a surrogate for metabolic rate. Portable, self-contained respirometry systems have been devised (Handy and Depledge, 2000).

8.4 Effects at Whole Organism Level

The effects of pollutants on whole organisms fall into three main classes: neurophysiological, behavioral, and reproductive effects. However, these effects can often be interrelated. Neurological changes can affect behavior, changes in behavior can affect reproduction, and so on.

8.4.1 Neurophysiological Effects

All animals except sponges have nervous systems, and the systems are most highly developed in vertebrates. Among the invertebrates, certain groups, for example, arthropods, have well developed nervous systems. Nerves provide communication between receptors (retinal cells, taste receptors) and effectors (muscles and glands); see Figure 8.8. Information travels through the nervous system via electrical impulses transmitted along the axons of neurons (nerve cells).

Information passes between neurons across synapses (nerve junctions) via chemical messengers (neurotransmitters). Acetylcholine, noradrenaline, and serotonin are examples of neurotransmitters released from nerve endings that cross the synaptic cleft and interact with receptors on the postsynaptic membranes of adjacent neurons (Figure 8.8). More advanced nervous systems (e.g., of vertebrates) are differentiated into two separate but interrelated parts: a central system in which information is integrated and a peripheral system through which impulses are transmitted.

Many natural and man-made toxic chemicals act on the nervous system. With a wide range of potential sites of action both on axonal membranes and at synapses, this is not altogether surprising considering the vital role of the nervous system for regulating the functions of organisms. It is noteworthy that five major groups of insecticides (organochlorine, organophosphorous, carbamate, pyrethroid, and neonicotinoid) act as neurotoxins. For descriptions of their modes of action, see Section 7.4.2, Eldefrawi and Eldefrawi (1990), and Walker (2008).

Chemicals can exert effects on the nervous system at different types of receptors and locations. The central nervous system, peripheral nervous system, or both may be affected by chemicals. Central nervous system effects can be monitored using an electroencephalogram (EEG); peripheral effects are detectable with an electromyelogram (EMG). Both techniques can detect changes in the passages of nervous impulses. The actual sites of action may be the synapses or axonal membranes.

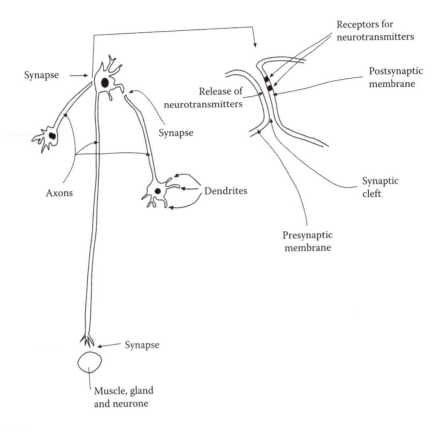

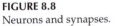

FIGURE 8.8
Neurons and synapses.

Anticholinesterases such as organophosphorous and carbamate insecticides are examples of poisons that act at synapses. They disturb synaptic transmission at cholinergic junctions by inhibiting the enzyme acetylcholinesterase that has the function of destroying the neurotransmitter acetylcholine. Inhibition of the enzyme lengthens the residence time of acetylcholine and prolongs the stimulation of cholinergic receptors on the postsynaptic membranes (see Section 7.4.2). Thus chemical messages are not rapidly terminated, as they should be, and normal transmission of impulses across synapses is disrupted.

If the disruption is sufficiently prolonged, the system for relaying impulses across the postsynaptic membrane and beyond will run down, leading to synaptic block, i.e., no synaptic transmission. Synaptic blockage causes tetanus (rigid fixation of muscle), and in vertebrates death is likely to follow quickly as a result of consequent paralysis of the diaphragm muscles and respiratory failure. Anticholinesterases act on cholinergic junctions throughout the central and peripheral nervous systems of vertebrates. The pattern depends on the compound in question and the species affected. Typical symptoms include tremors, convulsions, respiratory and circulatory failure, coma, dizziness, and depression.

Since 2000, the neonicotinoid group of neurotoxic insecticides has gained prominence. These compounds act on nicotinic receptors for acetylcholine (Section 1.2.10). Evidence shows that they can cause sublethal neurotoxic effects on honeybees. The literature on this subject has been reviewed and the data from experiments with bees subjected to meta-analysis (Cresswell, 2011). Interestingly, this statistical analysis conducted over an extended data set indicated that these insecticides produced sublethal effects that were not evident from statistical analysis of the individual studies.

DDT, an organochlorine insecticide, acts on the sodium channels of axonal membranes (Section 7.4.2) and disturbs the action potential of the nerves and thus the transmission of impulses along the nerves. DDT can exert both central and peripheral effects on the nervous system, ranging from tremors and twitches due to peripheral effects to more coordinated disturbances from central effects. Dieldrin, on the other hand, acts on GABA receptors located primarily on pre- and postsynaptic membranes adjoining synapses. In insects, dieldrin affects the inhibitory synapses of the central nervous system. Symptoms of poisoning are the consequences of effects upon the central nervous system, e.g., tonic convulsions.

Neurotoxic compounds can have diverse and wide-ranging effects on the whole animal including endocrine and behavioral disturbances. Organophosphorous insecticides, for example, have behavioral effects on birds, including reduced mobility and singing. Endocrine function can be affected by disturbances in the nervous system and vice versa.

8.4.2 Effects on Behavior

The effects of pollutants on behavior are discussed in Chapter 16 of Walker (2009). Atchison et al. (1996) focused on aquatic species. Prominent among the pollutants that can cause behavioral effects are neurotoxic compounds described in Section 7.4.2. Fish have proved to be sensitive organisms for the detection of behavioral effects from pollutants. Several studies have demonstrated effects of organophosphorous insecticides on fish behavior at concentrations well below lethal levels (Beauvais et al., 2000; Sandahl et al., 2005).

These studies measured reaction to stimuli, swimming stamina, swimming performance, prey detection, prey capture, and predator avoidance. Although all behaviors are potentially vulnerable to alteration by pollutants, particular attention has been given to foraging, with some attention to vigilance. Impaired foraging results in reduced acquisition of resources and subsequent reduced production. Impaired vigilance results in increased vulnerability to predators and subsequent increased mortality. Clearly the effects of pollution on behavior may cause lowered production and increased mortality.

The components of foraging behavior are illustrated in Figure 8.9. All these components may be adversely affected by pollution. Although little work has been undertaken on the effects on appetite, a common effect of chemical stressors is the cessation of feeding. Prey encounter rates depend on many factors including search strategy, learning, and sensory systems. All can be affected by chemical stressors that reduce the efficiency of searching for prey. A few studies of prey choice indicate that it too may be affected. Most importantly, capture rates are known to be affected by toxicants in some species, at least when larger or more evasive prey are hunted. The time from capture to ingestion (handling time) has been shown to be increased in the presence of toxicants, e.g., copper in bluegill feeding on *Daphnia*, zinc and lead in zebrafish (*Brachydanio rerio*), and alkyl benzene sulfonate detergent in flagfish (*Jordanella floridae*) (Atchison et al., 1996). However, these increases in handling times seem to be due to repetitive rejection and recapture, perhaps caused by blockage of the gustatory senses by contaminants. These predators use vision to identify, pursue, capture, and start processing their prey. Rejection may result from lack of gustatory confirmation of the edibility of a captured animal.

Another example of the effects of pollutants on foraging behavior is the disturbance of the foraging activities of bees by neurotoxic insecticides (Thompson, 2003). Pyrethroid insecticides, for example, at realistic exposure levels, appear to affect the ability of foraging

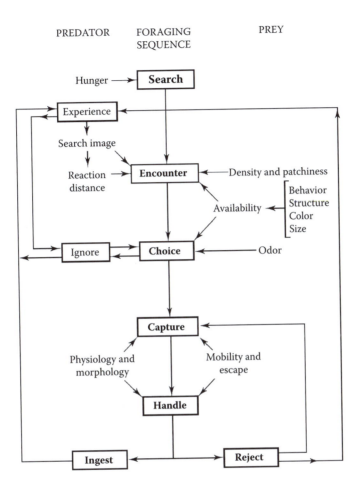

FIGURE 8.9
Components of foraging behavior. (*Source:* Atchison, G.J. et al. (1996). In *Quantitative Ecotoxicology: A Hierarchical Approach.* Lewis Publishers. With permission.)

honeybees to return to their hives. The neonicotinoid insecticide imidacloprid was shown to affect the dance of honeybees that conveys information about the locations of nectar. This may explain the negative effect of low levels of this chemical on foraging by bees.

Foraging is not the only goal, however, because as mentioned above an animal must also avoid being eaten. "Few failures in life are as unforgiving as the failure to avoid predation" (Lima and Dill, 1990). Most studies of vulnerability to predation have been simple experiments in which prey were first exposed to a pollutant after which a predator was introduced. The predators were generally not exposed to the pollutant. Such studies have shown, for example, that ionizing radiation and mercury both increased the risk to mosquitofish (*Gambusia affinis*) of predation by largemouth bass (*Micropterus salmoides*; Atchison et al., 1996). In other species, the risk of predation has been shown to increase with exposure of prey to thermal stress, insecticides, pentachlorophenol, fluorine, and cadmium. Beitinger (1990) found that vulnerability of prey to predation was increased in twenty-three of twenty-nine experiments he reviewed.

The relationship between depression of acetylcholinesterase (AChE) activity by OPs and carbamates and behavior has been studied in many species ranging from invertebrates to mammals. The mechanism whereby these compounds prevent the normal functioning of the nervous system has been considered in Section 7.3.

Impaired burrowing activity in earthworms (*Pheretima posthuma*) related to AChE inhibition caused by the carbaryl carbamate was studied by Gupta and Sundararaman (1991). They found that the ability of the earthworm to burrow in soil decreased with increased dose over the entire range of concentrations of carbaryl studied. The activity of AChE and locomotory behavior in carabid beetles (*Pterostichus cupreus*) exposed to dimethoate was studied by Jensen et al. (1997). In the case of males, all locomotory indices (path length, time during which the organism is active, average velocity, stops per walked meter, and turning rate) were affected at the lowest dosage. Females were less affected, although the degree of inhibition of AChE was similar.

Behavioral responses of mammals and birds after depression of AChE activity were reviewed by Grue et al. in 1983 and 1991 and Grue and Heinz in Dell'Omo (2000). One difficulty of these behavioral studies is to follow the changes in AChE activity nondestructively, i.e., in blood. Unfortunately for investigators, the recovery of activity of AChE in the blood is rapid and the coefficient of variation is three times greater than the activity in the brain (Holmes and Boag, 1990). Thus plasma measurements can be used only over a much shorter time (up to 12 h) and with considerably less precision than brain measurements. In practice, most investigators have sacrificed subsamples during the experiment and relied on measurements of brain AChE activity.

Detailed quantitative analyses at low levels of intoxication have also been attempted. Hart (1993) investigated the relationship between behavior and AChE activity in the starling (*Sturnus vulgaris*) and found that, although most behavioral effects occurred when brain levels fell below 50%, some subtle effects on behavior such as effects on posture (time spent standing on one leg while resting) were found at relatively high levels of AChE activity (88% of normal) and may reflect impaired balance or coordination.

Another example of a quantitative study is shown in Figure 8.10. House sparrows (*Passer domesticus*) were dosed with the organophosphate chlorfenvinphos, and their feeding behavior was then recorded during four successive 1.5-hour periods. The percentage of seeds dropped was plotted against dose (assayed at the end of the day). Birds whose brain AChE activities were most reduced initially dropped some 30% more seeds than they dropped on control days (Figure 8.10A). However, the effect wore off 3 hours after dosing (Figure 8.10C and D). Figure 8.11 shows that the birds that dropped the most seeds lost the most weight.

Extrapolating from the laboratory to events in the wild is notoriously difficult (Hart, 1993). Animals may be able to detect and avoid toxicants in their diets. If they ingest toxicants, they may be able to compensate for disabilities by retreating to safe places. Alternatively, disabilities may make them more vulnerable to predation or less able to maintain their territories or to care for offspring. Some field studies succeeded, but others failed to find effects on behavior when brain acetylcholinesterase activity was depressed by 40 to 60% (Hart, 1993). This may be consistent with the existence of a general threshold for behavioral effects at about that level.

A good example of a field study is that of Busby et al. (1990), who examined the effects of the organophosphorous insecticide fenitrothion on white-throated sparrows (*Zonotrichia albicollis*) in spruce fir forest in New Brunswicka, Canada (see Box 8.2).

The extrapolation to nature has also been tackled by similar studies both in the laboratory and in the field. In one study, the effects of the OP dimethoate on behavioral parameters

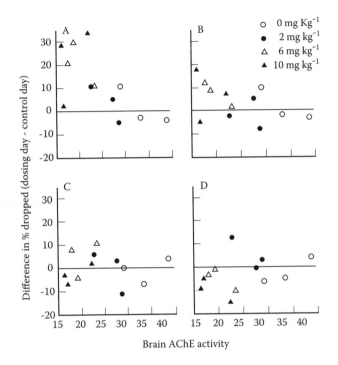

FIGURE 8.10
Differences in percentages of seeds dropped between dosing days and control days plotted against brain acetylcholinesterase activity in house sparrows. A through D show results in successive 1.5-hour time bins after
dosing. Doses are shown in the key. (*Source:* Fryday, S.L. et al. (1994). *Bulletin of Environmental Contamination and
Toxicology* 53, 864–876. With permission.)

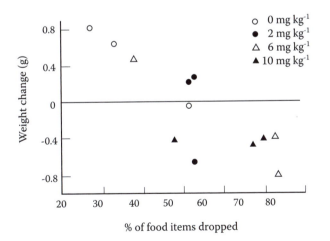

FIGURE 8.11
Body weight loss in relation to percentage food items dropped by birds in Figure 8.10. (*Source:* Fryday, S.L. et al.
(1994). *Bulletin of Environmental Contamination and Toxicology* 53, 864–876. With permission.)

BOX 8.2: FOREST SPRAYING IN NEW BRUNSWICK, CANADA

The forest has been sprayed aerially each year since 1952 to control spruce budworm (*Choristoneura fumiferana*). Fenitrothion is the preferred insecticide. In the study of Busby et al. (1990), fenitrothion was applied aerially twice: once at 420 g/ha and again 8 days later at 210 g/ha. An earlier study indicated that spraying at these levels would have resulted in brain acetylcholinesterase inhibition of 42% and 30%, respectively. A team of experienced field workers ringed and were able to identify thirteen breeding pairs of white-throated sparrows in the sprayed area and seven in a nearby control area. Each pair was followed until it abandoned its territory or its offspring fledged.

The results showed that the adult population of white-throated sparrows in the sprayed area was reduced by one-third, primarily as a consequence of mortality and territory abandonment after the first spray. Other behavioral responses of breeding birds included inability to defend their territories, disruption of normal incubation patterns, and clutch desertion. Pairs that managed to hatch at least one chick produced only one-third of the normal number of fledglings. Overall, the reproductive success of the pairs in the sprayed area was only one-quarter of that of the pairs in the control area.

of the wood mouse (*Apodemus sylaticus*) were studied in the laboratory. The parameters included frequency and duration of sniffing, rearing, grooming, and general activity. Similar responses were demonstrated with radiotagged mice in the field (Dell'Omo and Shore, 1996a and b).

8.4.3 Reproductive Effects

Some of the best documented cases of effects of pollutants on reproduction are considered elsewhere in this volume. These are the effect of tributyl tin (TBT) on mollusks (Section 16.3), DDE-induced eggshell thinning in raptorial and fish-eating birds (Section 16.1), and reproductive failures of fish and birds in the Great Lakes caused by a mixture of pollutants (Section 16.2). The mechanisms involved in these reproductive failures are diverse. TBT causes females to develop male characteristics (imposex) that can lead to sterility. DDE causes eggshells to be thinner than normal, leading to egg breakage. In the Great Lakes studies (apart from eggshell thinning in some species), the mechanisms were less clearly defined.

Disruptors of the endocrine system received considerable attention in recent years. Some of the basic biochemistry has already been covered in this book. Here we consider some of the evidence for effects on reproduction in wildlife. The mechanisms whereby endocrine disruptors affect the reproductive health and survival of wildlife are shown in Figure 8.12.

The decline of populations of ringed seals (*Phoca hispida*) and grey seals (*Halichoerus grypus*) in the Baltic even after hunting pressures were reduced sparked an investigation. The main reason was that females were unable to reproduce; miscarriages increased, and many females were sterile because of damage to their uterine walls Other findings were indications of metabolic disorders, immunosuppression, and hormonal imbalance, a condition known as hyperadrenocorticism. Although it is generally considered that persistent organochlorines and their metabolites are the cause, "it is not possible to make conclusive statements about the relation between disease and poor reproduction among Baltic seals and the concentration of organohalogen compounds. Even though PCB and DDT together

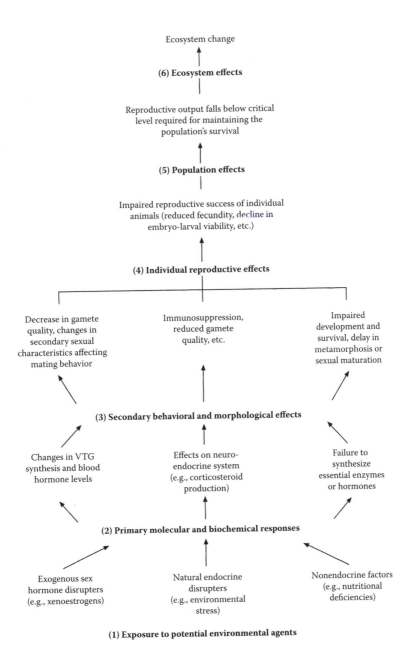

FIGURE 8.12
Mechanisms by which endocrine disruptors affect the reproduction and survival of wildlife. (*Source:* Campbell, P.M. and Hutchinson, T.H. (1998). *Environmental and Toxicological Chemistry* 17, 127–135. With permission.)

with metabolites are the dominating xenobiotics in the seals, synergistic and additive effects of other compounds cannot be excluded" (Ollson et al., 1992).

Jaw bone damage was also found, and this effect made it possible to assemble a historical record from museum specimens. The increase in the lesions (less than 10% to over 50%) started in the late 1950s and the 1960s, which correlates with the increase of organochlorine pollution (Bergman et al., 1992).

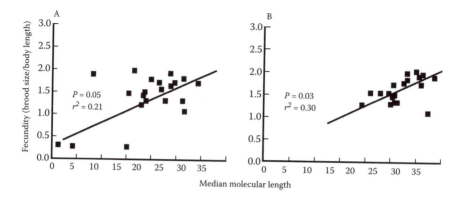

FIGURE 8.13

Correlation between median molecular length (inversely proportional to strand breaks) and fecundity of mosquitofish collected from a radionuclide-contaminated pond. (A) Total strand breaks (liver). (B) Double-strand breaks (liver). (*Source:* Theodorakis, C.W. et al. (1997). *Ecotoxicology* 6, 205–218. With permission.)

Endocrine disruptors have been shown to cause effects on fish. The most prominent disruptor is ethynylestradiol (EE2), a component of oral contraceptives (Walker, 2009, Chapter 15; Harries et al., 1999; Allen et al., 1999). Lifelong exposure of zebrafish to 5ng/l EE2 can cause reproductive failure (Nash et al., 2004). Other environmental chemicals that may act as endocrine disruptors include *o,p′*-DDT, PCBs, TCDD, and bisphenol A.

Vitellogenin is a yolk precursor, the synthesis of which is triggered by gonadotropin hormones in females. The induction of vitellogenin synthesis by endocrine disruptors in male fish is covered in Chapter 10. Studies linking DNA damage with reproduction have been carried out on mosquito fish (*Gambusia affinis*) in streams contaminated with radionuclides (Theodorakis et al., 1997). The researcher were able to relate the number of strand breaks with the number of embryonic abnormalities and fecundity and discovered an interesting seasonal variation. In the spring, the percentage of broods with at least one abnormality was 14% in one contaminated pond and 57% in another; no abnormalities were found in the control ponds. In summer, although the percentages were still significantly higher in the contaminated ponds, the percentages of abnormal broods in the control ponds were 20% and 40%. The relationship of fecundity (brood size and length of female) to strand breaks is shown in Figure 8.13.

8.5 Effects on Plants

The effects of chemicals in vertebrates are much better understood than effects in plants, reflecting the much larger investment in human toxicology than in ecotoxicology. Some important effects in plants will now be briefly reviewed.

Photosynthesis is a complex process occurring in the chloroplasts of green plants; it can be inhibited by a number of herbicides (Section 7.4.9). Organic compounds are synthesized by the reduction of carbon dioxide using light energy absorbed by chlorophyll, as described by the empirical equation

$$6\,CO_2 + 6\,H_2O \rightarrow C_6H_{12}O_6 + 6O_2$$

The trapping of solar energy and its utilization to biosynthesize organic compounds are essential to the lives of plants and to the animals that feed on them. Also, the generation of oxygen by photosynthesis ensures the maintenance of an aerobic atmosphere and the operation of aerobic processes that are essential to most life forms. The evolution of photosynthesis provided the essential basis for the evolution of aerobic processes to follow the anaerobic activities of the earliest life forms.

Herbicides such as substituted ureas, triazines, and paraquat inhibit photosynthesis, but their exact modes of action are somewhat uncertain. Two forms of chlorophyll and a number of carrier proteins are involved in the complex electron carrier systems known as photosystems I and II; a detailed discussion of how and where these compounds may act lies outside of the scope of this text. The critical point is that the effects of herbicides and other environmental chemicals on photosynthesis can be measured in living plants or in isolated chloroplasts.

Respiration, which is simply the reverse of the empirical equation shown for photosynthesis, can also be inhibited by environmental pollutants including nitrophenolic herbicides such as dinoseb that act as uncouplers of oxidative phosphorylation by mitochondria. The actions of respiratory poisons can be determined by measuring changes in the rate of oxygen consumption in whole plants or in isolated mitochondria.

Plant growth rates are relatively easy to measure and often provide indices of the toxic effects of pollutants. A reduction in the rate of growth can lead to stunted development and dwarfism, a response that can be observed in the leaves of trees exposed to sulfur dioxide and other aerial pollutants. Another issue concerns the actions of chemicals that can affect growth regulation. The so-called plant growth regulator herbicides (Sections 1.2.10 and 7.4.10) can cause characteristic growth disturbances in plants. The herbicides include the widely used phenoxyalkanoic acids MCPA, 2,4-D, CMPP, and 2,4-DB. They cause distorted growth, malformed leaves, severe epinasty (downward curvature) of stems and petioles, and unequal growth of young, rapidly developing tissues near meristems. Effects of this kind are familiar to gardeners who have used these compounds to control broad-leaved weeds (dicots) in lawns.

8.6 Energy Costs of Physiological Change

In this chapter, we have seen cases in which the machinery of resource acquisition or uptake is damaged. In other cases, damage is avoided, but organisms may still be affected because detoxification generally consumes energy and other resources that thus cannot be used for production. Both damage and detoxification are likely to lead to reduced production.

The effect of pollution on production is usually measured by its effect on scope for growth (SFG), defined as the difference between energy intake and total metabolic losses (Warren and Davis, 1967; Widdows and Donkin, 1992; Figure 8.14). Figure 8.15 shows the effects of tributyl tin (TBT) concentration on SFG in the mussel *Mytilus edulis*. Note that above a threshold of 2 µg/g, SFG declines as TBT concentration increases, indicating a loss of production. In the field, this decline could translate to smaller abundance of animals (see Chapter 16). No effect of TBT at low levels was noted on SFG. However, in cases of essential nutrients (e.g., some metals), decreases in SFG at very low levels of nutrient were found. SFG is very useful for assessing the effects of pollution on aquatic animals (e.g., fish, Crossland, 1988; Maltby et al., 1990a). Field and microcosm studies confirmed

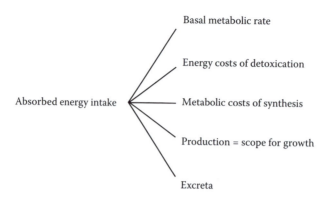

FIGURE 8.14
Energy–nutrient allocation diagram illustrating scope for growth.

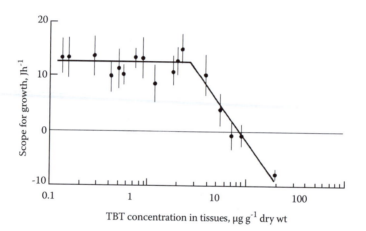

FIGURE 8.15
Effects of TBT (tributyl tin) on scope for growth (SFG) in *Mytilus edulis* mussels. (*Source:* Widdows, J. and Donkin, P. (1992). In *The Mussel* Mytilus: *Ecology, Physiology, Genetics, and Culture.* Elsevier Applied Science. With permission.)

that the long-term effects on growth and survival of individuals can be predicted from measured effects on energy balance observed at the individual level (Widdows and Donkin, 1991).

Widdows and Donkin (1991) described how reductions in SFG in *Mytilus edulis* in contaminated sites can be apportioned between specific pollutants. In a study in Bermuda, Widdows et al. (1990) showed that the overall reduction in SFG of *Mytilus edulis* could be proportional such that, at the most contaminated sites, tributyl tin accounted for 21% and hydrocarbons for 74% of the observed effects.

A decline in SFG in the freshwater amphipod *Gammarus pulex* was due to increased stress; a decline in feeding rates led to decreased offspring weight and increased numbers of abortions, with important consequences for the long-term viability of affected populations (Maltby and Naylor, 1990; Maltby et al., 1990a and b).

Quantifying SFG relies on measuring parameters in organisms exposed to pollutants and comparing measurements with those of unexposed individuals. Nonsedentary organisms can be caged in micro- and mesocosms to prevent their migration away from the pollutants. However, advances in microelectronics may soon enable some indicators of environmental stress to be measured remotely.

Turning now to the effects of chemicals on growth, a vast literature covers a wide variety of species. Virtually every class of pollutants—OCs, OPs, heavy metals, PAHs—have been shown to retard growth. Two examples are as follows. First, during an investigation of the effects of high levels of boron and selenium in the Central Valley of California, the growth rates of ducklings were found to be reduced (Stanley et al., 1996), although duckling survival was not affected. Second, on the Great Lakes, the growth rates of young ospreys (*Pandion halieatus*) inversely correlated with the concentrations of dioxin in eggs (Woodford et al., 1998), although overall breeding success was not affected. It appears that growth can be a sensitive but nonspecific indicator of pollution.

In many cases, an organism can compensate for the effects of the pollutant on its physiology, but only at a price. Along with direct damage is the energy cost of detoxification mechanisms. In this case, the resources that the organism invests in detoxification reduce its chances of dying but at the cost of lost production. In other words, the organism trades off loss of production for reduced mortality.

The concept of trade-off is important in modern evolutionary ecology and is described in detail in Chapter 13. In this section we note only that the genetic possibilities for species are limited by trade-offs and are concerned with the trade-off between production rate and mortality rate. This can also be considered a trade-off between production and defense because mechanisms that reduce mortality rates serve to defend the organisms.

A trade-off between production and defense could be achieved in several ways (shells, spines, vigilance, etc.; Sibly and Calow, 1989), but here we are concerned with defenses against toxins. Possible methods of defense include relatively impermeable exterior membranes (Oppenoorth, 1985; Little et al., 1989), more frequent molts (e.g., in Collembola, in which metals stored in the gut cells are voided during ecdysis; Bengtsson et al., 1985), and more comprehensive immune systems and detoxification enzymes (Terriere, 1984; Oppenoorth, 1985; Bass and Field, 2011). Many examples are given in this book.

Although it is clear that such defenses generally have energy costs (Sibly and Calow, 1989; Hoffman and Parsons, 1991), these could be small, e.g., in the case of inducible enzyme responses. Even here, however, there must be a cost, because amino acids are required at every stage of the genetic response, and because all genetic mechanisms involve overheads (molecular checking, DNA turnover, and disposal of waste). However, in environments that are naturally highly stressful (wide fluctuation in temperature, humidity, etc.), the costs of coping with pollution may be only a tiny proportion of the total stress that the organism has to cope with and may consequently be difficult to quantify.

To illustrate what can be achieved in this area, we consider two studies in more detail. In a classic study of insecticide detoxification in *Myzus persicae* aphids, Devonshire and Sawicki (1979) used strains that differed in insecticide resistance. The strains were genetically different and their mortality and growth rates were examined in a standard environment. More resistant strains contained more detoxifying enzyme (phosphatase E_4, up to 1% of total protein) because they contained more duplications of a structural gene (Figure 8.16A). LD_{50} values were measured for three strains from which mortality rates could be calculated (Figure 8.16B). Note that strains containing more copies of the structural gene (that presumably spent more on detoxification) achieved reductions in mortality rate.

Bengtsson et al. (1983, 1985) studied detoxification of metals by the springtail *Onychirus armatus*. The springtails were fed fungi grown on a nutrient broth contaminated with 0,

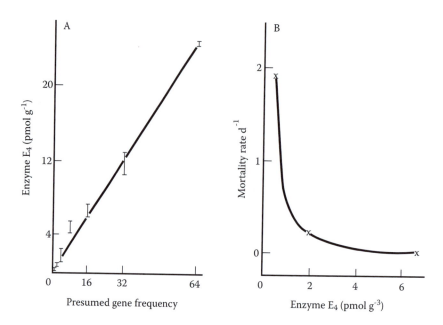

FIGURE 8.16

(A) Concentration of detoxifying enzyme E_4 in seven strains of the *Myzus persicae* aphid in relation to the number of copies of a structural gene hypothetically present in each strain. (B) Mortality rates of three strains in relation to E_4 concentration. Aphids were placed on potato leaves dipped in an organophosphorous insecticide (Demeton-S-methyl). Data from Sawicki and Rice (1978) and Devonshire and Sawicki (1979). (*Source:* Sibly, R.M. and Calow, P. (1989). *Biological Journal of the Linnean Society* 37, 110–116. With permission.)

30, 90, or 300 µg/g of copper and lead in equal proportions. Concentrations of these metals within the animals reached high levels initially but were then reduced by detoxification and reached steady state after a few weeks (Figure 8.17A). Detoxification was achieved by more frequent molting of the lining of the gut (where most metals are stored), and this reduced the growth rate as shown in Figure 8.17B. Detoxification was not complete, however, and body metal levels were elevated in more contaminated environments even at steady state as shown in Figure 8.17A. Mortality rates were higher in contaminated environments (Figure 8.17B), possibly as a result of the increased metal levels in the animals (Figure 8.17C). Note, however, that a certain amount of copper is needed physiologically, so that in a completely deficient environment growth rate must be reduced and mortality rate increased (Figure 8.17B). In this study, growth rate was reduced at higher levels of pollution, presumably because the animals molted more frequently, but metals were not completely eliminated and mortality rates still increased.

Where there is a trade-off between production and defense, Figure 8.18 illustrates the likely form of the relationship. The trade-off curve has the following features: (i) even when growth rate is maximal (zero allocation to defense), growth rate is finite—equal to the limiting production rate in the study environment; (2) even if all resources are allocated to defense, the mortality rate is still greater than zero, equivalent to the extrinsic mortality rate in that environment. The simplest curve having these general features is convex viewed from below and has the general form of the curve shown in Figure 8.18. The evolutionary implications of this trade-off are considered in Section 13.3.

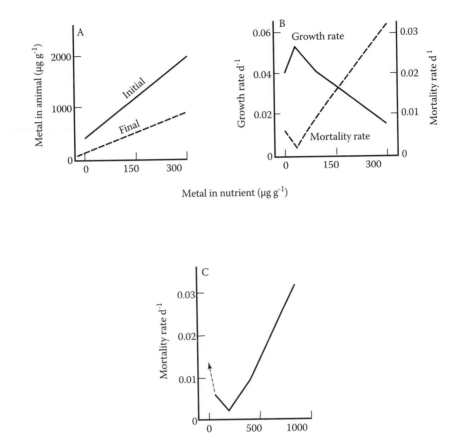

FIGURE 8.17
(A) Initial peak levels and final steady-state levels of Cu and Pb in the bodies of springtails in relation to levels in nutrient broth. (B) Growth rate (reciprocal of time to first reproduction) and mortality rate (calculated from survivorship over first 10 weeks of life). (C) Mortality rate in relation to final steady-state levels of metals in the bodies. Mortality must increase at very low levels as the animals then suffer from copper deficiency. Data from Bengtsson et al. (1983 and 1985). (*Source:* Sibly, R.M. and Calow, P. (1989). *Biological Journal of the Linnean Society* 37, 110–116. With permission.)

8.7 Summary

In this chapter, the effects of pollutants on physiological processes at the cellular, organ, and whole organism level were examined. The range of effects is diverse, including effects on the metabolism of cells and the behaviors of individuals. We examined the effects of single pollutants; in the next chapter the interactive effects of pollutants are examined.

Pollutants may damage organisms directly by increasing their mortality rates or interfering with resource acquisition and uptake, thereby reducing production rates. These effects on individuals can result in slower population growth or even population decline. These effects are discussed in Chapter 12.

Alternatively, organisms may avoid or restrict damage via detoxification (induction of monooxygenases) or repair mechanisms such as DNA repair. However, all these activities

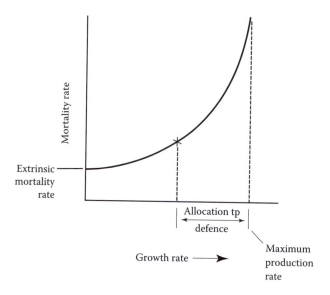

FIGURE 8.18

Likely form of trade-off between production rate and mortality rate that may constrain the operation of detoxi-fication mechanisms. Allocation of resources to defense reduces mortality rate and simultaneously cuts growth rate. (*Source:* Sibly, R.M. and Calow, P. (1989). *Biological Journal of the Linnean Society 37*, 110–116. With permission.)

consume energy and resources that are thus diverted from production. It follows that the population effect will still be detrimental. The evolutionary implications of this trade-off are discussed in Chapter 13.

Further Reading

Atchison, G.J., Sandheinrich, M.B., and Bryan, M.D. (1996). Review of the effects of pollution on the behaviors of aquatic animals.

Dallinger, R. and Rainbow, P.S., Eds. (1993). *Ecotoxicology of Metals in Invertebrates.* Several relevant papers.

Dell'Omo, G. (2000) *Behavior in Ecotoxicology.* Important text.

Donker, M.H. et al., Eds. (1994). *Ecotoxicology of Soil Organisms.* Several relevant papers.

Grue, C.E. et al. (1991). Effects of acetylcholinesterase on behaviors of mammals and birds.

Hopkin, S.P. (1989). *Ecophysiology of Metals in Terrestrial Invertebrates.* Comprehensive review of the distribution and effects of metals in terrestrial invertebrates.

Langston, W.J. and Bebianno, M.J. (1998). *Metal Metabolism in Aquatic Environments.* Useful papers describing current ideas on the roles of metals in aquatic organisms.

Widdows, J. and Donkin, P. (1992) Reviews the effects of pollution on scope for growth in *Mytilus edulis.*

9

Interactive Effects of Pollutants

In a natural environment, organisms are frequently exposed to complex mixtures of pollutants, and it is relatively uncommon to find any single pollutant dominant over all others. Due to limitations of time and resources, nearly all regulatory toxicity testing is carried out using single compounds. It is not feasible to test the toxicity of more than a very small proportion of the chemical combinations present in terrestrial, marine, or freshwater ecosystems or that may arise from the release of new chemicals into the environment. Figure 9.1 illustrates the complexity of the situation and shows analytical data for residues of PCBs in tissues of organisms from a polluted area. A number of different PCB congeners are found in both species, with a wider selection in mollusks than in harbor seals. When effluents or contaminated environmental samples are subjected to testing, toxicity is often caused by more than one chemical component of a mixture, and questions arise concerning possible potentiation. This issue was discussed in Section 6.2.

9.1 Introduction

When regulatory authorities consider the toxicities of mixtures, it is usually assumed (without definite evidence to the contrary) that the toxicity of a combination of chemicals will be approximately additive. In other words, the toxicity of a mixture of compounds will approximate to the total toxicities of its individual components. This is usually a correct assumption.

However, in a relatively small yet very important number of cases, toxicity may be substantially more than additive—i.e., when organisms are exposed to a combination of two or more chemicals, potentiation of toxicity may occur. Sometimes the term *synergism* is used to describe this phenomenon. However, many scientists restrict the use of this term to situations in which only one of two components is present at a level that can cause a toxic effect and the other compound (synergist) would have no effect if applied alone. This practice will be followed here. Table 9.1 illustrates some examples of synergism. The effectiveness of a synergist is usually measured as a synergistic ratio (SR):

$$\frac{\text{MLD (or conc.) for chemical alone}}{\text{MLD (or conc.) for chemiocal + synergist}} = \text{SR}$$

Thus where synergism is present, the SR will be greater than 1; in effect, the synergist will lower the median lethal dose (MLD) or concentration of the chemical. The following section covers additive toxicity before moving on to consider potentiation and synergism.

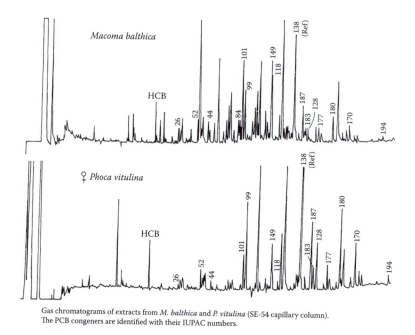

Gas chromatograms of extracts from *M. balthica* and *P. vitulina* (SE-54 capillary column).
The PCB congeners are identified with their IUPAC numbers.

FIGURE 9.1

PCB congeners in tissues of marine organisms [mussels (*Macoma balthica*) and harbor seals (*Phoca vitulina*)] from the Dutch Wadden Sea. The compounds were separated, identified, and quantified by capillary gas chromatography. Each of the numbered peaks represents a PCB congener. HCB (hexachlorobenzene) served as an internal standard. (*Source:* Boon, J.P. et al. (1989). *Marine Environmental Research* 27, 159–176. With permission.)

TABLE 9.1

Examples of Synergism

Organisms	Pesticide	Detoxifying Enzyme System	Inhibitor (Synergist)	Increase in Toxicity
Insect strains resistant to pyrethroids	Cypermethrin	Monooxygenase	Piperonyl butoxide	<40×
Insects	Carbaryl	Monooxygenase	Piperonyl butoxide	<200×
Mammals and some resistant insects	Malathion	Carboxyl (B) esterase	Organophosphorous compounds	<200×

Note: See Chapter 7 of Hodgson and Levi (1994).

9.2 Additive Effects

The toxicity of a mixture is often roughly equal to the total toxicity values of its individual components. In other words, each chemical expresses roughly the same toxicity in a mixture as it would when tested alone. Where no evidence indicates potentiation or antagonism, estimates of the toxicity of a mixture can be made by adding together the expected contributions from each component.

The toxicity of each component in a mixture depends on its concentration and can be estimated from a dose–response curve, e.g., a percentage of mortality from an LD_{50} plot. Thus a mixture containing three components at concentrations that would, if tested individually, cause 5, 10, and 15% mortality would be expected to cause 30% mortality overall if toxicity were simply additive. Toxicity data working to end points other than mortality can be treated similarly, e.g., reduction of photosynthetic rate expressed as a percentage. Clearly, where potentiation or antagonism exist, the estimated toxicity will differ markedly from the measured toxicity.

Where several compounds share a common mechanism of action and interact with the same site of action, it is probable that they will show additive toxicity when present as mixtures unless the picture is complicated by toxicokinetic factors (Sections 9.3 to 9.5). Differences in affinities for the site of action (receptor) and corresponding differences in the relationship between concentration (or dose) and toxic effect are bound to arise within such groups of compounds. However, toxicity is frequently closely related to the percentage of receptor sites to which the toxic molecules bind. In this case, the concentrations of individual compounds are corrected by affinity factors, and all toxicity data are fitted to a single dose–response curve. In practice, this has been carried out by calculating toxic equivalency factors (TEFs) in relation to the most toxic component of a group.

The calculation of dioxin equivalents for mixtures of polychlorinated aromatic compounds illustrates this approach (Safe, 1990; Ahlborg et al., 1994). A standard toxic response is defined (e.g., 50% mortality), and a comparison is then made between the concentrations required to produce this effect by (i) the most toxic compound of the group and (ii) another chemical of the group. The ratio of (i) to (ii) is the TEF. The contribution of each compound to the overall toxicity of the mixture (toxic equivalent or TEQ) is then determined by multiplying its concentration in an environmental sample (e.g., water or tissue) by its TEF. The TEQs for individual compounds can then be added up to yield a toxic equivalent for the mixture, i.e., equivalence relative to a reference compound.

An important example of this approach is the calculation of dioxin equivalents for PCDDs and coplanar PCBs, in which TCDD (dioxin) is the reference compound. See Section 15.2 for details. Recently, cellular systems have been developed to measure dioxin equivalents in environmental samples. They depend on the principle that toxicity is a consequence of the binding of these compounds to the Ah receptor (Ah-receptor-mediated toxicity). The receptor is present in, for example, hepatocytes, and the extent to which polychlorinated compounds bind is indicated by a characteristic response such as light emission (see Section 6.7). Other examples in which members of a group of compounds share a common mode of action will now be briefly discussed.

Warfarin and related anticoagulant rodenticides share a common binding site in hepatic microsomes, where they inhibit the operation of the vitamin K cycle (Sections 1.2.1 and 7.4.4). Evidence suggests that they express additive toxicity when present as mixtures. Several metabolites of PCBs act as thyroxine antagonists, competing with it for binding sites on the protein transthretin (Section 7.4.5). Finally, estrogen and androgen binding sites serve as binding sites of certain environmental endocrine disruptors (Section 7.4.7).

9.3 Potentiation of Toxicity

The phenomenon of potentiation of toxicity requires further explanation. In Figure 9.2 two compounds (A and B) are given consideration. The maximum dose of either one will yield the same degree of toxic response (X, representing a percentage reduction in respiration or mortality). Between the two maximum doses are different mixtures of the two

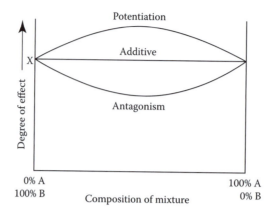

FIGURE 9.2

Potentiation of toxicity. The vertical axis indicates the degree of toxic effect of the compound, and the horizontal axis represents the composition of the mixture. The maximum doses of compounds A and B both yielded the same degree of toxic response X. Potentiation occurs when the toxicity of a mixture of two compounds exceeds the total toxicities of the individual components. (*Source:* Moriarty, F. (1999). *Ecotoxicology*, 2nd ed. Academic Press. With permission.)

compounds. Doses for either compound run from 0 to 100% of the maximum. The total contributions of the two components of the mixture will always be 100%, e.g., 30% of the maximal dose of A + 70% of the maximal dose of B. If toxicities are simply additive, all these combinations should give the same toxic response (X) as the maximal dose of A or B—a straight line represents the combined effect in Figure 9.2. If, however, potentiation or synergism occurs, the toxic effect should be greater than expected; antagonism will produce a smaller toxic effect.

Care is needed when deciding whether toxic effects of combinations of chemicals are truly greater than additive. In the first place, because of errors in measurement, interest is confined to situations where toxicity is *substantially* greater than additive (e.g., where SRs exceed 2). Smaller differences usually reflect no more than the compounding of errors. Also, consideration should be given to the relationship between dose and toxic effect for each individual component of a mixture (Figure 9.3). It is vital to know whether a straight-line relationship exists between dose (or log dose) and toxic response (Figure 9.3). If it does, then increases in toxicities of combinations of chemicals that are substantially greater than additive must be regarded as examples of potentiation. If, on the other hand, the results are not linear, for example, where the increase in toxicity of an individual compound is proportionately greater than the corresponding increase in dose, this conclusion does not necessarily follow (Figure 9.3).

An enhancement of toxicity above a simply additive level may merely reflect the result when the dose of an individual chemical (or chemicals) is increased and may not therefore represent potentiation from interactions between chemicals. Note that this chapter is mainly concerned with cases in which potentiation represents considerable enhancement of toxicity—frequently of at least one order of magnitude (i.e., 10-fold or more).

The question of interactive effects is also complicated by the determination of the end point used for measuring toxicity. For example, in tests with the springtail *Folsomia candida*, Van Gestel and Hensbergen (1997) found that mixtures of cadmium and zinc were antagonistic for growth but additive with respect to reproduction when comparison was made with the effects produced by the metals administered singly.

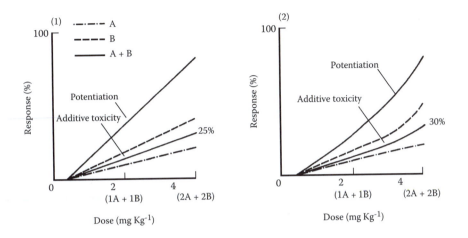

FIGURE 9.3
Additive toxicity and potentiation. In (1), compounds A and B produce a linear response over the dose range 0 to 4. If toxicity is simply additive, the response to 1 mg/kg A + 1 mg/kg B is intermediate between the responses to 2 mg/kg A and 2 mg/kg B. If potentiation occurs, the response to the combination will be higher than is shown for the additive response. In (2), the response to A is linear and the response to B is nonlinear; the additive response to 2 mg/kg A + 2 mg/kg B is greater than in (1). This demonstrates that linearity cannot be assumed for individual response curves. If, in this case, dose–response curves were known only up to 2 mg/kg for A and for B and the response to 2 mg/kg A + 2 mg/kg B were as shown, it might be wrongly assumed that potentiation occurred. To establish that potentiation occurred, the dose–response curves for compounds A and B must be determined for doses above those used in combination.

The identification of combinations of pollutants that give rise to problems of potentiation might seem an impossible task. However, there are guidelines that aid the recognition of such combinations. In particular, recent rapid advances in biochemical toxicology have given more insight into the potentiation of toxicity due to interactions at the toxicokinetic level (Chapter 5). When one compound (A) causes a change in the metabolism of another (B), two types of interaction are recognized which lead to potentiation of toxicity:

1. Compound A inhibits an enzyme system that detoxifies compound B. Thus the rate of detoxication of B is slowed down because of the action of A.

2. Compound A induces an enzyme system that activates compound B. Thus the rate of activation of B is speeded up because of the action of A.

The two types of interactions will now be discussed separately.

9.4 Potentiation Due to Inhibition of Detoxification (Box 9.1)

In terrestrial vertebrates and invertebrates, the effective elimination of lipophilic xenobiotics depends on their enzymatic conversion to water-soluble products that are readily excreted (Chapter 5).

The inhibition of enzymes concerned with detoxification can lead to an increase in the toxicity of the compounds they metabolize. This phenomenon is illustrated by the effects

BOX 9.1 TWO EXAMPLES OF POTENTIATION FROM INHIBITION OF DETOXIFICATION

Pyrethroid insecticides, for example, cyhalothrin, are considerably toxic to bees, but they do not usually cause much damage when properly used in the field. Bees are repelled by pyrethroids, perhaps as the consequence of low-level sublethal effects. However, pyrethroids can become far more toxic in the presence of certain ergos- terol biosynthesis inhibitor (EBI) fungicides. Synergistic ratios of 5 to 20 have been reported for bees when EBI fungicides are added to pyrethroid insecticides. This potentiation of toxicity has been attributed to the inhibition of detoxification of pyre- throids by the MO system. Several reports from France and Germany have cited field deaths of hive bees attributed to this type of potentiation (Figure 9.4).

In a second example, organophosphorous insecticides containing thion (P=S) groups can inhibit microsomal MOs of vertebrates when the enzyme converts them to oxons (P=O; Figure 9.5). This process is oxidative desulfuration and leads to the binding of sulfur to cytochrome P_{450} with consequent losses of MO activity. Exposure to relatively low levels of organophosphorous compounds can make birds more sensitive to the toxicity of carbamate insecticides; carbamates are detoxified by MOs, so a reduction of MO activity can cause an increase in toxicity (Figure 9.5).

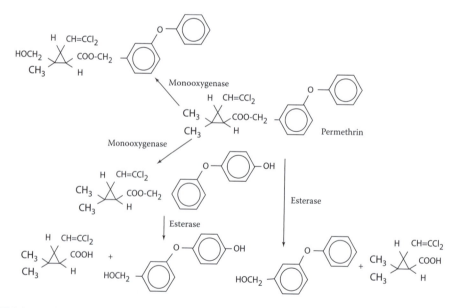

FIGURE 9.4

Metabolism of permethrin. Permethrin is detoxified by two systems. Monooxygenase attacks both acid and alcohol moieties. The B esterase system breaks the carboxyester bonds. Inhibitors of monooxygenase can increase the toxicities of permethrin and other pyrethroids.

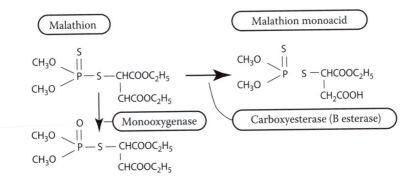

FIGURE 9.5
Metabolism of malathion. Malathion is detoxified by the action of a carboxyesterase but activated by monooxygenase. Toxicity depends on the relative importance of these competing enzymes. Chemicals that induce the monooxygenase system can make malathion more toxic.

of synergists on insecticide toxicity (Table 9.1). Some of the most striking examples of synergism involve inhibition of monooxygenases by piperonyl butoxide and other methylenedioxyphenyl compounds. Synergists of this type can increase the toxicity of pyrethroid and carbamate insecticides by as much as 60- and 200-fold, respectively. (The synergistic ratios provide measures of increases in toxicity.)

9.5 Potentiation from Increased Activation

Metabolism of lipophilic xenobiotics brings a reduction of toxicity in most cases, but some important exceptions have been observed. Oxidation by the monooxygenase (MO) systems sometimes generates highly reactive metabolites that can cause cellular damage. In principle, therefore, the induction of MO by a nontoxic dose of one compound can increase the toxicity of a second compound that is subject to oxidative activation. It does not follow that the induction of MO will automatically lead to increases in the rate of activation (or degree of consequent cellular damage). Induction may also lead to the induction of other enzymes that have detoxifying functions and can compensate for increased activation.

9.6 Field Detection of Potentiation

Although potentiation is a well recognized phenomenon in the laboratory, the extent to which it occurs in the field is virtually unknown. There are good reasons for suspecting that it may occur in heavily polluted areas. In some marine areas, for example, a wide range of organochlorine compounds (PCBs, TCDDs, and others; Figure 9.1) are found (Malins and Collier, 1981). Elevated levels of $P_{450}1A1$ have been found in fish and birds from such areas, and relatively high rates of activation of mutagenic PAHs have been suspected.

BOX 9.2   TWO EXAMPLES OF POTENTIATION
FROM INCREASED ACTIVATION

Benzo(a)pyrene and certain other carcinogenic polycyclic aromatic hydrocarbons (PAHs) are activated by an inducible form of cytochrome P_{450} known as P_{450}1A1 (Chapter 5). A range of planar organic compounds, not in themselves carcinogens or mutagens, can cause the induction of P_{450}1A1. Examples are certain PAHs, coplanar PCBs, and 1,2,7,8-tetrachlorodibenzodioxin (TCDD; Chapter 7). Such compounds can act as promoters that potentiate the carcinogenic actions of other compounds. By increasing the rate of activation of carcinogens, they can also increase the rate of formation of DNA adducts; this may lead to an increased rate of chemically induced mutation.

In marine environments, evidence indicates that (i) fish, birds, and mammals sometimes have elevated levels of P_{450}1A1, and (ii) that the levels of P_{450}1A1 relate to degree of exposure to pollutants such as coplanar PCBs. It is suspected but not proven that individuals with elevated 1A1 will experience higher levels of DNA damage caused by environmental carcinogens and mutagens.

In a second example, OP insecticides containing P=S groups are activated by P_{450} forms of the MO system (Figure 9.4). Thus the induction of MO by a nontoxic dose of a xenobiotic can lead to enhanced activation and consequently increased toxicity of the OP. Partridges exposed to EBI fungicides can show increased levels of hepatic MO due to induction. In the induced state, they show increased susceptibility to OP insecticides such as malathion and dimethoate (Walker et al., 1993; Johnston, 1995).

The time dependence of this potentiation must be emphasized. After exposure to the EBI, several hours will elapse before the rate of activation of the OP increases. Furthermore, the MO activity will return to its normal level after a few days if exposure to the EBI does not recur. Thus birds may become more susceptible to certain OPs between 6 hours and 5 days after exposure to an EBI. Potentiation of this kind has so far been demonstrated only in the laboratory. It has not been shown to occur in the field after normal approved use of pesticides (see further discussion in Section 9.6).

In intensive agriculture and horticulture, animals, birds, and nontarget invertebrates are exposed to a variety of insecticides, herbicides, and fungicides. Combinations of pesticides are sometimes used as seed dressings and sprays (e.g., tank mixes) and raising the possibility of potentiation of toxicity when animals are exposed to such mixtures. Mobile species (e.g., birds, flying insects) can be exposed sequentially to various compounds as they move from field to field.

As explained earlier, evidence indicates that pyrethroid sprays containing concentrations of an insecticide that do not show toxicity to bees become toxic when mixed with EBIs. Potentiation here has been attributed to EBIs inhibiting detoxification (Box 9.2; Pilling et al., 1995).

Other combinations of pesticides cause concern with regard to possible potentiation of toxicity under field conditions. In laboratory trials, Schmuck et al. (2003) demonstrated that the toxicity of the thiacloprid neonicotinoid is markedly potentiated by the EBIs prochloraz and tebuconazole. Concentrations of either the insecticide or the EBI administered individually caused less than 10% mortality of honeybees within 24 hours of exposure. However, if one of the EBIs was mixed with the insecticide at the same concentrations, mortality at 24 hours was 80% for the prochloraz mixture and 70% for the tebuconazole mixture.

In another laboratory investigation, Iwasa et al. (2004) reported a very high synergistic ratio of 1141 when triflumazole was mixed with thiacloprid. It is noteworthy that it is not only the lethal effects of neurotoxic insecticides such as these that cause concern. Evidence shows that sublethal effects, for example, on foraging, may also adversely affect bee populations. However, the significance of these observations related to effects of pesticides on bee populations will remain speculative until thorough field studies have been carried out. The effects of neurotoxic insecticides on bees are discussed in more detail in Section 8.4.2 and in Walker (2009).

The difficulty of identifying harmful effects caused by pollutants in the field is a recurring theme of this book. The potential of biomarker strategies to aid the resolution of this problem was emphasized in Chapter 8 and is discussed further in Chapters 15 and 16. This point can be illustrated by a hypothetical example in which the interaction of two pesticides is investigated in a field trial. One or more biomarkers of toxic effect would be chosen to measure responses to a pollutant in an appropriate indicator species. These responses would then be determined in three different field situations:

1. Where only compound A was applied
2. Where only compound B was applied
3. Where both A and B were applied

If potentiation occurred, the response in the case of 3 would be substantially greater than the total responses for 1 and 2. This approach has the advantage that it can measure potentiation at a sublethal level and thus can give early warning of enhancement of toxicity before lethal effects occur.

Nondestructive biomarkers exhibit several advantages in situations of this kind. In particular, they make it possible to conduct serial sampling in individual animals or birds; any changes caused by chemicals can then be measured in relation to internal controls. In other words, biochemical and physiological parameters can be measured in individuals before and after exposure to pesticides, thereby overcoming the serious difficulty of inter-individual variation. Serial sampling is possible, for example, in the case of nestling birds in nest boxes and in animals or birds living in limited areas (e.g., enclosures) from which they can readily be recaptured and sampled.

9.7 Summary

In the field, living organisms are exposed to mixtures of pollutants, and questions arise about possible interactions between the components of mixtures. When chemicals are tested during an environmental risk assessment, they are usually tested individually. Very rarely are mixtures tested. In the absence of evidence to the contrary, it is normally assumed that the toxicity of mixtures of compounds will be additive, i.e., the toxicity of a mixture will approximate to the total toxicities of its individual components.

This is often the case, and several types of environmental chemicals share a common mode of action which behave this way (e.g., Ah receptor-mediated toxicity caused by mixtures of coplanar PCBs and dioxins). However, in a small but significant number of cases toxicity is potentiated when organisms are exposed to mixtures and the level of toxicity is substantially greater than additive. The best known cases involve interactions at the toxicokinetic level, where one compound inhibits the detoxification of another or one increases the rate of activation of another. With increasing knowledge of the biochemical toxicology of pollutants, it becomes increasingly possible to predict such interactions and test for them.

Further Reading

Boon, J.P. et al. (1989). Complex pollution patterns caused by PCBs.

Malins, D.C. and Collier, T.K. (1981). Discusses effects of mixtures on marine organisms.

Moriarty, F. (1999). *Ecotoxicology*, 3rd ed. Discusses interpretation of data purporting to show potentiation.

Walker, C.H. et al. (1993). Laboratory evidence for potentiation in birds.

Wilkinson, C.F. (1976). Potentiation of toxicity of insecticides.

10

Biomarkers

The term biomarker has been gaining acceptance in recent years, albeit with some inconsistency in definition. Here we define a biomarker as any biological response to an environmental chemical at the individual level or below demonstrating a departure from the normal status. Thus biochemical, physiological, histological, morphological, and behavioral measurements may be considered biomarkers. Table 10.1 illustrates examples of biomarkers at different organizational levels. Biological responses at higher organizational levels—population, community, and ecosystem—are considered bioindicators. Notwithstanding the importance of changes at these higher levels (Chapters 12, 15, and 16), they are too general to be considered specific biomarkers.

The relationship between biomarkers and bioindicators based on specificity and ecological relevance is shown in Figure 10.1. In general, it is difficult to relate biochemical changes to ecological changes (although eggshell thinning caused by p,p'-DDE and imposex in dog whelks caused by TBT, discussed in Chapter 16, show that a physiological change can be related to a massive population change). It is also difficult to relate ecological changes to specific chemical causes.

10.1 Classification of Biomarkers

A number of classifications of biomarkers have been proposed. The most widely used is division into biomarkers of exposure and biomarkers of effect. Biomarkers of exposure indicate exposure of an organism to chemicals but do not indicate the degree of adverse effect the change causes. Biomarkers of effect, or more correctly toxic effect (because all biomarkers by definition show effects) demonstrate adverse effects on organisms.

Figure 10.2 depicts a biomarker approach based on changes in physiological parameters. The change in the health status of an individual with increasing exposure to a chemical is shown by a smooth curve running from healthy through reversible to irreversible changes leading to death. The important transition points along the way are: (i) first stress (h), i.e., physiology is no longer normal and the organism, although stressed, is able to compensate for this stress; (ii) the organism is no longer able to compensate (c) but the changes are still reversible and removal of the stress enables the organism to recover; and (iii) the point r beyond which the changes are irreversible and death ensues. The second part of Figure 10.2 shows the responses of five biomarkers used to measure health status (Depledge et al., 1993; Depledge, 1994).

TABLE 10.1

Biomarkers at Different Organizational Levels

Organizational Level	Example of Biomarker
Binding to receptor	TCDD binding to Ah receptor
	Nonylphenols binding to estrogen receptor
Biochemical response	Induction of Cytochrome $P_{450}1A$
	Inhibition of ACh-ase
	Vitellogenin formation
Physiological alteration	Eggshell thinning
	Feminization of embryos
Effect on individual	Behavioral changes
	Scope for growth

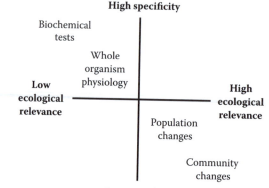

FIGURE 10.1

Specificity and ecological relevance of biochemical effects measurements. (*Source:* Addison, F. (1996). *Environmental Reviews* 4, 225–237. With permission.)

10.2 Specificity of Biomarkers

Biomarkers range from those that are highly specific—an enzyme of the heme pathway known as aminolevulinic acid dehydratase (ALAD) is inhibited only by lead—to those that are nonspecific. Effects on the immune system can be caused by a wide variety of pollutants. Table 10.2 lists biomarkers by degree of specificity.

Both highly specific and highly nonspecific biomarkers have value in hazard assessment. Taking blood samples of waterfowl and measuring ALAD activity can determine the percentage of the waterfowl at risk from lead poisoning without further measurements. However, determination of ALAD does not indicate what other pollutants may be present.

Inhibition of AChE can be used to provide legal proof of death by organophosphorous and carbamate pesticides (Hill and Fleming, 1982), and such inhibition is considered specific to these classes of chemicals. However, in recent years, evidence indicates that the inhibition of AChE is not caused solely by OPs and carbamates. Payne et al. (1996) found depression of AChE activity as high as 50% in fish in Newfoundland remote

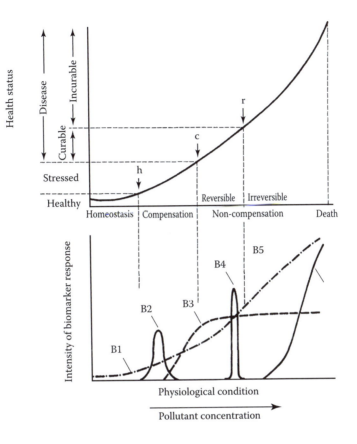

FIGURE 10.2

Relationship of exposure to pollutants, health status, and biomarker responses. Upper curve shows the progression of the health status of an individual as exposure to pollutant increases. *h* = Point at which departure from normal homeostatic response range is initiated; *c* = limit at which compensatory responses can prevent development of overt disease; *r* = limit beyond which the pathological damage is irreversible by repair mechanisms. The lower graph shows responses of five hypothetical biomarkers used to assess the health of the individual. (*Source:* Depledge. M.H. et al. (1993). In *Biomarkers: Research and Application in the Assessment of Environmental Health.* Springer-Verlag. With permission.)

from pesticide use and considered that a complex mixture of pollutants may have been involved. Studies by Guilhermino et al. (1998) noted that both detergents and metals can inhibit the activity of AChE and suggest that use of this enzyme as a biomarker could be extended.

The induction of monooxygenase is caused by a wide variety of chemicals and is often a sensitive indicator of pollutants (Besselink et al., 1997). Thus it is a useful indication that organisms are affected by pollutants, although it seldom reveals a specific cause. An exception is the induction of cytochrome $P_{450}1A1$ and $1A2$ by planar compounds such as dioxins and coplanar PCBs. Here, the degree of induction of the enzyme can give a measure of Ah receptor-mediated toxicity that may be caused by these pollutants. See Walker (2009) for further discussion. The induction of other P_{450}s that have less substrate specificity can provide the basis for useful biomarker assays of exposure without providing much evidence for identifying pollutants causing the response.

TABLE 10.2

Biomarkers Listed by Decreasing Specificities to Pollutants

Biomarker	Pollutant	Comments and References
ALAD inhibition	Lead	Sufficiently reliable to replace chemical analysis (Wigfield et al., 1986)
Metallothionein induction	Cadmium	More difficult to measure than cadmium levels (Hamer, 1986)
Eggshell thinning	DDT, DDE, dicofol	Degree of eggshell thinning is easily measured (Ratcliffe, 1967)
AChE inhibition	OPs, carbamates	Easier and more reliable than chemical analysis (Fairbrother et al., 1991)
Anticoagulant clotting proteins	Rodenticides	Measurements similar in complexity to chemical analysis (Huckle et al., 1989)
Monooxygenase induction	OCs, PAHs	Dioxin equivalent more easily measured than chemical analysis (Murk et al., 1997)
Porphyrin profiles	Several OCs	Separation by high-performance liquid chromatography well developed (Kennedy and James, 1993)
Retinol profiles	OCs	Demonstrates exposure to specific chemicals (Shugart, 1994); considerable natural variations; ratios more reliable than absolute values (Spear et al., 1986)
DNA and hemoglobin adducts	Largely PAHs	Several tests available; complicated by repair mechanisms
Vitellogenin induction	Estrogenic chemicals	Induction in male fish is sensitive indicator of estrogenic chemicals (Harries et al., 1997)
Other serum enzymes	Metals, OCs, PAHs	Several enzyme systems have been studied (Fairbrother, 1994)
Stress proteins	Metals, OCs	Wide range of stress proteins have been studied (Sanders, 1993)
Immune responses	Metals, OCs, PAHs	Many tests available (Wong et al., 1992)

10.3 Relationship of Biomarkers to Adverse Effects

The ability to relate a degree of change of a biological response to the harm it causes is useful, primarily for defending the cost of a proposed remedial action. Tabled 10.3 lists the same biomarkers shown in Table 10.2 in the order of the adverse effects they measure. A quick examination of the two tables will reveal differences in ranking order.

The first biomarker in Table 10.3 is eggshell thinning. It is possible to define a critical degree of eggshell thinning; it has been found for a variety of species that eggshell thinning in excess of 16 to 18% is associated with population declines. This phenomenon is discussed in more detail in Chapter 16.

The fact that the relationship between the biomarker response and an adverse effect is not clear does not invalidate the use of the biomarker. First, it demonstrates that the organism has been sufficiently exposed to a pollutant or pollutants to undergo a physiological change. In cases such as the induction of metallothionein, the change is protective (Chapter 7); a knowledge of how much of the possible protective mechanism

TABLE 10.3

Biomarkers Listed by Decreasing Specificities of Adverse Effects

Biomarker	Organizational Level	Comments and References
Eggshell thinning	Intact animal—population	Wide species variation in sensitivity; related to reproductive success (Peakall, 1993)
Inhibition of AChE	Organ—intact animal	Degree of inhibition related to mortality and sublethal effects (Grue et al., 1991)
Inhibition of ALAD	Organ—intact animal	Degree of inhibition related to mortality (Scheuhammer, 1989)
Clotting proteins	Intact animal—population	Related to mortality; risk assessed from blood protein levels (Hegdal and Blaskiewicz, 1984)
Induction of monooxygenases	Organ—population	Analysis of dioxin equivalents related to reproductive success; induction of P_{450} enzymes related to specific chemicals (Bosveld and van den Berg, 1994)
Depression of plasma retinol and thyroxine	Organ	Binding to specific protein; relation to adverse effects tenuous (Brouwer and van den Berg, 1986)
DNA integrity	Organ	DNA damage indicates serious harm; relationship to effects often tenuous (Everaarts et al., 1998)
Immune responses	Organ	Proper functioning critical to health, but system has considerable reserve (Richter et al., 1994)
DNA and hemoglobin adducts	Organ	Good monitor of exposure, especially for PAHs; relation to effects is tenuous (Varanasi et al., 1989)
Other enzymes	Organ	Relationship to effects not clear (Fairbrother, 1994)
Porphyrin profiles	Organ	Levels in environmental samples well below those causing adverse effects (Fox et al., 1988)
Induction of vitellogenin	Organ—intact animal	Clear link between induction of vitellogenin and presence of estrogenic chemicals; biological significance is speculative (Arcand-Hoy and Benson, 1998)
Induction of metallothionein	Organ	Protective mechanism, not related to mechanism of toxicity (Hamer, 1986)
Stress proteins	Organ	Difficult to separate effects from nonchemical stresses (Pyza et al., 1997)

has already been induced is valuable to assess the risk to individuals. Second, in the case of vital systems, it is an indication that further investigations should be undertaken. For example, few would take damage to the integrity of DNA lightly, even though in many cases the damage is repaired and no adverse effects occur (Chapter 7). In other cases, such as changes in porphyrin levels, it is clear that the levels are much lower than those shown to cause harm. Nevertheless, these biomarkers can be used to demonstrate exposure. The use of these various types of biomarkers in hazard assessment is considered later in this chapter.

10.4 Specific Biomarkers

Some of the biomarkers listed in Tables 10.2 and 10.3 are covered elsewhere in this book. Specifically, eggshell thinning is discussed in Chapter 16, and some information about the inhibition of AChE and induction of monooxygenases appears in Chapter 7. The next sections discuss some of the available biomarkers. More complete coverage is given in the books on biomarkers cited in the reading list at the end of this chapter.

10.4.1 Inhibition of Esterases

From the point of view of ecotoxicology, AChE is particularly useful because it represents the site of action, and its degree of inhibition relates to toxic effects. Butyrylcholinesterase (BuChE) is sometimes studied in parallel with AChE, but its physiological role is unknown and its degree of inhibition is not simply related to toxic effect. The study of neuropathy targets esterase, the interaction of which with organophosphorous compounds (OPs) can lead to organophosphorous compound-induced delayed neurotoxicity, has been confined to laboratory studies.

The mode of action is well established and has been considered in some detail in Chapter 7. Two classes of compounds, the OPs and carbamates, inhibit AChE, causing an accumulation of acetylcholine at the nerve synapses and disruption of nerve function that produces obvious effects: tremors, motor dysfunction, and death. The assay for AChE is more straightforward, quicker, and cheaper than chemical analysis for OPs or carbamates. The degree of inhibition of AChE has been related to the symptoms observed.

With vertebrates, the inhibition of brain AChE has often been used to establish that death has been caused by OP or carbamate pesticides (Mineau, 1991). Under ideal conditions, inhibition in the range of 50% to 80% can be taken as proof of mortality from the pesticide (Hill and Fleming, 1982). In practice, the degree of denaturation is often unknown and adequate controls are often difficult to obtain. Also, inhibition by carbamates is readily reversible (cf. OPs) and can be quickly lost after death. However, the measurements of OPs and carbamates are made more difficult because they are rapidly metabolized and eliminated; in fact, chemical analysis for residue levels has not been widely used for diagnosing poisoning by these compounds. The use of AChE inhibition for diagnosing damage caused by pesticides as a consequence of forest spraying in eastern Canada is discussed in Chapter 16.

The usual wild vertebrate organ studied is the brain—the principal site of action of OPs and carbamates. Although using inhibition of blood AChE or BuChE would be more acceptable, the relationship to inhibition of brain AChE is complex. Studies have shown that variability of esterase activity is much greater with plasma than with brain and that the recovery of plasma AChE activity is much more rapid than recovery of brain AChE. Also, plasma AChE does not represent a site of action of OPs and carbamates, and there is no simple relationship between degree of inhibition and toxic effect. Thus diagnosis based on plasma AChE activity is difficult.

10.4.2 The Induction of Monooxygenases

The heme-containing enzymes known as cytochromes P_{450} are major components of the defenses of organisms against toxic chemicals in their environment. Originally evolved, perhaps as long as 2,000 million years ago, to handle naturally occurring toxic compounds

(Nebert and Gonzalez, 1987), they now play an important role in the detoxification of man-made chemicals.

The monooxygenase system is a coupled electron transport system composed of two enzymes—a cytochrome and a flavoprotein (NADPH-cytochrome reductase). The system is found in the endoplasmic reticulum of most organs, but the activity is far greater in the liver than in most other tissues (Chapter 5). Recent work has shown the complexity of the system. One review identifies more than 750 isoforms belonging to 74 different gene families (Nelson, 1998). Cytochrome P_{450} I-dependent reactions include N-oxidation and S-oxidation; widely studied enzyme activities include ethoxyresorufin O-deethylase (EROD), benzo(a) pyrene hydroxylase (BaPH), and aryl hydrocarbon hydroxylase (AAH). Cytochrome P_{450} II-dependent oxidations include aromatic hydroxylation, acyclic hydroxylation, dealkylation, and deamination. Monooxygenases are induced by a wide variety of compounds (Section 7.2). Inducers of environmental interest include some organochlorine, organophosphorous, and pyrethroid insecticides, polycyclic aromatic hydrocarbons (PAHs), PCBs, and TCDDs.

Using the induction of monooxygenase activity in fish as a monitor of pollution of the marine environment by oil was proposed in the mid 1970s. Since then, a wide variety of studies has been published, ranging from the induction of AHH in fish near the Los Angeles sewage outlets to EROD induction by pulp mill effluent in Sweden. Induction of monooxygenases by paper mill effluent has proven one of the most sensitive biomarkers for tracking this type of pollution.

Monooxygenase activity is shown by a wide range of species (see Chapters 5 and 7), and some studies on fish-eating birds in the Great Lakes are detailed in Chapter 16. The value of monooxygenases for biological monitoring has been clearly demonstrated in the case of hydrocarbon pollution in fish and aquatic invertebrates and for both PAH and OC contamination in a wide range of organisms. As the response is caused by a very wide variety of chemicals, the system is capable of detecting exposures sufficiently high to cause biological responses to many xenobiotics. Conversely, it has only limited value for identifying causative agent(s) but can be used to delimit an area that may later warrant more detailed investigations.

From a practical view, the considerable variation within a specific population means that the sample size usually must be fairly large. If a system is induced by a large variety of natural compounds and xenobiotics and is affected by a wide variety of other parameters—temperature, diet, etc.—great care must be taken to ensure there are reliable control levels.

10.4.3 Studies of Genetic Materials

The fundamental role of DNA in the reproductive process is well known and will not be discussed further. The end points used to assess the damage to DNA by environmental pollutants are specific genotoxic effects, especially increased carcinogenesis, rather than effects on the reproductive process.

The sequence of events between the first interaction of a xenobiotic with DNA and consequent mutation may be divided into four broad categories. The first stage is the formation of adducts. At the next stage, secondary modifications of DNA such as strand breakage or an increase in the rate of DNA repair occur. The third stage is reached when the structural perturbations to the DNA become fixed. At this stage, affected cells often show altered functions. One of the most common assays to measure chromosomal aberrations is sister chromatid exchange. Finally, when cells divide, damage caused by toxic chemicals can lead to the creation of mutant DNA and consequent alterations in gene function.

The covalent binding of reactive metabolites of environmental pollutants to DNA—adduct formation—is a clear demonstration of exposure to these agents and an indication of possible adverse effects. The relationships of environmental levels, degrees of adduct formation, and ultimate effects are complex. For example, although a direct relationship between the extent of cigarette smoking and the number of DNA–BaP adducts has been clearly shown, the relationship between DNA–BaP adducts and the onset of lung cancer is less well defined. In the field of wildlife toxicology, the establishment of the sequence of events from the initial DNA lesion to harm is even more difficult. Nevertheless, it is reasonable to conclude that the reactions of chemicals with DNA cause harmful consequences such as tumor formation.

Three approaches have been used to study DNA damage caused by genotoxins and the formation of DNA adducts, after exposure to pollutants:

1. Radioactive postlabeling (usually with ^{32}P), leading to the separation of a range of adducts by two-dimensional thin-layer chromatography

2. Techniques to identify specific adducts, including fluorescence spectrometry, chromatographic techniques, and enzyme-linked immunosorbent assay (ELISA)—techniques that are sensitive if properly used and can detect one adduct among 10^8 normal nucleotides

3. Random amplified polymorphic DNA (RAPD) assays to detect alterations to DNA caused by chemicals

The three techniques yield different data. Radioactive postlabeling reveals the degree of total covalent binding. The techniques for identifying specific adducts provide information about the degrees of binding for a few specific compounds. RAPD gives evidence of genotoxin- induced DNA damage and mutations (Atienzar and Jha, 2004).

Monitoring of adduct formation provides one of the best means of detecting exposure to polycyclic aromatic hydrocarbons (PAHs). The stability of DNA and hemoglobin adducts formed by this class of compounds means that evidence of exposure to them remains after the compounds are cleared from the body. The ability to study adduct formation by hemoglobin means that nondestructive testing is possible.

In studies at Puget Sound, Washington, fish and sediments were sampled. The levels of PAHs in sediment and gut were determined, and the extent of DNA–xenobiotic adducts in the liver measured by ^{32}P labeling. Additionally, concentrations of PCBs and the degree of induction of MFOs were determined. The various indices enabled workers to discriminate sites that exhibited considerable differences in chemical contamination by both PCBs and PAHs (Stein et al., 1992).

DNA fingerprinting using the polymerase chain reaction has also been used as a biomarker for the detection of genotoxic effects of environmental chemicals (Savva, 1998; Atienzar et al., 1999). The differences between control and experimental fingerprints are thought to be caused by DNA adducts. In studies of bivalves in the lagoon of Venice, good correlation was found between DNA fingerprinting and the degree of strand breakage measured by the alkaline unwinding assay (Castellini et al., 1996).

Breakage in chromosomes can be examined directly under a microscope or by the alkaline unwinding assay (Peakall, 1992). This technique is based on the DNA strand separation that takes place where there are breaks. The amount of double-stranded DNA remaining after alkaline unwinding is inversely proportional to the number of strand breaks, provided that renaturation of the DNA is prevented.

Chromosome breaks in the gills of mud minnows were used to study pollution of the Rhine. Also, increased chromosomal aberrations were found in rodents collected from areas near a petrochemical waste disposal site. Although damage to chromosomes can lead to serious effects, it must be remembered that repair mechanisms are capable of preventing them.

Sister chromatid exchange (SCE) is the reciprocal interchange of DNA during the replication of chromosomal DNA. Chromosomes of cells that have gone through one DNA replication in the presence of labeled thymidine or the nucleic acid analogue 5-bromodeoxyuridine and then replicated again in the absence of the label are generally labeled in only one of the chromatids. The label is exchanged from one chromatid to the other. The SCEs can be easily visualized using a light microscope or differential staining.

Chromosomes that have undergone SCE should not be regarded as damaged in the conventional sense because they are morphologically intact. Nevertheless, SCE occurs at sites of mutational events including chromatid breakage. Good correlations have been observed between the number of induced SCEs per cell against the dosage of x-rays and the concentration of a number of chemicals known to cause chromosomal aberrations.

A relationship between SCE level (Nayak and Petras, 1985) and distance from an industrial complex was demonstrated in wild mice in Ontario, Canada. A variety of chromosomal aberrations were found in cotton rats living near hazardous waste dumps in the US. Fish exposed to water from the Rhine showed marked increases of SCE levels (van der Gaag et al., 1983).

Flow cytometry can detect chromosomal aberrations in a large number of cells rapidly and accurately. It has been shown to detect mutagenic and clastogenic effects in a wide variety of species (Bickham et al., 1998; Whittier and McBee, 1999).

A number of changes in genetic materials can be used to monitor for pollution by specific assays. Monitoring has been carried out on the incidence of tumors in fish. Fish are frequently sampled in considerable numbers and many of their tumors are visible externally. Such data could not be readily collected from other classes of organisms (muskrats from trappers or ducks from hunters), even when large numbers of samples are available, because of the cost of dissection.

In the North American Great Lakes, surveys to determine the incidence of tumors were conducted as part of a surveillance program. The levels of occurrence of tumors in brown bullheads (*Ictalurus nebulosis*) and white suckers (*Catostomus commersoni*) were highest in the most polluted areas. Because of the large numbers of contaminants in the Great Lakes, it is virtually impossible to link carcinogenesis to a specific chemical, but circumstantial evidence for a chemical origin is strong in many cases, although it should be cautioned that viral agents and parasites can cause neoplasms.

The occurrence of liver neoplasms in brown bullhead and the levels of PAHs in sediment have been monitored for twenty years on the Black River, Ohio (Baumann and Harshbarger, 1998). In the early 1980s, the prevalence of liver cancer was high (22% to 39% in fish over 3 years old). The coke factory was closed in 1983, and by 1987 levels of PAHs in sediment had decreased by two orders of magnitude and cancer rates had fallen to one-quarter of the previous figure. Dredging of the most contaminated sediments was carried out in 1990. A subsequent marked rise in cancer rates was followed by a decrease over the next few years.

Overall, the studies involving DNA have reached an interesting stage. A great deal of medical research indicates that this information can be used to assess the impacts of pollutants, especially PAHs, on wildlife.

10.4.4 Porphyrins and Heme Synthesis

Porphyrins are produced by the heme biosynthetic pathway—a vital system for most of the animal kingdom. Two major disruptions of heme biosynthesis by environmentally important agents have been studied. These are the formation of excess porphyrins after exposure to some organochlorines (OCs) and the inhibition by lead of the aminolaevulinic acid dehydratase (ALAD) enzyme.

Heme biosynthesis is normally closely regulated, and levels of porphyrins are ordinarily very low. Hepatic porphyria is characterized by massive liver accumulation and urinary excretion of uroporphyrin and heptacarboxylic acid porphyrin. Although the mechanism of OC-induced porphyria has not been completely elucidated, several researchers consider inhibition of the uroporphyrinogen decarboxylase enzyme the proximal cause. The two OCs most involved in inducing porphyria are hexachlorobenzene (HCB) and the PCBs. Although HCB has been shown in both mammals and birds to induce porphyria, the dosages required are high compared with environmental levels. PCBs have also been shown to be potent inducers, although their various congeners act quite differently.

Studies of the Rhine River showed that the patterns of hepatic porphyrins were markedly different and that the total porphyrin levels were much higher in pike collected there than in those from the cleaner River Lahn. The levels of organochlorines were up to 40-fold higher in the fish from the Rhine.

The variation in the means of the hepatic levels of highly carboxylated porphyrins (HCPs) in seven species (five orders) of birds was only twofold, and the total range was 4 to 22 pmol/g (Fox et al., 1988). No similar study appears to have been carried out on other classes of organisms. Baseline data collected from areas of low contamination show only small variations, but in view of the variability of responses to OCs in experimental studies, variability in areas of high contamination is a problem. Nevertheless, the levels of hepatic HCPs were markedly elevated in herring gulls (*Larus argentatus*) collected from the North American Great Lakes when compared with those from the Atlantic coast (Fox et al., 1988).

Aminolaevulinic acid dehydratase (ALAD) is an enzyme in the heme biosynthetic pathway. Inhibition of ALAD was first studied over thirty years ago as a means of detecting environmental lead exposure in humans and has since become the standard bioassay for this purpose. It has also been used in wildlife investigations. The assay is highly specific for lead because other metals are 10,000 times less active in causing inhibition. ALAD inhibition is rapid, but the effect is only slowly reversed. ALAD values return to normal only after about four months.

Inhibition of ALAD has been used as an indicator of lead exposure in general situations such as urban areas and along highways, and also specifically to study lead shot in waterfowl. A threefold difference in blood ALAD activity was found between rats in a rural area and rats in an urban site in Michigan (Mouw et al., 1975). The main physiological indications of lead toxicity in the urban rats were increases in kidney weight and the incidence of intranuclear inclusions. Both effects could be correlated with lead levels. Similarly, marked differences in ALAD activity were found among feral pigeons (*Columbia livia*) from rural, outer urban, suburban, and central London areas (Hutton, 1980).

The lead levels, ALAD activity, and reproductive success of barn swallows (*Hirundo rustica*) and starlings (*Sturnus vulgaris*) along North American highways with different traffic densities were studied (Grue et al., 1986). They found a significant increase of the lead levels in the feathers and carcasses of both adults and nestlings, and a 30% to 40% decrease in plasma ALAD activity. However, the number of eggs laid, the number of young fledged, and prefledgling body weights were not affected, indicating that lead from automotive emissions does not pose a serious hazard to birds nesting near motorways.

Mortality of ducks and other waterfowl caused by the ingestion of lead shot has been a serious concern for many years. The issue was first raised in North America over seventy years ago. Ducks and geese ingest spent lead shot during the course of feeding. A nation-wide survey found that 12% of the gizzard samples examined contained at least one lead shot and noted that 2% to 3% of all waterfowl in North America died from lead poisoning. Secondary poisoning of bald eagles feeding on waterfowl is another concern. National surveys of eagles found dead in the US showed that about 5% died from lead poisoning. The inhibition of ALAD has been shown to be sensitive enough to detect the effect of a single pellet. Many researchers found a strong negative correlation between blood lead concentration and log ALAD activity. The ALAD assay is simple and involves no expensive equipment or lengthy training. ALAD inhibition represents one end of the biomarker spectrum. It is a sensitive, dose-dependent measurement that is specific for a single environmental pollutant: lead.

10.4.5 Induction of Vitellogenin

The discovery of hermaphrodite roaches in stretches of rivers near sewage outlets triggered research on the effects of estrogenic disruptors in fish because natural hermaphroditism is assumed to be minimal. Examination of these fish showed that males contained high levels of vitellogenin—an egg yolk protein usually produced only by females. Thus increased levels of vitellogenin in male fish provided the basis for a biomarker assay for studies of endocrine disruptors (Box 10.1).

10.4.6 Behavioral Biomarkers

The behavioral effects of pollutants were discussed in Section 8.4.2. Use of behavioral effects in the development of biomarker assays represents a higher organizational level than others considered to date. One of the early proponents of the value of behavioral toxicology stated that the behavior of an organism represents the final integrated result of a diversity of biochemical and physiological processes. Thus a single behavioral parameter is generally more comprehensive than a physiological or biochemical parameter. Although much interesting work has been carried out in recent years, behavioral biomarkers are still not accepted as components of formal testing procedures.

Two fundamental difficulties surround the use of behavioral tests in wildlife toxicology. First, the best studied and most easily performed and quantified tests exhibit the least environmental relevance. Second, the most relevant behaviors are the most strongly conserved against change (Peakall, 1985).

Operant behavior such as conditioning to respond to a colored key to obtain food is too remote from real life to relate to survival. It can merely be presumed that a decrease in learning ability is an unfavorable response. Avoidance behavior is more directly related to survival, although the relationship has not been quantified. The ability to capture food is clearly important to predatory species but is difficult to measure under field conditions.

Field observations are difficult to quantify. It was suggested that behavioral changes possibly led to the decline of the peregrine falcon. However, observations by time-lapse photography at the eyries of highly contaminated peregrines revealed little abnormal behavior (Enderson et al., 1972). This study was based on seven peregrine eyries in

BOX 10.1   INDUCTION OF VITELLOGENIN IN FISH

A survey of five rivers in England was conducted in the summer of 1994 (Harries et al., 1997). Caged male rainbow trout were deployed at five sites in each river, one upstream from the suspected sources (waste treatment plants), one at the point of effluent discharge, and the other three at different distances downstream.

In four cases, the fish placed in the neat effluent showed very marked and rapid increases in vitellogenin levels. In two cases, none of the downstream sites showed estrogenic activity; in another, activity was detected 1.5 km downstream. The fourth river was quite different: the effluent was extremely estrogenic and so was water sampled at all other sites (maximum distance 5 km). The effluents contained much higher levels of alkylphenolic compounds. On the fifth river, even the neat effluent did not cause increased levels of vitellogenin. In this case, the small waste treatment plant did not receive industrial waste.

A disturbing finding is that some UK estuaries show high degrees of estrogenic contamination. Flounder (*Platichthys flesus*) in the Tyne and Mersey estuaries had vitellogenin levels four to six orders of magnitude higher than controls (Scott et al., 1999). Elevation of vitellogenin was less marked in the Crouch and the Thames.

The implications of these findings at the population level are not known and are difficult to determine. Sewage discharges may well contain other compounds that cause reproductive toxicity by other means. The effect on lowland coarse fish may well be different from effects on trout, which usually do not occur in these waters. It would be important to examine estuaries for effects on fish that reproduce in these areas rather than the flounder that breeds at sea. Myriad other factors also affect wild populations of fish. For further discussion, see Section 15.2. and Chapter 15 in Walker (2009).

Alaska, using battery-powered time-lapse motion picture cameras that took pictures about every 3 min. The film cartridges were replaced every 6 or 7 days. Replacement was difficult because the eyries were widely separated and the terrain was difficult. In two of the nests, the eggs broke, but no evidence of abnormal behavior was observed. The other five nests were successful. In all, some 70,000 pictures covering 4,200 hours were obtained. One of the drawbacks of this type of experiment is the time required to analyze the data obtained.

In another study (Nelson, 1976), 300 hours of observations of twelve peregrine eyries from a hide were analyzed. Four clutches lost single eggs, probably by breakage, but no abnormal behavior was observed. Although 300 hours is a lot of observation time, the average observation per clutch was only 25 hours among 400 total hours of daylight during the incubation period. These two studies illustrate the difficulties of observing behavioral changes in the field. Even if behavioral changes are documented, it is difficult to relate them to specific chemicals.

The best documented studies that can be extended to real life situations are those involving the organophosphorous pesticides and the subsequent inhibition of AChE. These were covered in Section 8.4, and their use in field conditions is discussed further in Section 16.4.

Studies of avoidance responses of fish to toxicants dates back over eighty years. From the first simple studies on acetic acid, the field has grown enormously, and the equipment has become highly complex, for example, sophisticated fish avoidance chambers with video monitors and computer-interfaced recording systems.

Recent studies include many on the effects of heavy metals. A comparison of the lowest observed effect concentration (LOEC) based on behavioral studies (avoidance, attraction, fish ventilation, and cough rates) with chronic toxicity studies indicates that some of the behavior tests are more sensitive than life cycle or early life stage tests. Other studies involving predator avoidance, feeding behavior, learning, social interactions, and locomotor behaviors have been insufficient to allow judgments about their sensitivity or utility.

At present, behavioral tests have not replaced conventional toxicity tests. However, they may provide ecological realism, e.g., the effects of pollutants on predator–prey relationships but they must be capable of field validation.

The behavioral tests that are the most advanced involve fish. The fish avoidance test is well established in the laboratory as a means of showing effects well below the lethal range, and highly automated procedures are available. Nevertheless, a note of caution should be injected. Pre-exposure to effluent reduced the avoidance behavior, and pre-exposed fish were observed more often in contaminated than in clean water (Hartwell et al., 1987). This desensitization caused by pre-exposure makes it likely that laboratory experiments will overestimate the responsiveness of fish to metal pollution in the wild.

These difficulties do not imply that behavioral effects caused by pollution are unimportant. Studies such as those examining the predation pressure on fiddler crabs (*Uca pugnax*) have shown that operational levels of pesticide use can cause population effects through behavioral changes (Ward et al., 1976). Substandard prey are more readily captured by predators, but field studies of the impact of chemicals on behavior are difficult.

The overall conclusion is that behavioral parameters are not especially sensitive to exposure to pollutants and that biochemical and physiological changes are usually at least as sensitive. Further, the variability of biochemical data is generally less and the dose–response relationship clearer than those obtained from behavioral studies. In general, physiological and biochemical changes are more readily measured and quantified.

10.4.7 Biomarkers in Plants

Plants have been widely used as biomonitors to localize emission sources and analyze the impacts of pollutants, especially gaseous air pollutants, on plant performance. One of the earliest studies was by Angus Smith, who examined the damage to plants around Manchester from what he called *acid rain*. However, biomarkers should go beyond the visible parameters of sentinel species. They should establish processes and products of plants that will allow recognition of environmental stress before damage is apparent. A biomarker must be able to predict an environmental outcome and consequential damage. Ideally, biomarkers should be selected from the events of biochemical or physiological pathways, but reaching this stage with plant biomarkers is difficult (Ernst and Peterson, 1994).

Specific biomarkers have been identified in sensitive plants. In a few cases, excess of a specific chemical will lead to the production of a metabolite that differs in tolerant and sensitive plants. In the presence of excess selenium, Se-sensitive plants fail to differentiate between S and Se. They incorporate Se in sulfur amino acids, leading to the synthesis of enzymes of lower activity that can lead to plant death. In contrast, Se-tolerant plants biosynthesize and accumulate nonprotein selenoamino acids such as selenocystathionein and Se-methyl-selenocysteine that do not cause metabolic problems. Thus selenoproteins in plants are excellent biomarkers for Se stress, although their use in the field has not been widely reported.

Another example is that after exposure to excess fluor (a mineral containing fluorine), plants synthesize fluoroacetyl coenzyme A and then convert it via the tricarboxylic acid cycle to fluorocitrate—a compound that blocks the metabolic pathway by inhibiting the aconitase enzyme. As a result, fluorocitrate accumulates and is a very reliable biomarker for fluor poisoning.

Some plant biomarkers reveal free metals. Phytochelatins are synthesized during exposure to a number of metals and anions such as SeO_4^{2-}, SeO_3^{2-}, and AsO_4^{3-}. Dose- and time-dependent relationships have been established under laboratory conditions for cadmium, copper, and zinc. For monitoring purposes, research is needed into the phytochelatin production of plants in the field.

General plant biomarkers that respond to a variety of environmental stresses may be useful to indicate that some component in the environment is a hazard to plant life. For example, the activity of the peroxidase enzyme has been used to establish the exposure of plants to air pollution, especially to SO_2.

Changes of enzyme systems during the development stage of plants and the effects of seasonal and climatic processes are not yet sufficiently known to demonstrate the reliability of enzyme activity as a monitoring device. At present, plant biomarkers are not as well advanced as animal biomarkers, but this is likely to change. The stationary nature of plants aids greatly in measuring exposure to pollutants; monitoring surveys for measuring levels of metals in lichens and organochlorines in pine needles are in place and would be more valuable if the measurements could be linked with biological changes in the plants.

10.5 Role of Biomarkers in Environmental Risk Assessment

The most compelling reason for using biomarkers in environmental risk assessment is that they yield information about the effects of pollutants. Thus their use in biomonitoring is complementary to the more usual monitoring by determining or predicting residue levels (see Section 6.5 for a discussion of risk assessment).

The first point in any assessment process is to decide what to assess. This may sound self-evident, but it is surprising how seldom precise objectives are defined. This is the case with many monitoring programs intended to determine levels of environmental chemicals. Take, for example, the International Mussel Watch Programme that measures metals in mussels in many parts of the world. The justification of such surveys is that we should know what pollutants are where and at what concentrations. However, what will be done with the information? Only if we know what concentrations are hazardous and what effective remedial action can be taken will such information be of practical use.

Similar considerations apply to biomarkers. Again, action levels must be determined if a monitoring program is to be effective. An advantage of the biomarker approach is that it may show that the physiology of an organism is within normal limits, indicating that no action is necessary. By contrast, zero levels are rarely found in analytical determinations. Ideally, both approaches (i.e., residues and biomarkers) should be utilized in an integrated manner (Chapter 11).

Before legislation to limit risks can be enforced, two fundamental questions must be answered: (i) the amount of damage we are prepared to tolerate and (ii) the amount of proof required. At one extreme, little concern will arise if a few aquatic invertebrates die within a few meters of the end of an outlet pipe. At the other end of the scale, events such as the destruction of most of the biota of the Rhine River (Deininger, 1987) set off a world-wide reaction. Our concerns are also species dependent and tend to increase as we move from algae to mammals and widespread differences exist even within classes of animals. It is easy to arouse concern about pandas and whales and difficult in the case of rats. Further, some types of damage are considered more serious than others; alteration of the genetic material that may be passed onto future generations is considered one of the most serious effects.

The question of how much damage we are prepared to tolerate is for society to answer; then it will be possible to design protocols to meet the standards required. Without an answer, scientists face the problem of trying to set up regulations without a goal. The question of how much proof is enough is largely scientific and it cannot be answered until the question about tolerable damage is answered. Despite the inertia, decisions must be made. Failure to make a decision constitutes making a decision.

Why use biomarkers in risk assessment? One important reason arises from the limitations of classical risk assessment. The basic approach is to measure the chemical present and then use animal experiment data to relate the chemical data to the adverse effects caused (Section 6.5). The limitation of this approach is that we have defined the levels critical to organisms for very few compounds. Under real life conditions, many organisms are exposed to complex and changing levels of mixtures of pollutants. Chemical monitoring works only if a material is persistent. Chemicals such as the PAHs and many pesticides have very short biological half-lives in most species but may nevertheless exert long-term effects. Biological and chemical monitoring systems should be complementary; we must know what chemicals are where and what their effects are.

The first question that biomarkers can help answer is whether environmental pollutants are present at sufficiently high concentrations to cause effects. If the answer is positive, further investigation to assess the nature and degree of damage and the causal agents is justified. A negative answer means that no further resources must be invested. In this way, biomarkers can act as important early warning systems.

The role of biomarkers in environmental risk assessment is determining whether organisms in a specific environment are physiologically normal. The approach resembles the study of biochemistry in humans. A suite of tests can be carried out to see whether an individual is healthy. It is necessary to select both the tests and the species to be tested.

In selecting tests, the specificity of a test to pollutants and the degree to which the change can be related to harm must be considered. Both specific and nonspecific biomarkers are valuable in environmental assessments. In an ideal world, we would have biomarkers to indicate exposures to and assess risks of all major classes of pollutants, and nonspecific biomarkers to assess accurately and completely the health of organisms and their ecosystems.

Clearly, the definition of harm must cover scientific and social aspects. Scientifically, it is important to demonstrate unequivocally that changes resulted from pollution, and biomarkers have an important role to play. Whether a change is sufficiently serious to warrant the cost of remedial action is for society to decide. These issues will be taken up further in Chapter 17 of this book.

10.6 Summary

A biomarker is defined as a biological response to an environmental chemical at the individual level or below that reveals a departure from normal status. The specificity of biomarkers to pollutants varies greatly. Highly specific biomarkers are valuable for detecting exposure to and possible effects of specific chemicals but yield no information about other pollutants. In contrast, nonspecific biomarkers show that exposure to pollutants has occurred without identifying the pollutant responsible. A number of specific biomarkers were considered in some detail. Biomarker assays are particularly useful when they relate to toxic effect (mechanistic biomarkers)—not just to exposure. The most important reason for using biomarkers in environmental risk assessment is that they demonstrate the effects of pollutants; their use is complementary to biomonitoring that involves the determination of levels of environmental chemicals.

Further Reading

Fossi, M.C. and Leonzio, C., Eds. (1994). *Nondestructive Biomarkers in Vertebrates*. Proceedings of a workshop devoted to the use of blood and other tissues that can be collected nondestructively to measure biomarkers.

Huggett, R.J. et al. (1992). *Biomarkers. Biochemical, Physiological, and Histological Markers of Anthropogenic Stress*. Techniques and extensive coverage of DNA alterations and immunological biomarkers.

McCarthy, J.F. and Shugart, L.R. (1990). Compilation of papers presented by American and European scientists at an American Chemical Society meeting.

Peakall, D.B. (1992). *Animal Biomarkers as Pollution Indicators*. Use of biomarkers of higher animals for environmental assessment.

Peakall, D.B. and Shugart, L.R. eds. (1993). *Biomarkers: Research and Application in the Assessment of Environmental Health*. NATO symposium.

Peakall, D.B. et al. (1999). *Biomarkers: A Pragmatic Basis for Remediation of Severe Pollution in Eastern Europe*. NATO workshop.

11

In Situ Biological Monitoring

11.1 Introduction

It is difficult to predict the effects of pollutants on organisms to an acceptable degree of accuracy by simply measuring concentrations of a chemical in the abiotic environment (Figure 11.1). Factors that affect bioavailability of chemicals to organisms include temperature fluctuations, interactions with other pollutants, soil and sediment type, rainfall, pH, and salinity. Even using biotic monitoring of chemical residue levels (see Section 11.3), there are considerable difficulties in knowing the effects of this level of chemical on the organism. This process is made more difficult by the presence of mixtures (see Chapter 9) and the considerable interspecies differences in response. In situ biological monitoring attempts to get around these problems by analyzing various parameters of natural populations that reflect the situation in the field rather than the standardized conditions of laboratory experiments.

There are four main approaches to in situ biological monitoring of pollution (Hopkin, 1993a). Each is examined in this chapter and is illustrated with examples of studies on a range of animals and plants from terrestrial and aquatic ecosystems. The four types are:

1. Monitoring the effects of pollution on the presence or absence of species from a site or changes in species composition, otherwise known as community effects (see also Chapter 14)

2. Measuring concentrations of pollutants in indicator or sentinel species (see also Chapters 4 and 5)

3. Assessing the effects of pollutants on organisms and relating them to concentrations in those organisms and other biotic and abiotic indicators (see also Chapters 7 and 8)

4. Detecting genetically different strains of species that have evolved resistance in response to a pollutant (see also Chapter 13)

11.2 Community Effects (Type 1 Biomonitoring)

The most frequent response of a community to pollution is that some species increase in abundance, others (usually the majority) decrease in abundance, and populations of others remain stable. The patterns of the species abundances reflect effects integrated over time and are widely used to monitor effects of pollutants on communities. Effects of pollution on communities and ecosystems are covered in greater detail in Chapter 14.

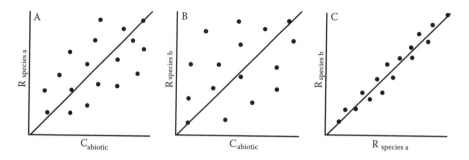

FIGURE 11.1

Schematic graphs to illustrate a principle of in situ biological monitoring of pollution. In this hypothetical example, the responses R (as measured by concentrations or effects of the pollutant on the *y* axis) of species "a" (A) and "b" (B) in sites with different levels of contamination are not closely related to concentrations of the pollutant (*x* axis) in abiotic samples (soil, air, water sediment) from the same sites. Because the relationship between species in the same sites is much closer (C), the responses of species "a" to the pollutant can be used to predict the responses of species "b" more accurately than similar predictions from abiotic samples (see Figures 11.5 and 11.6 for data that support this hypothesis). Reproduced from Hopkin, S.P. (1993). *In situ* biological monitoring of pollution in terrestrial and aquatic ecosystems. In Calow, P., ed., *Handbook of Ecotoxicology*, Vol. 1, pp. 397–427. Oxford, Blackwell. With permission from Blackwell Scientific.

11.2.1 Terrestrial Ecosystems

To be able to recognize an unusual assemblage of species, it is necessary to monitor changes over time or to have sufficient background knowledge of the normal ecology of similar but unpolluted sites (Van Straalen, 2004). In the former approach, sites must be monitored for several years to distinguish effects of pollutants from natural fluctuations. Perkins and Millar (1987) showed that emissions from an aluminum works in Anglesey, North Wales, were responsible for the almost complete elimination of lichens within 1 km of the factory soon after its opening in 1970. Some recovery has taken place since 1978, when new emission controls were introduced, but the lichen flora was still much impoverished in 1985 in comparison with pre-1970 populations (Figure 11.2). In the background knowledge approach, one or more sites are examined in the contaminated area and are compared with at least one reference site (e.g., Figure 11.3). However, if the differences between polluted and reference sites are subtle, the investigator is left with the difficult problem of deciding when a change indicates a toxic effect or natural between-site variations.

11.2.2 Freshwater Ecosystems

In Britain, three main approaches have been adopted to assess the effects of pollution on communities of freshwater organisms (British Ecological Society, 1990, on which this section is based). These are:

1. The biotic approach, based on the differential sensitivities of species to pollutants
2. The diversity approach, based on changes in community diversity
3. River invertebrate prediction and classification (RIVPACS), which combines an assessment in terms of both the types of species present and the relative abundances of families

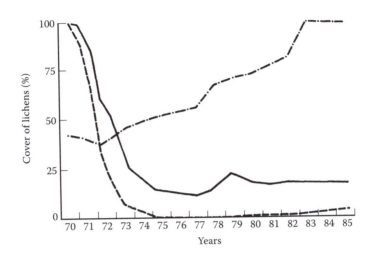

FIGURE 11.2

Cover (as percentage of type maximum) of foliose (—), fruticose (– – –), and crustose (–·–) corticolous lichens in eight permanent quadrats set up on broad-leaved trees within 1 km of the aluminum works in Anglesey, North Wales, from 1970 to 1985. Reproduced from Perkins, D.F. and Millar, R.O. (1987). Effects of airborne fluoride emissions near an aluminium works in Wales. 1. Corticolous lichens growing on broad-leaved trees. *Environmental Pollution* 47, 63–78. With permission from Elsevier Applied Science.

The most frequently used biotic indices have been the Trent Biotic Index (TBI), the Chandler Biotic Score (CBS), and the Biological Monitoring Working Party (BMWP). All three are based largely on relative tolerances of macroinvertebrates to organic pollution. TBI and CBS require identification of species, whereas BMWP requires only family level identification (see Box 11.1). Only the CBS takes abundance into account. However, these scores are generally assumed rather than experimentally derived. Furthermore, sensitivity rankings are assumed to apply across a range of toxicants, even though laboratory experiments have shown that this is not necessarily the case.

BOX 11.1   THE USE OF THE BMWP SCORE TO ASSESS THE HEALTH OF FRESH WATERS

A system for rapid appraisal of the health of freshwater ecosystems was developed in the late 1970s by the Biological Monitoring Working Party (BMWP). Scores ranging from 1 to 10 are given to families of invertebrates depending on their sensitivity to organic pollution. Families with a low tolerance to pollution are given a high score (e.g., the mayfly family *Heptageniidae* is allocated a score of 10), whereas those tolerant of severe pollution are given the lowest score of only 1 (e.g., tubificid oligochaete worms). At a particular site, scores for all families are added up to give the total BMWP score. The average score per taxon (APST = total BMWP score divided by the number of taxa) is a particularly valuable index because it is less sensitive to sampling effort and can be predicted with greater reliability than the number of taxa or the BMWP score.

Sites can be graded into four biological classes on the basis of these scores (see Table 11.1). Thus the highest quality of site gets a grade of A, whereas the grades B, C, and D are indicative of progressive loss of water quality.

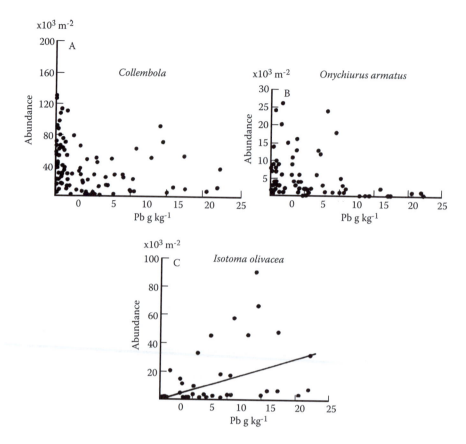

FIGURE 11.3

Abundance of (A) total Collembola, (B) *Onychiurus armatus* (Collembola), and (C) *Isotoma olivacea* (Collembola) in the 0- to 3-cm layer in lead-contaminated soils in the vicinity of a natural metalliferous outcrop in a Norwegian spruce forest. The concentrations of lead represent metal extracted from soil over 18 h in 0.1 M buffered acetic acid. Note that *O. armatus* is sensitive to lead pollution, whereas *I. olivacea* reached higher population densities in contaminated soils. Reproduced from Hågvar, S. and Abrahamsen, G. (1990). Microarthropoda and Enchytraeidae (Oligochaeta) in naturally lead-contaminated soil: A gradient study. *Environmental Entomology* 19, 1263–1277. With permission from the Entomological Society of America.

TABLE 11.1

Biological Banding of Average Score per Taxon (ASPT), Number of Taxa, and Biological Monitoring Working Party Score (BMWP) Based on Sampling for Three Seasons

Biological Class	Observed/Expected ASPT	Observed/Expected Number of Taxa	Observed/Expected BMWP Score
A	≥ 0.89	≥ 0.79	≥ 0.75
B	0.77–0.88	0.58–0.78	0.50–0.74
C	0.66–0.76	0.37–0.57	0.25–0.49
D	< 0.66	< 0.37	<0.25

Source: Wright, J.F., et al. (1993). A good summary of the RIVPACS technique by some of the workers who developed it. *European Water Pollution Control* 3, 15–25.

The most frequently used diversity indices have concentrated on species richness and the distribution of individuals among species. The Shannon–Weiner diversity index is the most commonly used. However, there are three main problems with diversity indices. First, many factors influence community structure, and as diversity indices do not take account of the species present, their usefulness as measures of water quality can be questioned. Second, there is an unresolved debate as to which index to use to measure diversity and as to which taxonomic group and level should be considered. Third, it is not clear how diversity responds to pollution. For example, diversity of plankton reduces continuously with organic enrichment, but for benthic invertebrates the response is bell-shaped, with the greatest diversity at intermediate pollution levels (British Ecological Society, 1990).

The RIVPACS approach is used to predict the fauna of a site using environmental variables (Wright et al., 1993). Hypothetical target communities are provided against which the combined effects of physical and chemical stresses can be assessed. Comparison of observed values with these predictions provides environmental quality indices. Although RIVPACS is the most widely used of the techniques described, it is still a broad-brush approach more useful for highlighting rivers and streams in need of more detailed study than for giving a final verdict on the level of environmental damage.

11.2.3 Marine Ecosystems

One of the simplest methods of detecting a pollution-induced change in communities of marine benthic organisms is to analyze the log-normal distribution of individuals per species in sediment samples (Gray, 1981). In many samples of benthic communities, the most abundant class is not that represented by one individual per species but often lies between classes with three and those with six individuals per species. Thus the curve relating numbers of individuals per species (x axis) to number of species (y axis) is often strongly skewed. This curve can be brought back to a normal shape by plotting the number of individuals per species on a geometric scale (usually × 2). Plotting the geometric classes on the x axis (class I = 1, class II = 2–3, class III = 4–7, class IV = 8–15, and so on) against the cumulative percentage of species on the y axis invariably gives a straight line. In polluted sites, there is often a break in the line, indicating a departure from an equilibrium community. If this persists over several sampling occasions, it is indicative of pollution-induced disturbance.

The responses of marine-fouling communities to pollution stress, in terms of changes in species composition, can be monitored in situ by reciprocal transplants. Climax communities are allowed to develop on submerged surfaces in a clean and a polluted site and are then moved between the sites. An experiment in Australia at Woolongong Harbour (uncontaminated) and Port Kemblar Harbour (polluted by discharges from heavy industry) demonstrated rapid changes in the community structure in response to pollution (Moran and Grant, 1991). Indeed, within two months, those communities on submerged plates that had been transferred from Woolongong to Port Kemblar were similar in structure to those that had developed entirely in Port Kemblar. Most changes occurred in short time periods when sensitive species were killed by periodic discharges (an effect difficult to predict by measuring levels of pollutants in the water). Space previously occupied by these species was quickly colonized by opportunists more tolerant to the pollutants, thus leading to changes in community structure.

11.3 Bioconcentration of Pollutants (Type 2 Biomonitoring)

Destructive measurement of levels of pollutants in organisms provides an indication of how much is present at a particular moment in time and may enable effects on predators to be assessed (see Chapters 12 and 16). Take, for example, a species of wading bird that feeds primarily on estuarine bivalve mollusks. "Critical" (safe?) concentrations for the birds could be set based on levels in bivalves rather than sediment or water or indeed their own tissues. The bivalves provide a critical pathway from the abiotic environment to the waders, the importance of which can be monitored biologically by analyzing the mollusks.

11.3.1 Terrestrial Ecosystems

Contamination of plants has been monitored either by collecting samples such as moss directly from the field (Ruhling and Tyler, 2004) or by exposing material (usually bags of Sphagnum moss) for a specified period and returning it to the laboratory for analysis. The decline in concentrations of lead in air in the UK following the reduction of permitted levels of lead in petrol was mirrored by a decline in lead concentrations in plants (Jones et al., 1991).

Biological monitoring of radioactive fallout in Italy derived from the Chernobyl disaster in 1986 showed that levels of ^{137}Cs in mushrooms increased after the accident (Borio et al., 1991). Basidiomycete hyphae in the soil accumulated the radioisotopes, but it was not until they produced sporophores that the cesium became available to above-ground fungivores. No correlation was found between the level of ^{137}Cs in the mushrooms and that in the soil in this study, emphasizing the importance of biological monitoring of radioactive pollution. Traditional risk assessment does not make allowances for effects such as this.

In Norway, where similar results have been obtained, the level of ^{137}Cs in the milk of goats increased fivefold in 1988 after the abundant growth of mushrooms in grazing land (Figure 11.4). The sporophores contained levels of radioactivity up to 100 times greater than those in green vegetation. Thus fungi provide an important critical pathway for the concentration and transport of radioactive isotopes along food chains. As far as invertebrates are concerned, it is clear that some groups, such as woodlice, snails, and earthworms, accumulate significant amounts of metals from their diet (Hopkin, 1989; Hopkin et al., 1986; Van Straalen, 1996), whereas most insects are able to regulate concentrations to relatively low levels (Hopkin, 1995). Thus species in the former three groups provide a significant route for the transfer of metal pollutants to their predators (Figure 11.5 and Figure 11.6).

Monitoring concentrations of pollutants in vertebrates may pose some difficulties. They are usually difficult to catch, population densities are lower than in invertebrates, and a license may be required (or may even be unobtainable for some species). Ways around this are to take blood samples or use a nonliving product of the animal that indicates previous exposure to a pollutant. Analysis of eggs has been widely used for organochlorines, and this is less destructive than the collection of adults. Feathers have been proposed as possible indicators. In Finland, feathers from nestlings of a range of birds provide a good indicator of mercury exposure of the adults (Solonen and Lodenius, 1990). Regurgitated pellets from owls can be used to monitor exposure to rodenticides. The metal content of teeth of bank voles has been used as an exposure indictor in Poland (Gdula-Argasinska et al., 2004).

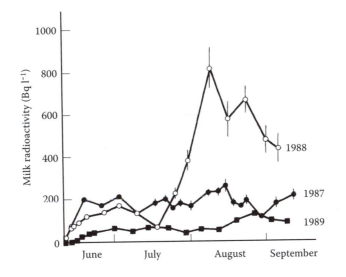

FIGURE 11.4
Radiocesium ([134]Cs and ^{137}Cs) activity from 1987 to 1989 in goat milk during the grazing seasons from 15 June to 15 September in the Jotunheimen mountain range (mean ± SE, three to eight animals in 1987 and eight in 1988 and 1989). Reproduced from Hove, K., Pederson, O., Garmo, T.H. et al. (1990). Fungi; a major source of radiocesium contamination of grazing ruminants in Norway. *Health Physics* 59, 189–192. With permission from the Health Physics Society and Williams & Wilkins, Baltimore.

11.3.2 Freshwater Ecosystems

Predicting bioaccumulation in aquatic systems is more difficult than in terrestrial ones because of the greater mobility of water and sediments in comparison with soils and because of the difficulty of knowing whether the main route of exposure is via water or food. Organisms can be collected directly from the field for analysis or caged in polluted and unpolluted sites to assess bioavailability. The freshwater amphipod *Gammarus pulex* has been used extensively for such work.

Monitoring concentrations of pollutants in fish is carried out all over the world. For example, a monitoring program for mercury in Brazil showed that levels in edible parts of fish from gold mining areas (where mercury is used extensively in gold extraction and refining) were five times greater than the "safe" level for human consumption (Pfeiffer et al., 1989). Global contamination with butyltin compounds has been monitored using skipjack tuna (Ueno et al., 2004). The long-term studies of Schmitt and Brumbaugh (1990) detected a decrease in the concentrations of lead in fish in the United States between 1976 and 1984. This coincided with regulatory measures that reduced the influx of lead to the aquatic environment. A similar downward trend has also been found in the levels of polychlorinated biphenyls (PCBs) in the eggs of herring gulls from the Great Lakes (Figure 11.7).

11.3.3 Marine Ecosystems

Marine mammals at the top of the food chain are particularly difficult to study as they are usually protected; therefore, many studies are based on beached specimens. This procedure is likely to be biased in favor of high residue levels. Some progress has been made in obtaining skin samples nondestructively (Aguilar and Borrell, 1994; Hobbs et al., 2003), but systematic biomonitoring of marine mammals is still at an early stage of development.

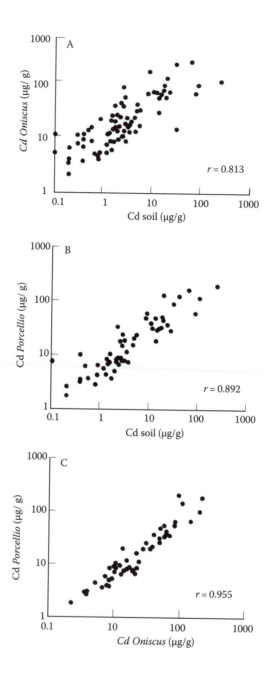

FIGURE 11.5
Relationships between concentrations of cadmium (dry weight) in the terrestrial isopods (woodlice) *Oniscus asellus* (A) and *Porcellio scaber* (B) and soil, and between the two species (C) collected from sites in Avon and Somerset, southwest England in 1998 and 1989. The region includes a primary zinc, cadmium, and lead smelting works and disused zinc mining areas. Each point represents the mean of twelve isopods and six samples of soil from each site. Note that the concentrations of cadmium in *P. scaber* in this region can be predicted more accurately from the concentrations in *O. asellus* (C) than from levels in soil (B). Reproduced from Hopkin, S.P. (1993). *In situ* biological monitoring of pollution in terrestrial and aquatic ecosystems. In Calow, P., ed., *Handbook of Ecotoxicology*, Vol. 1, pp. 397–427. Oxford, Blackwell. With permission from Blackwell Scientific.

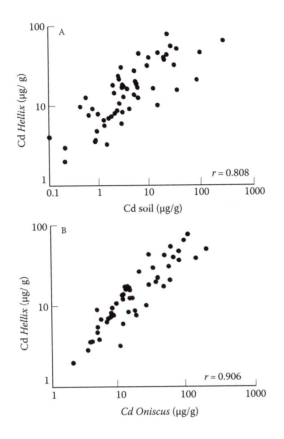

FIGURE 11.6
Relationships between concentrations of cadmium (dry weight) in the snail *Helix aspersa* and soil (A) and *Oniscus asellus* (B) collected from the same region as in Figure 11.5. Each point represents the mean of seven snails, twelve woodlice, or six samples of soil from each site. Note that the concentrations of cadmium in *H. aspersa* can be predicted more accurately from the concentrations in *O. asellus* (B) than from levels in soil (A). Reproduced from Hopkin, S.P. (1993). *In situ* biological monitoring of pollution in terrestrial and aquatic ecosystems. In Calow, P., ed., *Handbook of Ecotoxicology*, Vol. 1, pp. 397–427. Oxford, Blackwell. With permission from Blackwell Scientific.

Colonial nesting marine seabirds have been studied with eggs being used for the determination of organochlorine (OC) levels and feathers for the levels of heavy metals. In the latter case, it is possible to extend the time backward by using museum specimens. For example, Applequist et al. (1995) were able to examine mercury levels in guillemots (*Uria aalge*) over a 150-year period and found that levels increased from 1–2 $\mu g/g^{-1}$ during the period 1830–1900 to 3–6 $\mu g/g^{-1}$ by the 1970s.

By far the greatest research effort has been directed toward bivalve mollusks. The reasons for this are fourfold. First, many species are a source of food for predatory vertebrates, particularly birds. Second, they are widespread and common, are easily collected in large numbers, and are sedentary. Third, because they are filter feeders, bivalves pass large volumes of water through their bodies, accumulate pollutants continuously, and act as integrators of exposure over long periods. Fourth, there is good background knowledge of their basic biology to allow results to be put into an ecotoxicological context.

Mytilus edulis has been analyzed most frequently as it is common and has a global distribution. Research on *Mytilus* has been so extensive that global mussel-watch schemes have

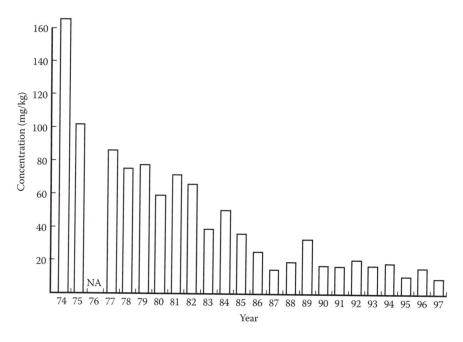

FIGURE 11.7
Concentrations of PCB in eggs of herring gulls from Muggs Island/Leslie Spit colonies, Lake Ontario, 1974–1997.
(Data from Environment Canada.)

been established. These have shown trends in pollutant levels that, in many cases, have declined. For example, Fischer (1989) was able to demonstrate that concentrations of cadmium in *Mytilus edulis* in Kieler Bucht in the western Baltic in 1984 had declined to about 30% of their level in 1975. Many of the schemes are ongoing and have been running for several years. This will ensure that long-term trends in the bioavailability of inorganic and organic pollutants to bivalves and their predators can be separated from natural fluctuations (for recent examples, see Cattani et al., 1999; Gunther et al., 1999).

Vertebrates are at the top of marine food chains and are vulnerable to poisoning from pollutants contained in their diets. This is particularly true of some organic pollutants that may accumulate in fatty tissues of marine mammals (Kannan et al., 2004) and birds (Hario et al., 2004). Residence times may be very long in organic pollutants that are highly lipophilic (Kow > 105) or that are only slowly metabolized to water-soluble products that can be excreted (see Chapter 5). Thus even after the complete removal of the source of pollution, contaminant levels in the tissues of vertebrates may take several years to decline to background levels. Levels of PCBs and dichlorodiphenyltrichloroethane (DDT) in open-ocean dolphins did not decline between 1978 and 1986, despite a reduction in organochlorine contamination of the marine environment during this period (Loganathan et al., 1990).

11.4 Effects of Pollutants (Type 3 Biomonitoring)

The primary aim of ecotoxicologists should be to describe and predict effects of pollutants on organisms and ecosystems. The basis of such studies is that biochemical, cellular, physiological, and morphological parameters can be used as screening tools or biomarkers in environmental monitoring (see Chapter 9).

11.4.1 Terrestrial Ecosystems

One of the simplest in situ indicators of air pollution is the Bel W3 variety of tobacco. The plants develop a mottling of the leaves in response to low levels of ozone pollution. The test is easy to perform and has been used in several surveys involving schoolchildren (Heggestad, 1991). Forest dieback is one of the clearest examples of the effects of pollution on ecosystems. Growth rates of trees can also be useful and can be retrospective if determined from the widths of annual growth rings.

Ecological and physiological effects due to pollution may have unexpected side effects on behavior. For example, air pollution from a copper smelter in Finland resulted in a decline in caterpillars in the vicinity of the factory. Great tits (*Parus major*) obtain the carotenoids that give them their yellow color by feeding on these caterpillars. Birds near the factory had paler plumage than did birds farther away, and it is thought that this will reduce fitness in terms of choice of mate and survival (Eeva et al., 1998). Similar results have been obtained for radioactively contaminated barn swallows (*Hirundo rustica*) near Chernobyl, in which there appears to be a trade-off between increased use of carotenoids for free radical scavenging and their role in sexual signaling (Camplani et al., 1999).

In some situations, effects can be related directly to a local source of pollution. Walton (1986) obtained direct evidence for the effects of fluoride emissions from an aluminum plant on small mammals. Moles and shrews collected less than 1 km from the plant had extremely high levels of fluoride in their bones and teeth. Several manifested the symptoms of fluoride poisoning, including chipped and broken teeth and brittle bones.

An in situ bioassay using earthworms for assessing the toxicity of pesticide-contaminated soils was developed by Callahan et al. (1991). At each site, five *Lumbricus terrestris* were placed in enclosures distributed in transects throughout areas of high and low contamination at a superfund site in Massachusetts formerly used for mixing pesticides. Mortality, morbidity (coiling, stiffening, swelling, lesions, etc.), and whole body concentrations of a wide range of organic pollutants in worms were related to levels in the soils. This in situ method does not require removal of highly contaminated soils from the site. It provides an accurate dose–response relationship that can be used to predict the soil levels below which worms will return to the site and the "safe" levels in soils at which worms will not accumulate sufficient concentrations of pollutants to be harmful to their predators.

11.4.2 Freshwater Ecosystems

At the organism level, most recent research has been directed toward developing in situ bioassays for detecting sublethal effects. The most widely used parameter is scope for growth (SFG) (see Section 8.5). SFG measures the difference between energy input to an organism from its food and the output from respiratory metabolism and, at least in principle, can be related to population and community processes. Animals that are stressed (expending energy on detoxifying and excreting pollutants) have less energy available for somatic growth and reproduction. This is manifested as lower reproductive and growth rates compared with unstressed controls. Most measurements on SFG in fresh water have been on amphipod and isopod crustaceans. The SFG test using *Gammarus pulex* for measuring the effects of zinc and low pH is at least an order of magnitude more sensitive than acute 24-h LC_{50} tests (Naylor et al., 1989).

Many rivers and streams are affected by acute episodic pollution rather than long-term chronic contamination. Bursts of pollutant runoff can occur after thunderstorms, rapid snow melt (see Figure 4.8), accidents at factories, or deliberate release. These effects of intermittent increases in pollutant concentrations can be detected by continuous monitoring of caged organisms. Seager and Maltby (1989) described such a system that used rainbow trout (*Onchorhynchus mykiss*). Fish were caged in situ in Pendle Water, a polluted urban river in Lancashire, England. The trout responded to sewer out-flow discharges by increasing their breathing rate. This was monitored by measuring the small oscillating voltage produced by the muscles involved in gill ventilation.

11.4.3 Marine Ecosystems

The effects of pollutants on marine ecosystems are difficult to study because of the vastness of the system and the difficulty of relating any effects seen to specific chemicals. The use of biological effects to monitor marine pollution has been reviewed by Addison (1996). He discusses the use of three biochemical responses (monooxygenase induction, metallothionein induction, and AChE inhibition), measurements of energy partitioning in mollusks, and analysis of benthic community structure to assess the impact of marine pollution.

One of best-studied examples of effects of pollutants on marine organisms is that of imposex in mollusks caused by tributyltin (TBT). This phenomenon is considered in more detail in Chapter 16. In the present chapter, only the biomonitoring aspects are discussed (see Box 11.2 and Figure 11.8).

Studies to determine the effects of pollutants on marine mammals are few in comparison to studies on invertebrates. Detailed studies on the effects of PCBs on seals have been carried out in the Netherlands (Brouwer et al., 1989, 1990). More recently, evidence has been found of renal lesions in seals that may be related to pollutant accumulation (Bergman et al., 2001). Usually, extrapolation has been made from unrelated species when assessing risks to marine mammals.

Work on the beluga whale (*Delphinapterus leucas*) in the St. Lawrence estuary in Canada illustrates the difficulties of field studies with marine mammals. This population has failed to increase despite the cessation of hunting, and toxic chemicals have been blamed for the problems of this isolated population. Here the collection of specimens for scientific study cannot be made, and thus material can only be obtained from animals found dead or by nondestructive sampling. Determination of the residue levels revealed high levels of PCBs, but as the animals had been dead for some time, measurements of the biomarkers

---

BOX 11.2   IMPOSEX IN DOG WHELK

The level of imposex in a population of dog whelks is the percentage of females that possess a penis, whatever the size. The mean size of the female penis relative to males in a population of dog whelks is represented by the relative penis size index (RPSI). The RPSI is calculated by dividing the cube of the mean length of the female penis by the cube of the mean length of the male penis (both in mm) and multiplying by 100. Thus if the RPSI is 100%, then the mean size of the female penis is the same as that of the males. This index provides an indicator of the exposure of dog whelks to TBT at a site (Figure 11.8).

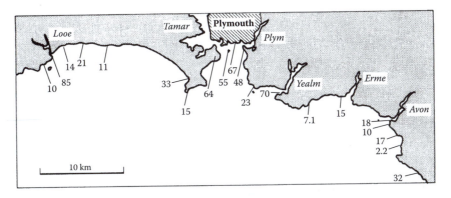

FIGURE 11.8
Levels of relative penis size index (RPSI) (percentage size of female penis relative to males) in dog whelks (*Nucella lapillus*) collected from southwest England in 1984–1985. Imposex develops in females in response to TBT leached from antifouling paints and is most prevalent in areas of high boating activity (e.g., the Looe and Yealm estuaries). Reproduced from Bryan, G.W., Gibbs, P.E., Hummerstone, L.G., and Burt, G.R. (1986). The decline of the gastropod *Nucella lapillus* around southwest England: Evidence for the effect of tributyltin from antifouling paints. *Journal of the Marine Biological Association of the United Kingdom* 66, 140–611. With permission from the Marine Biological Association of the United Kingdom.

studied by the Dutch workers were not possible. At present, there is no firm evidence that PCBs are causing harmful effects in whales.

It should be pointed out that although both pinnipeds and cetaceans are termed marine mammals, they are not closely related. Studies on DNA adducts (Sections 7.4.1 and 10.4.3) from stranded whales in the St. Lawrence have revealed high levels of benzo(a)pyrene adducts that were not detected in the brains of belugas killed by native hunters in the Arctic (Martineau et al., 1988). The levels of adducts have been correlated with high incidence of tumors in the St. Lawrence beluga whales.

The possibility that the widespread mortality of seals in various parts of the world caused by viruses is linked to effects of chemicals (especially PCBs) on the immune system has been put forward, but the Scottish verdict of "not proven" seems to be as far as one can go at the moment. Detailed immunological studies on beluga whales have been started based on nondestructive sampling of skin, and it will be interesting to learn whether the high levels of PCBs in the St. Lawrence population is affecting the immune system.

11.5 Genetically Based Resistance to Pollution (Type 4 Biomonitoring)

Strains of plants that possess genetically based resistance to high concentrations of metals in soils have been recognized for many years (Baker and Walker, 1989). Resistance is also well documented in insects. Such resistance is inheritable and should be distinguished from phenotypic tolerance that all members of a species may possess (preadaptation). The latter may consist of avoidance strategies, high excretion ability, or possession of enzymes that break down organic pollutants. Phenotypic tolerance can be induced (e.g., increased synthesis of metal-binding proteins). However, genetically distinct pollution-resistant strains will evolve only if the selection pressure persists for several generations.

Terrestrial invertebrates that have been shown by breeding experiments to develop genetic resistance to high concentrations of metals include races of earthworms (Spurgeon and Hopkin, 2000), Collembola, terrestrial isopods (woodlice), and *Drosophila* (Posthuma and Van Straalen, 1993). However, the selective advantage that this conveys is quite small. Resistant animals typically survive concentrations of metals only some 30–50% higher than controls. Freshwater oligochaetes and marine polychaetes have also evolved resistance to metals. In some of these cases, the basis of the increased tolerance is an increase in the copy number or transcription rate of the gene coding for detoxifying proteins. In the case of organophosphate insecticides, up to a 256-fold amplification of the genes coding for nonspecific esterases that break down the insecticide have been found in mosquitoes (see also Figure 13.6). For metals such as copper and cadmium, the gene that codes for the metal-binding protein metallothionein may be duplicated up to four times. The metals are bound more rapidly by resistant animals after ingestion. Such amplification has been found in wild *Drosophila* and has probably evolved in response to the spraying of fruit trees with copper-containing fungicides. Metallothioneins are ancient proteins known from such diverse groups as Collembola (Hensbergen et al., 1999) and snails (Dallinger et al., 1997). Thus the ability to develop resistance in sites naturally enriched with metals such as copper and zinc has been around for many millions of years.

11.6 Conclusions

In situ biomonitoring organisms should satisfy the five Rs if they are to be used successfully (Hopkin, 1993a). These are:

1. Relevant—To be ecologically meaningful, ecotoxicological tests should use species that play an important role in the functioning of the ecosystem.

2. Reliable—Species should preferably be widely distributed, common, and easily collected to facilitate comparison between sites separated by large distances.

3. Robust—Bioindicators should not be killed by very low levels of pollutants (with the exception of type 1 community structure monitoring where sensitivity is important) and should be robust enough to be caged in polluted field sites.

4. Responsive—The organisms should exhibit measurable responses to pollutant exposure by having greater concentrations of the contaminant(s) in the tissues (type 2 biomonitoring), by exhibiting effects such as reductions in scope for growth and fecundity, by increased incidence of disease or induction of a biochemical response (type 3 biomonitoring), or by possession of genetically based resistance (type 4 biomonitoring).

5. Reproducible—The species chosen should produce similar responses to the same levels of pollutant exposure in different sites.

11.7 Summary

Numerous factors influence the bioavailability of pollutants. This makes it very difficult to accurately predict the extent to which chemical residues are assimilated by animals and plants and their biological effects in the field. Consequently, many researchers have used in situ biological monitoring to gather such information. Four main approaches have been adopted. First, the presence or absence of taxa from clean and contaminated sites can be assessed (community effects). Second, concentrations of pollutants are measured in organisms collected from the field; this approach also allows field deployment of sentinel species to monitor pollutant availability (e.g., the mussel *Mytilus edulis* as part of the global mussel-watch program). The third type of monitoring quantifies the effects of chemicals on organisms and includes a wide range of biochemical and physiological parameters that come under the general heading of biomarkers. One of the most successful biomarker assays is the measurement of imposex in dog whelks. This phenomenon is caused by tributyltin (TBT), which has been widely used as an antifouling coating on boats. TBT imposes male characteristics on females, leading to a reduction in reproductive performance and local extinction of dog whelk populations in the most contaminated sites. The fourth type of monitoring detects the evolution of genetic resistance. This is most obvious in insects where, for example, resistant strains of mosquitoes may tolerate concentrations of organophosphate (OP) insecticides more than 200 times greater than those that will eliminate nonresistant populations. Natural selection has favored those mutations able to produce greater quantities of nonspecific esterases with a detoxifying function in response to the toxic action of the insecticides.

Further Reading

Baker, A.J.M. and Walker, P.L. (1989). Concise review of metal tolerance in plants (see also Ernst et al., 1992).

British Ecological Society. (1990). A useful summary of the background to monitoring water quality.

Bryan, G.W., et al. (1986). A classic paper on the effects of TBT on dog whelks.

Callahan, C.A., et al. (1991). Excellent paper describing in situ monitoring with earthworms at a superfund site in the United States.

Hove, K., et al. (1990). Interesting study on radiocesium in goats that emphasizes the importance of long-term monitoring.

Lowe, V.P.W. (1991). Use of laboratory data to interpret unexplained mass mortality of seabirds near the Sellafield nuclear reprocessing plant.

Naylor, C., et al. (1989). Study on scope for growth in *Gammarus pulex*.

Posthuma, L. and Van Straalen, N.M. (1993). Comprehensive review of metal tolerance and resistance in terrestrial invertebrates.

Wright, J.F., et al. (1993). A good summary of the RIVPACS technique by some of the workers who developed it.

Section III

Effects of Pollutants on Populations and Communities

12

Changes in Numbers: Population Dynamics

So far, this book has mainly described the effects of pollutants on individual organisms. In this chapter, we consider how populations should be described, how they should be analyzed, and the interpretations that may then be made as to the causes of population changes. The first six sections of the chapter are an exposition of population ecology theory, and some readers will find this quite heavy going. An attractive route for those encountering this material for the first time may be to begin by reading the case studies in Sections 12.7 and 12.8.

When pollutants enter an ecosystem, the species within it may be affected in any one of the ways shown in Figure 12.1. The numbers of some species will decline, perhaps to zero (Figure 12.1, curve i) if the species becomes locally extinct. Alternatively, numbers may decline but level out lower than before (curve ii), and the population may persist at this level if the pollution endures (chronic pollution). A third possibility is that population size initially increases (curve iii). If the pollution is transient, then the population may eventually recover, either rising from the level to which it was depressed by pollution (curve iv) or returning through immigration/recolonization if pollution had rendered the population extinct (curve v). In these last two cases, the population does not necessarily return to its original level. We shall see examples of some of these curves in the case studies described later in this chapter. A last possibility to consider is that in the case of chronic pollution, resistance may evolve within the population, allowing population numbers eventually to increase to a new equilibrium. The evolution of resistance is considered in Chapter 13. In this chapter we introduce the ecological theory needed to interpret and understand the processes that may operate to produce curves such as those in Figure 12.1.

Before we can describe populations, we need a method that determines which organisms are in the population and which are not. Often geographic boundaries are used. The choice of these boundaries has two very important consequences. In the first place, if the area chosen to define a population is too small, there will be extensive migration in and out of the population. This produces problems in population analysis because the migration rates have to be estimated, and this is far from easy. Second, population analyses are specific to the environment in which the population occurs. Change the environment, and the population parameters and the processes that determine them may all change. In these two ways, the choice of geographic boundaries is likely to have large effects on the results of population analyses.

The simplest description of a population is in terms of its abundance. This is discussed in the next section.

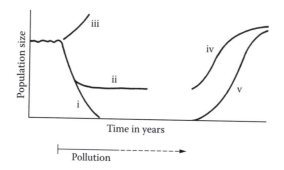

FIGURE 12.1
Possible responses of population size to pollution.

12.1 Population Abundance

Population abundance measures the size of a population, generally as the number or density of individuals within it. Abundance is the most widely used descriptor of populations and the easiest to understand, and the effects of pollutants on populations are most easily understood in terms of their effects on abundance, as in Figure 12.1. Most studies of the effects of pollution plot abundance against time; for examples, see Figures 12.10, 12.14, 12.17, 12.22, and 12.24. Note that some of these also show the duration of pollution events that may have been responsible for population changes.

What happens to abundance is of central concern in assessing the effects of pollution, because stressors and disturbances generally depress population size. After a disturbance, however, abundance may recover and eventually reach an equilibrium representing the carrying capacity of the assessed environment, i.e., the population size that can be sustained by the environment in the long term. Changes in this equilibrium abundance are generally of greater ecological significance than are changes in abundance alone, because "a change in population size without a change in equilibrium population size indicates an effect that may well be temporary, whereas a change in a population's equilibrium probably indicates something more long term" (Maltby et al., 2001). We shall return to the topic of carrying capacity and the factors that affect it throughout this chapter.

Measuring total abundance requires the counting of all the individuals in the population, but this is often not practicable and is not always the best approach. Counting only the breeding individuals will sometimes be easier and more relevant. In other cases, it may be necessary to estimate population density, for example, by taking a sample of a small area and then multiplying up. Alternatively one may have to use population indices, for example, densities of fecal droppings or numbers sighted along transects. Guidance on sampling and censusing methodology can be found in Krebs (1999) and Sutherland (2006).

Further population analysis is based on analyzing the causes of changes in population abundance. Methods by which this can be achieved are now described.

12.2 Population Growth Rate

A key feature of population growth or decline, as shown in Figure 12.1, is the rate at which it occurs. The curve i population, for example, declines fairly rapidly and becomes extinct. The curve ii population initially declines but then steadies and neither grows nor declines—its population growth rate is zero. The curve iii population initially increases in size—its population growth rate is positive. Thus population growth rate is positive, zero, or negative according to whether the population increases, is stationary, or decreases.

Population growth rate is the most important characteristic of the population, and population ecologists spend much time and effort measuring it and establishing the factors that affect it. Some of these factors are described in Section 12.3. Right at the outset, however, it is worth noting that there is one set of factors that is particularly important and particularly difficult to study in practice. These are density-dependent factors, i.e., factors that are affected by population density.

Typically, the effect of density dependence is that population growth rate reduces as population density increases. In the most straightforward case, the population density is then stabilized at a level characteristic of the environment in which the population exists. This level is referred to as the carrying capacity of the environment.

It will be immediately apparent that density dependence is an unwelcome complication for ecotoxicologists. It means that for a full understanding we need to know not only the way that pollution affects the population growth rate but also the way that density dependence affects it.

12.3 Population Growth Rate Depends on the Properties of Individual Organisms

Growth rates of a population can be measured in different ways, as shown in Box 12.1.

It is intuitively clear that population growth rate depends on individuals' birth and death rates and on the timing of their breeding attempts. Together, these characterize the individuals' life histories. In general, however, mortality, birth, and growth rates may change with age. For a complete description of the life history, therefore, we need a record of age-specific growth, birth, and death rates. Collectively, these are sometimes referred to as the organism's vital rates. At this point, the reader may wish to look at the real-life example provided in the case study in Section 12.3.1.

BOX 12.1   DIFFERENT WAYS OF ASSESSING POPULATION GROWTH

The measure we favor is population growth rate, r. This is a per capita growth rate, so r is defined as the population increase per unit time divided by the number of individuals in the population. The definition of population growth rate can be understood mathematically as follows. If population size $N(t)$ is plotted as a function of time t, as in Figure 12.1, then dN/dt represents mathematically the population increase per unit time, in units of animals per unit time. Population growth rate puts this on a per capita (i.e., per animal) basis by dividing by the number of individuals in the population. Mathematically, $r = 1/N \, dN/dt$. This is measured as animals per capita per unit time. Thus if population size is 1000 and is increasing by one animal per year, the population growth rate is 0.001 per capita per year.

If r is constant, then $N(t) = N(0)e^{rt}$. This shows that exponential population increase occurs if population growth rate is constant.

Another measure that is sometimes used is net reproductive rate, usually given by the symbol λ, defined as

$$\lambda = e^r \tag{12.1}$$

Conversely,

$$\lambda = \log_e \tag{12.2}$$

λ is the factor by which the population is multiplied each year. Thus if the population doubles each year, then $\lambda = 2$. For example, if $\lambda = 2$ and initial population size is 10, then in subsequent years population size is 10, 20, 40, 80 …. In this case, $r = \log_e 2 = 0.693$.

We now consider a method of recording the life histories of organisms with discrete breeding events—for example, they might breed annually. Suppose the organism breeds for the first time at age t_1, for the second time at age t_2, for the third time at age t_3, and so on, as shown in Figure 12.2. Suppose the number of offspring produced by each female is n_1 at her first breeding attempt, n_2 at her second, and so on. Last, suppose that the probability of a female surviving from birth to age t_i is l_i. The population growth rate, designated r, can be calculated from the formula

$$1 = \frac{1}{2}n_1 l_1 e^{-rt_1} + \frac{1}{2}n_2 l_2 e^{-rt_2} + \frac{1}{2}n_3 l_3 e^{-rt_3} + \kappa \tag{12.3}$$

Equation 12.3 is known as the Euler–Lotka equation. In deriving the equation, it is assumed that the proportion of females in each age class is invariant (the assumption of stable age distribution). However, estimates of population growth rate calculated from Equation (12.3) are still useful in practice even if the stable age distribution assumption does not hold.

It is usual to use a computer to calculate r from measurements of life history parameters (values of n_i, l_i, and t_i). The simplest method is to calculate the right side of Equation 12.3 for

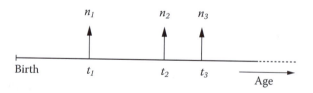

FIGURE 12.2

General life history. $t_1, t_2, t_3, \ldots$ represent the ages at which the organism breeds; $n_1, n_2, n_3, \ldots$ are the number of offspring then produced by each breeding female.

each of a number of trial values of r (e.g., try $-0.5, -0.4, -0.3, \ldots, 0.3, 0.4, 0.5$). Just one value of r makes the right-hand side of Equation 12.3 equal to 1. This value of r satisfies Equation 12.3 and so measures the population's growth rate.

The life history may be simpler than that shown in Figure 12.2. For example, the adult mortality rate may be constant, or the birth rate may be constant, or breeding attempts may be at regular intervals. Such simplifications often allow Equation (12.3) to be written in a simpler, more tractable form (Calow et al., 1997).

One other very important way of describing life histories is to record the number of organisms surviving and the number of offspring produced at regular intervals, e.g., daily. These records are conveniently tabulated in a matrix, which is referred to as a population projection matrix. Powerful methods of matrix algebra have been developed and applied to the analysis of these matrices (Caswell, 2001).

In the following case study, life histories were recorded by daily counting of the numbers of animals surviving and of the numbers of offspring produced.

12.3.1 The Life History and Population Growth Rate of the Coastal Copepod *Eurytemora affinis*

The life history and population growth rate of the coastal copepod *Eurytemora affinis* at different concentrations of dieldrin was studied by Daniels and Allan (1981). A population of *E. affinis* was obtained from Chesapeake Bay, Maryland, where it undergoes annual population expansions between February and May when water temperatures increase from 5 to 15–20°C. Animals were kept in the laboratory for 2 months (three or four generations) at 18°C before experimentation began.

The experiment consisted of recording the life histories of animals subjected to different concentrations of dieldrin. There were seven treatments, 0, 1, 2, 3, 4, 5, and 10 µg l^{-1} dieldrin, together with an acetone control because acetone was used as the carrier of dieldrin. Sixty newly hatched larvae (nauplii) were allocated to each treatment and maintained in groups of six in dishes containing 20 ml of bay water. The water was changed every other day, when the animals were fed a fixed number of algal cells. The numbers of survivors and births were counted daily.

The survivorship curves of animals undergoing the various treatments are shown in Figure 12.3A. Survival to day 20 was worse at concentrations of 5 to 10 µg/l than at lower concentrations. Reproduction began around day 18, and the birth rate was variable over time (two representative birth rate curves are shown in Figure 12.3B).

The survivorship curves of animals undergoing the various treatments are shown in Figure 12.3A. Survival to day 20 was worse at concentrations of 5–10 µg l^{-1} than at lower concentrations. Reproduction began around day 18, and the birth rate was rather variable over time (two representative birth rate curves are shown in Figure 12.3B).

Any of these effects on its own would produce a reduction in population growth rate. The reductions in population growth rate that would be produced by each effect on its own are shown in Figure 12.4B. The reduction due to birth rate depression is similar both to that due to increased mortality and to that due to increased development period. Thus the reduction in population growth rate with increasing dieldrin concentration is due equally to birth rate, death rate, and age at first reproduction effects (Figure 12.4B).

This case study shows how population growth rate varied with dieldrin concentration (Figure 12.4B). Population growth rate also depends on many physical aspects of the environment. For example, population growth rate of the rice weevil *Sitophilus oryzae* depends on the temperature and moisture content of the grain stores in which it lives. The dependence

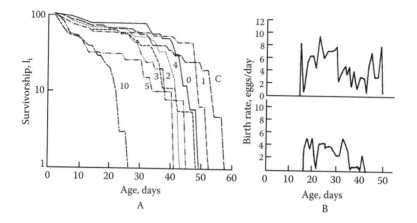

FIGURE 12.3

Effect of dieldrin on the life history of *Eurytemora affinis*. (A) Survivorship curves. These show for each treatment the proportion of animals that survive from birth to each age. Numbers indicate treatments, i.e., concentrations of dieldrin, in µg l⁻¹. C is the acetone control. (B) Birth rate in relation to age. Birth rate was measured at all concentrations, but only two representative concentrations, 2 and 4 µg l⁻¹, are shown here (upper and lower graphs, respectively). Modified from Daniels, R.E. and Allan, S.D. (1981). Life table evaluation of chronic exposure to pesticide. *Canadian Journal of Fisheries and Aquatic Sciences* 38, 485–494. With permission from the National Research Council of Canada.

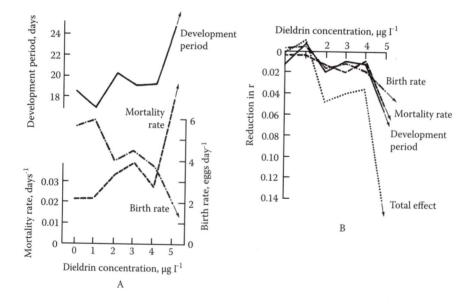

FIGURE 12.4

Effects of dieldrin concentration in *E. affinis*. (A) Effects on mortality rate, birth rate, and development period (age at first reproduction) estimated as described in the text. (B) Reductions in population growth rate, *r*, caused by the effects shown in (A). From Sibly, R.M. (1996). Effects of pollutants on individual life histories and population growth rates. In Newman, M.C. and Jagge, C., Eds. *Quantitative Ecotoxicology: A Hierarchical Approach*. Boca Raton, FL, Lewis. With permission.

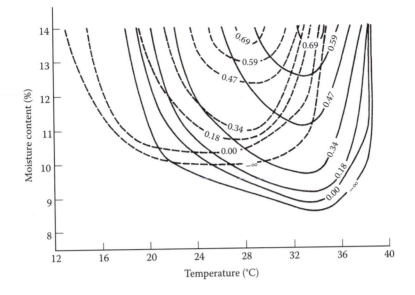

FIGURE 12.5
Contour plots of population growth rate for two species of grain beetle, *Sitophilus oryzae* (– – –) and *Rhizopertha dominica* (———). Contours are here labeled in terms of population growth rate; net reproductive rate was used in the original. Reproduced from Andrewartha, H.G. and Birch, L.C. (1954). *The Distribution and Abundance of Animals*. Chicago, University of Chicago Press. With permission.

is shown in a contour plot in Figure 12.5 (dashed lines). The axes represent the temperature and moisture content of the grain. The dashed lines are contours of equal population growth rate. Within the range of environments represented in Figure 12.5, there are some in which the population flourishes (high values of r). In general, the higher the value of r, the faster the population grows. Ecologists refer to those conditions for which $r \geq 0$ as the species' ecological niche.

In general, different species have different niches. The grain beetle *Rhizopertha dominica*, for example, flourishes at higher temperatures than *S. oryzae* (Figure 12.5).

In Figure 12.5, population growth rates were not affected by interactions with other animals. When interactions within and between species are taken into account, population growth rates are reduced. In the long term, population growth rates do not exceed zero (i.e., long-term population explosions do not occur).

As we have seen, many factors affect population growth rate. If the long-term population growth rate is zero, however, some factors must operate more strongly when the population is large but only weakly when the population is small, a phenomenon known as density dependence. We turn to these factors next.

12.4 Density Dependence

As mentioned earlier, factors that vary in their effect with population density are said to be density dependent. Their net result is that population growth rate is affected by population density. In the simplest case, population growth rate is a negative linear function of

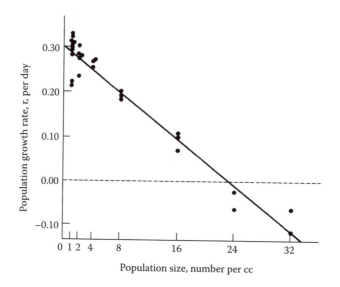

FIGURE 12.6
Density dependence in *Daphnia pulex*. From Frank, P.W., Boll, C.D., and Kelly, R.W. (1957). Vital statistics of laboratory cultures of *Daphnia pulex* DeGeer as related to density. *Physiological Zoology* 30, 287–305. With permission.

population density, as in Figure 12.6. Note that when the population is small (left-hand side of Figure 12.6), the population increases (population growth rate is positive). When the population is large (right-hand side of Figure 12.6), the population decreases (population growth rate is negative). In between, there is an (equilibrium) population density for which population growth rate is zero. This population density is called the carrying capacity of the environment in which the population lives. Ecologists give it the symbol K. The effect of density dependence here is to push the population density toward the equilibrium density if other factors have increased or decreased it. The equilibrium is therefore a stable equilibrium.

Suppose population growth rate reduces with population density in a straight-line relationship, as in Figure 12.6. Let the equation of the straight line be

$$\text{population growth rate} = r_0 - bN, \tag{12.4}$$

where r_0 and b are constants. r_0 represents population growth rate at low population density. We can write b in terms of r_0 and carrying capacity K as follows. When the population is at the carrying capacity of the environment, the population density is K and population growth rate is 0. Substituting these values in Equation 12.4, we obtain

$$0 = r_0 - bK, \tag{12.5}$$

Hence $b = r_0/K$, and substituting this value of b into Equation 12.4 we get:

$$\text{population growth rate} = r_0 - r_0 N/K. \tag{12.6}$$

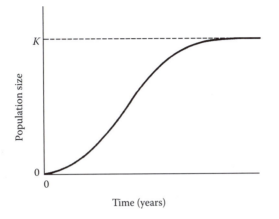

FIGURE 12.7
The sigmoidal growth curve that results from the logistic Equation 12.7. K is the carrying capacity of the environment.

As population growth rate $= 1/N \, dN/dt$ (from Box 12.1), Equation 12.4 can be written as

$$\frac{dN}{dt} = r_0 N \left(1 - \frac{N}{K} \right) \tag{12.7}$$

Ecologists call this the logistic equation. It produces a sigmoidal pattern of population growth (Figure 12.7). When small, the population grows exponentially (left-hand side of Figure 12.7). As population density approaches carrying capacity, population growth rate declines, resulting in a slow approach to the final equilibrium value (right-hand side of Figure 12.7).

12.5 Identifying Which Factors Are Density Dependent: *k*-Value Analysis

Population growth rate depends on the life history traits of the individuals in the population, as described in Section 12.3. In particular, population growth rate depends on individuals' age-specific birth and death rates and on the timing of their breeding attempts. Any or all of these may be density dependent. The analysis of which traits are density dependent is referred to as *k*-value analysis by population ecologists. This is because age-specific mortalities are known as *k*-values. Mortality at age (or stage) i is given the symbol k_i. In practice, it is usually mortalities that are analyzed.

k-value analysis assesses how the *k*-values vary as population density varies. Usually, natural variation in population density is used. Population density is measured repeatedly over a period of years together with the *k*-values. Mortality at age i, k_i, is then plotted against population density. The most appropriate measure of population density is usually that of individuals of age i. An example of a *k*-value analysis is shown in Figure 12.8. It was obtained from a 25-year study of sea trout by J. M. Elliott and collaborators (Elliott, 1993). By electric fishing at fixed times of year, population density *k* was established at various points in the life history, as shown in Figure 12.9.

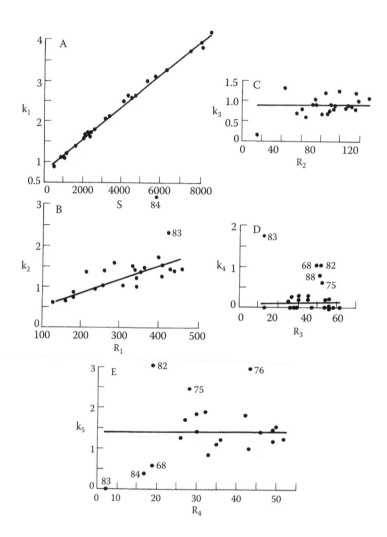

FIGURE 12.8

Sea trout mortalities $k_1–k_5$, in relation to population density in each of the five periods depicted in Figure 12.9. Thus (A) refers to alevin, (B) to young parr, and so on. Population densities $S, R_1, ..., R_4$ are defined in Figure 12.9; $k_1–k_5$ were calculated for the periods shown in Figure 12.9 using Equation 12.8. Each datum point refers to a single year. Reproduced from Elliott, J.M. (1993). A 25-year study of production of juvenile sea-trout, *Salmo trutta*, in an English Lake District stream. *Canadian Special Publication of Fisheries and Aquatic Sciences* 118, 109–122. With permission from the National Research Council of Canada.

k-values, which are measures of mortality, were calculated using the formula

$$k_i = \log_e R_{i-1}/R_i \qquad (12.8)$$

where R_i represents the population density of stage i, as indicated in Figure 12.9. These age-specific mortalities are plotted against population density for five phases of the sea trout life history in Figure 12.8. In the first two phases (Figures 12.8A and 12.8B), there are clear positive relationships between population density and age-specific mortality, but there is no relationship in the later phases of the life history. The effect of a positive relationship as shown in Figures 12.8A and 12.8B is to stabilize the population because higher mortality

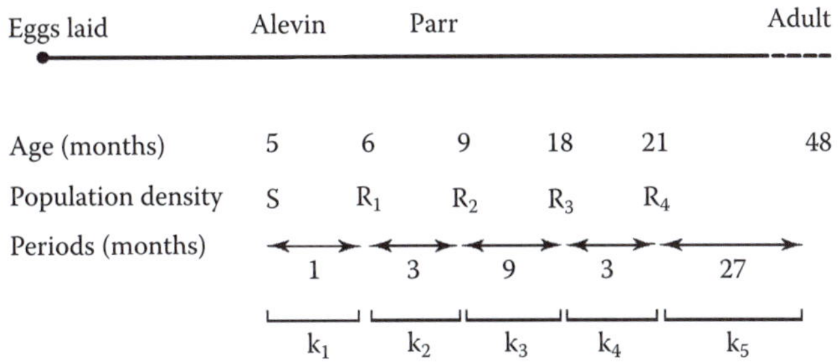

FIGURE 12.9
The life history of the sea trout at a stream in northwest England. The eggs hatch after about 5 months. The young trout are known as alevin from hatching until they resorb their yolk sacs; after this they are known as parr. Population density at each age was measured by electric fishing and designated S, R_1, R_2, ... as shown.

occurs at higher population density, making the population decrease when population density is high. Conversely, at low population density, mortality rate is relatively low, and this allows the population to increase.

Although population ecologists generally work with k_i-values, as in Equation (12.8), it is sometimes better to use mortality rates. These can be calculated as

$$\text{mortality rate} = k_i/t_i \tag{12.9}$$

$$= 1/t_i \log_e R_{i-1}/R_i \tag{12.10}$$

This example shows how population density affects mortality rate, and it may also affect somatic growth rate and birth rate. Moreover, as noted before, when considering the effects of pollutants, mortality, somatic growth, and birth rates together determine population growth rate.

The effects of population density on mortality rate are central to population ecology and have been reviewed by Sinclair (1989, 1996). Considering the importance of the topic and the attention it has received over the years, it is perhaps disappointing to record Sinclair's conclusion that "we still have a poor understanding of where density dependence occurs in the life cycle of almost every group of animals."

12.6 Interactions between Species

So far we have considered some of the factors that affect the numbers of a single population. If understanding a single population is difficult, untangling interactions between species is more than twice as hard. Here population growth rate depends not only on the population's own density but also on the population density of the other species. To establish these dependencies in the field is generally prohibitively expensive. For this reason, detailed study has usually been restricted to simple laboratory systems or to mathematical models.

Despite these difficulties, a number of general points can be made.

Interactions between species can be logically classified as being of one of three types, as follows.

1. Competition, in which the population growth rate of each species decreases the more there are of the other species. Generally, species compete for common resources (e.g., food, space, breeding sites), so the more competitors there are, the lower the average success of each.

2. Mutualism, in which the population growth rate of each species increases the more there are of the other species. Thus under mutualism, the species in effect help each other. Mutualism is the opposite of competition.

3. Predator–prey, in which the population growth rate of species A increases the more there is of species B, but the population growth rate of species B decreases the more there is of species A. These asymmetries are also evident in host–parasite, host–parasitoid, and plant–herbivore interactions, and these can therefore be treated under the same heading as predator–prey.

Although logically all interactions must fit into one of the above categories, because population growth rates are density dependent, the interaction of two species may not be in the same category at all population densities.

A species whose numbers vary substantially with time can persist in its environment only if it has a refuge that protects it and allows it to increase when its numbers have been reduced to low levels. The term *refuge* is here used in a very broad sense. It may refer to a physical refuge. Examples of physical refuges include defendable holes or cracks in rocks, as used by whelks to escape predation by crabs, or areas accessible to only one species, such as the splash zones on upper shores where barnacles live but their whelk predators cannot follow them.

Environmental patchiness can also result in refuges for prey species. For instance, examples are known in which infested patches go extinct, but prey are able to escape to uninfested patches. If there is a sufficient time lag before the predators find the uninfested patches, this effectively creates a refuge for the prey in which, for a time, they can increase.

Refuges can occur as a result of predators' foraging behavior. There are many cases known in which foraging effort decreases as the density of prey decreases, and this may result in a reduced mortality rate of prey at low prey densities, allowing prey populations to grow. Reduced foraging effort at low prey densities may result if predators switch to a different type of prey.

It follows from the definition of predator–prey interactions that removal of predators can lead to an increase in the population density of the prey species. Many cases are known in which pollutants are known to have had these two related effects (Dempster, 1975). For example, the red spider mite, *Panonychus ulmi*, appeared as a pest on outdoor fruit trees after the elimination of the slow-breeding predatory insects that previously controlled the mites. Fruit farmers used to apply pesticides to orchards in Britain as many as 20 times in a season, and this killed the mite's predators, upsetting the natural balance that had kept the mite population under control (Mellanby, 1967). More recently, Inoue et al. (1986), investigating the effects of spraying Kanzawa spider mites, *Tetranychus kanzawai*, with six kinds of insecticide and three kinds of fungicide, showed that the population density of the Kanzawa spider mites increased with the application of certain insecticides and fungicides, probably because of their adverse effects on the natural predators (three species of predatory mite, three species of predatory insect, and a spider).

Interactions between species can therefore explain otherwise anomalous increases in population abundance when the environment becomes polluted, as in curve iii in Figure 12.1. Let us call the increasing species A. If species A was the only one adversely affected by pollutants (e.g., decreased reproductive success, increased mortality rate), its increase would constitute a paradox. The only way species A can increase is as a result of complex interactions with other species. Perhaps some other species, call it species B, exists that depresses species A, as in the case of the insects predating spider mites described above. If pollution were to depress species B, the consequence for species A could be beneficial. Thus the red spider mite increased in numbers when its predators were killed by pesticides.

12.7 Field Studies: Three Case Studies

Ecological studies of the effects of pollutants are generally based on circumstantial evidence, although experimental studies are possible (see below). Commonly, the time course of pollution is monitored and compared with observed changes in species numbers. Negative correlations between pollution and changes in species numbers do not, however, necessarily imply causal links, and there is always a worry that such patterns could have been coincidental. This makes it attractive, as in all science, to carry out replicated experiments in which a treatment is applied to experimental areas but not to control areas. The use of proper replication allows an assessment of whether the variation between treated and untreated areas is significantly greater than natural variation between control areas. The effect of the treatment can then be measured as the difference between the two types of areas, and this can be assessed statistically provided that a suitable experimental design has been used.

12.7.1 The Decline of the Partridges

The decline of the partridge was due in part to the indirect effects of pesticides killing the insect prey necessary for chick survival. Eighteen years of intensive study of the partridge (*Perdix perdix*) in Sussex, England, are summarized and worldwide trends reviewed in a book by G. R. Potts (1986), from which the following account is taken.

The decline in numbers in Britain and worldwide is shown in Figure 12.10. In seeking the reasons for the decline, it is sensible to start by examining the trends in mortality rates (*k*-values). Table 12.1 summarizes data from 34 populations, each studied for 2–29 years between 1771 and 1985. This suggests that the main reason for the population decline lies in the increase in chick mortality k_3. This raises the question of what factors cause increases in chick mortality rate.

The causes of partridge chick mortality have long been matters of controversy among gamekeepers and ecologists. Both weather and the availability of insect food could be important. The advocates of weather are struck by the fact that small partridge chicks can produce only about one third of their own body heat; the rest comes from brooding parents. In cold weather, chicks therefore cannot afford to spend too long away from their parents, so feeding time is limited. This might lead to chick starvation. Arguing on this basis, there have been many attempts to correlate chick production with summer weather, but none of these have been particularly successful.

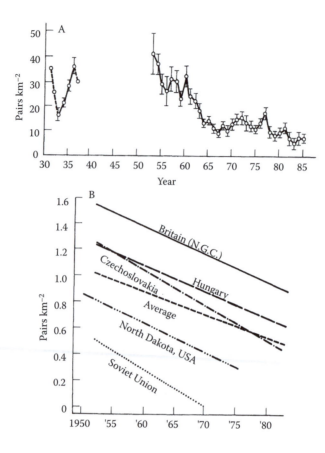

FIGURE 12.10

(A) The UK Game Conservancy's National Game Census March pair counts for the partridge from 1933 to 1985, with estimated minimum densities for the early 1930s ± 2 standard errors. (B) The trend in density of breeding pairs km^{-2} over the period 1952–1985 in various regions of the world range. Reproduced from Potts, G.R. (1986). *The Partridge*. London, HarperCollins. With permission.

TABLE 12.1

Comparison of Mortality Rates (*k*-Values) in Stable and Declining Populations ± Standard Errors[a]

		Populations		
		Stable (21)	**Declining (13)**	**Significance**
Nest loss	$k_1 + k_2$	0.26 ± 0.02	0.21 ± 0.03	ns
Chick mortality	k_3	0.29 ± 0.02	0.44 ± 0.02	$P < 0.01$
Shooting mortality	k_4	0.07 ± 0.01	0.08 ± 0.02	ns
Winter loss	k_5	0.38 ± 0.03	0.41 ± 0.05	ns
Total loss		1.00	1.14	

Source: From Potts, G.R. (1986). *The Partridge*. London, HarperCollins. With permission.

[a] The last column gives the results of statistical tests comparing stable and declining populations. ns = not significant.

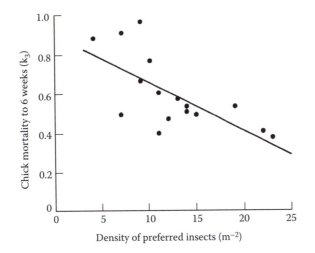

FIGURE 12.11
Annual chick mortality in relation to the density of preferred insects in the third week of June. Data from the Sussex study 1969–1985. Reproduced from Potts, G.R. (1986). *The Partridge*. London, HarperCollins. With permission.

The other school of thought holds that insects must be an important food for chicks because chicks go to a great deal of trouble to find insects, and they eat them in large quantities. Laboratory studies have shown that insects are nutritionally necessary for growing chicks; also in the Sussex study, there was a good relationship between chick mortality k_3 and the density of preferred insects (Figure 12.11).

Distinction between the effects of insect availability and weather was achieved by multiple regressions. This showed that 48% of the variation in chick mortality was explained by the density of preferred insects, and an additional 10% was explained by average daily temperature in the critical period 10 June to 10 July. Both these were statistically significant ($P < 0.001$), but it seems that insect density was much more important than temperature in the Sussex study.

Given that reduced availability of insects is the key to the partridge population decline, it is natural to ask whether the insects declined as a result of pesticide usage. The increase in the use of herbicides is shown in Figure 12.12, and chick mortality rates k_3 are shown in relation to pesticide usage in Table 12.2. Note that chick mortality rates appear to be considerably increased by the use of herbicides, and increased again when some insecticides were also used. The effect of herbicides was presumably indirect, removing the food necessary for the survival of the insects eaten by the partridge chicks. The effects of insecticides on insects appear less important. It seems reasonable to conclude that the decline of the partridge population occurred largely because of increased chick mortality, and that this in turn was the result of a decrease in the density of insects consequent on increased use of pesticides.

The above are not the only factors known to affect partridge populations. Nest losses are known to be density dependent, the form of the relationship depending on whether gamekeepers are present (Figure 12.13). Dispersal and adult mortality due to shooting are also density dependent. These density-dependent effects would return populations to carrying capacity within a few years were it not for the high levels of chick mortality. Not surprisingly, equilibrium levels are considerably higher if gamekeepers are present.

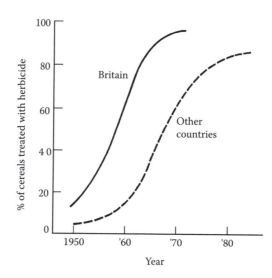

FIGURE 12.12
Trend in herbicide use on cereals. Reproduced from Potts, G.R. (1986). *The Partridge*. London, HarperCollins.
With permission.

Because fewer gamekeepers are employed in the UK now than formerly, it is reasonable to ask whether the population decline could be better ascribed to reduced gamekeepers than to increased pesticide use. The Sussex study attempted to answer this by entering the key known relationships into a simple model of partridge population dynamics. The model consisted of four basic equations showing how the four *k*-values listed in Table 12.1 were affected by population density and some environmental features. The model was used to see how the population would have reacted if gamekeepers had been employed and if pesticides had not been used. Figure 12.14 shows that, according to the model, use of gamekeepers would have reduced the rate of population decline but would not have prevented it. Only when herbicides are not used, restoring chick mortality rates to their former levels, is the population decline prevented altogether.

TABLE 12.2

Summary of Estimates of Partridge Chick Mortality Rates Grouped According to Herbicide and Insecticide Use

	Mean k_3 ± SE		
	Up to 1952 (No Herbicide)	**1953–1961 (Some Herbicide)**	**1962–1985 (Herbicide + Some Insecticide)**
National game census	0.33 ± 0.02	0.45 ± 0.05	0.51 ± 0.02
Damerham	0.32 ± 0.05	0.50 ± 0.06	0.50 ± 0.05
Lee Farm	–	0.44 ± 0.06	0.67 ± 0.07
North Farm	–	0.45 ± 0.05	0.63 ± 0.04
Sussex study	–	0.40 ± 0.05	0.61 ± 0.04
Mainland Europe	0.29 ± 0.03	0.37	0.45 ± 0.03
North America	0.25 ± 0.07	0.28	0.36 ± 0.04

Source: From Potts, G.R. (1986). *The Partridge*. London, HarperCollins. With permission.

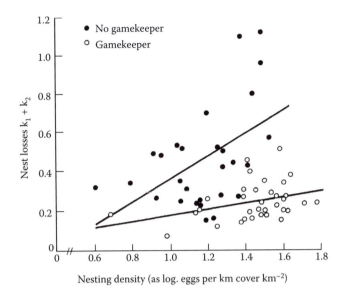

FIGURE 12.13
Dependence of nest losses on nesting density with and without gamekeepers. Reproduced from Potts, G.R. (1986). *The Partridge*. London, HarperCollins. With permission.

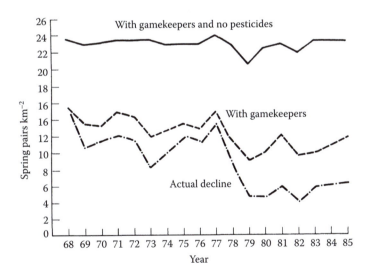

FIGURE 12.14
Simulations of the Sussex partridge population showing the actual decline, the less severe decline that would have occurred if gamekeepers had been employed at 1968 levels, and how no decline would have occurred if gamekeepers had been employed and pesticides had not been used. Reproduced from Potts, G.R. (1986). *The Partridge*. London, HarperCollins. With permission.

TABLE 12.3

Population Trends of Selected Species of Birds in the UK

Species	Year Decline Started	Population Trends 1972–1996 (%)	
		Farmland	Countrywide
Tree sparrow (*Passer montanus*)	1978	−87	−76
Turtle dove (*Streptopelia turtur*)	1979	−85	−62
Grey partridge (*Perdix perdix*)	1978	−82	−78
Spotted flycatcher (*Muscicapa striata*)	Before 1969	−78	−78
Skylark (*Alauda arvensis*)	1981	−75	−60
Song thrush (*Turdus philomelos*)	1975	−66	−52
Lapwing (*Vanellus vanellus*)	1985	−46	−42

Source: Data from Crick, H.Q.P., Baillie, S.R., Balmer, D.E., Bashford, R.I. et al. (1998). Breeding birds in the wider countryside; their conservation status. Research Report 198. British Trust for Ornithology. Thetford.

The final message of Potts's (1986) book was that, as far as could then be seen, density dependence alone would not be sufficient to save partridge populations from extinction. Partridge preservation could be achieved only by increasing the supply of insects to chicks, thereby reducing the mortality rate of chicks. At the time of writing, little has changed except locally where appropriate management has been introduced; overall the partridge continues to decline (Potts, 2000). This huge loss of a valuable natural resource is arguably worth over £500 million annually in Europe alone.

Although the decline of the grey partridge is the best-documented case, many other species of birds have declined markedly, especially on farmland, in the UK over the last 20 to 30 years. A list of some of these species is given in Table 12.3. For them, there is much less information upon which to base an analysis of causation. However, using an epidemiological approach (Peakall and Carter, 1997), intensive agriculture does appear to be the primary cause.

The databases of the British Trust for Ornithology (BTO) (Table 12.4) have proved valuable in examining the problem. The population declines appear to have been caused by intensive farming practices, and in western Europe these changes have been fueled by the European Union Common Agricultural Policy. This massive ecotoxicological experiment has been reviewed in detail by Pain and Pienkowski (1997). Although it is to be expected that several factors are involved, the present evidence suggests that indirect effects of herbicides and insecticides are the most important causes of the declines shown in Table 12.3.

A recent concern given much prominence by the media has been over the development of herbicide-resistant crops. There is the possibility that farmers could treat fields with high levels of herbicides, thereby effectively eliminating food essential to wildlife. The contrary view is that farmers could let the weeds grow alongside crops for a longer period, secure in the knowledge that weeds can be eradicated later in the season. Crops engineered to be resistant to insect pests could cause a problem in that they are designed to produce a steady stream of natural insecticides that could harm beneficial predators. It is possible that this could be more harmful than the intermittent use of chemicals that could allow insect populations to recover. The recent case of the monarch butterfly (*Danaus plexippus*) affected by transgenic pollen (Losey et al., 1999) has focused attention on this problem. Research is needed to see whether these effects are seen under realistic conditions of use.

TABLE 12.4

Some of the Surveys Undertaken by the British Trust for Ornithology (BTO)

Survey	Technique	Characteristics	Outcome
Bird ringing	Rings (or bands) are placed on leg of birds. Subsequent recoveries reported to BTO	Recoveries allow calculation of annual mortality besides giving information on migration	Started in 1909; 22 million birds ringed; 450,000 subsequently recovered
Constant effect sites	Standardized mist netting of birds on 12 visits	Monitors change in abundance and productivity	Started in 1983; 120–130 sites annually
Nest record scheme	Using special cards, the contents of nests are recorded at each visit	Provides data on clutch size and nesting success	Started in 1939; 30,000 records per year, over a million total
Common birds census (CBC)	Detailed census of plot on 10 visits during the breeding season	Detailed data on the population changes of common species	Started in 1962; 220–230 sites covered per year
Breeding bird survey	Randomly selected sites, transects carried out on three visits per year	Breeding population data, less detailed but larger sample size than CBC	Started 1994; 2169 squares covered in 1997
Garden bird watch	Recording numbers of 10 common species, presence or absence of others	Population data from gardens. Lack of fixed protocol balanced by large sample size	Started 1995; now has 10,000 observers

12.7.2 Population Studies of Pesticides and Birds of Prey in the UK

The study of the effects of pesticides on birds of prey in the UK has a special place in eco-toxicology. This is partly because some pesticide effects were reported first in these species and partly because of the intensive nature of the studies, which have been conducted over 50 years. Despite this, our knowledge of some of the population processes is less complete than in the case of the partridge. It may appear surprising that even where pesticides are known to affect individual vital rates (e.g., mortality rate or breeding success) this does not necessarily lead to population decline. Bear in mind, however, that density dependence can compensate for the effects of the pesticides on particular age classes, and the net result could be that carrying capacity is unchanged.

The peregrine falcon, *Falco peregrinus*, was the first bird of prey to be analyzed in detail from the point of view of pesticide impact. Very similar, and in some instances fuller, data have been obtained for another bird of prey, the sparrowhawk, *Accipiter nisus*, by I. Newton. Some of these data will be referred to where appropriate. The following account is based mostly on books by Ratcliffe (1993) and Newton (1986). We begin by reviewing the properties of the pesticides involved and their known effects on mortality, breeding success, and behavior and then go on to consider their effects on populations.

The pesticides that have been shown to affect the birds of prey are insecticides belonging to the organochlorine group (Section 1.2.6). They include DDT, introduced into agricultural use in the late 1940s, and the cyclodiene insecticides aldrin, dieldrin, and heptachlor, introduced in the mid-1950s. Both these types are extremely stable in their original form and/or as metabolites and so persist in the environment for many years. In addition, they are readily dispersed in wind and water and in the bodies of migratory birds and insects.

The organochlorine pesticides have properties that make them particularly hazardous to birds of prey and their predators. Because of their high fat solubility and resistance to metabolism, they have long biological half-lives in many species (see Chapter 5). Consequently, they tend to bioaccumulate as they move up food chains, reaching their highest concentrations in predators. It follows that top carnivores are especially useful indicators of the environmental effects of these compounds.

In the case of both sparrowhawks and peregrines, organochlorine insecticides have two distinct types of effect upon wild populations. First, the cyclodiene insecticides aldrin, dieldrin, and heptachlor (used, for example, as seed-dressing chemicals) can cause lethal toxicity and so may increase mortality rates. Some sparrowhawks and peregrines found dead in the field contained lethal concentrations of these insecticides in tissues such as liver, brain, or muscle. Evidence from both laboratory and field studies indicate that tissue concentrations of dieldrin and heptachlor epoxide above 10 ppm will cause death of predatory birds because of their direct neurotoxic action. The cyclodiene insecticides affect the central nervous system by inhibiting GABA receptors, and it is known from laboratory studies that they can have sublethal effects on the function of the nervous system (e.g., irritability, disorientation, and convulsions) before lethal effects are produced. A high level of skill and coordination is required in predators if they are to catch their prey, so disturbances of the nervous system can seriously affect their hunting skills (Chapter 8, Section 8.4.2). There can be no doubt that serious sublethal effects as well as lethal ones were produced in the field during and after the 1950s, but there is no means in retrospect of quantifying this.

A second type of effect of organochlorine insecticides is eggshell thinning, caused by p,p'-DDE, a persistent metabolite of p,p'-DDT. The association between eggshell thinning in British peregrines and sparrowhawks and exposure to DDT residues was first noted by Ratcliffe (Figure 12.15). Initially, it was not clear whether this was a causal relationship, but dosing predatory birds with p,p'-DDE in the laboratory has established that low levels of

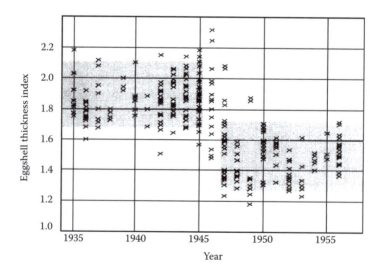

FIGURE 12.15
The decline in peregrine eggshell thickness that commenced in the UK in 1947. Shaded areas represent 90% confidence limits. The eggshell thickness index is defined as index = weight of eggshell (mg)/length × breadth (mm). From Peakall, D.B. (1993). DDE-induced eggshell thinning: an environmental detective story. *Environmental Reviews* 1, 13–20. With permission.

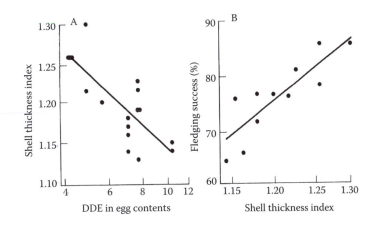

FIGURE 12.16

In sparrowhawks, (A) DDE is negatively correlated with shell thickness, and (B) eggshell thickness is positively correlated with the percentage of young raised per brood (fledgling success). Data from different areas, modified from Newton, I. (1986). *The Sparrowhawk*. Calton, Poyser. With permission from Academic Press. Data analogous to (A) for peregrine can be found in Figure 15.2.

the chemical do cause eggshell thinning. It is now known that p,p'-DDE reduces transport of Ca_2^+ to the developing eggshell, probably because of inhibition of calcium ATPase in the shell gland (Chapter 16). Thus there are good scientific grounds for regarding the close negative correlation between residue of p,p'-DDE and eggshell thickness as a causal relationship (Figure 12.16A). Furthermore, because eggshell thinning is correlated with fledging success in wild sparrowhawks (Figure 12.16B), it is probable that DDT affects hatching success directly.

Although it is well established that lethal toxicity and eggshell thinning have occurred in the field in the UK, difficulties arise in quantifying them and relating them to population change. Consider p,p'-DDE and shell thinning first. In both sparrowhawk and peregrine, substantial reductions in shell thickness occurred during 1946 and 1947 in the UK, coincident with the large-scale introduction of DDT as an insecticide. There was, however, no evidence of a general decline in these species at this time, or for several years after (see Figure 12.17 for the peregrine). As sparrowhawk and peregrine breed at ages 1–2 and 2 years, respectively, and then have relatively short life expectancies (about 1.3 and 3.5 years respectively), the very sharp decline in populations in the late 1950s cannot be directly attributed to eggshell thinning, which commenced in the period 1946–1947. Thus although eggshell thinning was occurring, and this was having some effect on fledging success (Figure 12.16B), it did not bring any reduction in population size. Indeed, there was an increase in numbers of peregrines at this time. Presumably, population sizes at this stage were still at or close to carrying capacity, despite the reduction in hatching success brought about by DDT. Probably some density-dependent process compensated for the reduction in fledging success brought about by DDT. Perhaps the smaller number of fledglings was individually more successful in obtaining food because they had fewer competitors and so grew faster and were more successful than they otherwise would have been.

Whereas the introduction of DDT did not lead to population decrease, the populations of both species declined rapidly in the mid- to late 1950s (see Figure 12.17 for peregrine, Figure 12.18 for sparrowhawk). These declines coincided in both time and space with the introduction of aldrin, dieldrin, and heptachlor as seed-dressing chemicals. As discussed

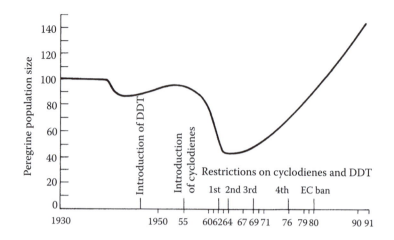

FIGURE 12.17

Peregrine population size in Britain (1930–1939 = 100) showing the 1961 population decline and subsequent recovery, together with an outline of pesticide usage. Reproduced from Ratcliffe, D. (1993). *The Peregrine Falcon*, 2nd ed. London, Carlton. With permission from Academic Press.

above, these compounds caused deaths in the field, and extensive sublethal effects were suspected but were not quantifiable.

A series of bans, placed between 1962 and 1975, led to the progressive removal of the cyclodienes and DDT from the UK (Figure 12.17). These bans were followed by a decline in organochlorine residues and population recovery in both these species. The pattern of recovery differed between areas. Particularly interesting was the late recovery of sparrow-hawk populations in eastern England (Figure 12.18). Sparrowhawk populations recovered when tissue concentration of *dieldrin* fell below 1 ppm (Figure 12.19 and Figure 12.20). (In this account, the term *dieldrin* is used synonymously with the abbreviation HEOD, which is the chemical term for the active constituent of the commercial insecticide diel-drin.) Interestingly, the same was true of the kestrel during this period. Closer inspection of the data for both sparrowhawks and kestrels showed that there was a wide range of tis-sue dieldrin levels in individuals whose liver samples showed concentrations of 1 ppm or more. Death could be attributed to direct dieldrin poisoning (on the grounds of high tissue residues ≥ 10 ppm and symptoms of toxicity) in only a small proportion of the kestrels or sparrowhawks found dead during the period 1976–1982. On the other hand, a substantial proportion had residues in the range 3–9 ppm, increasing the suspicion that sublethal effects were more important than lethal ones (Sibly et al., 2000; Walker and Newton, 1999). Population modeling for the sparrowhawk strengthened the case for suggesting that sub-lethal effects of dieldrin were important in causing population decline in Britain over this period. There is consequently a strong suspicion that sublethal effects of dieldrin on adults were widespread and may have been important in causing population decline; it is also possible that embryotoxicity was a contributing factor.

The recovery of the sparrowhawk shown in Figure 12.20 was rapid between 1980 and 1990 but leveled off in 1993—a sigmoidal population increase of a type frequently encoun-tered as populations approach carrying capacity (compare Figure 12.7) (Newton, 1988; Newton and Haas, 1984; Newton and Wyllie, 1992). Peregrines also have now recovered (Figure 12.17).

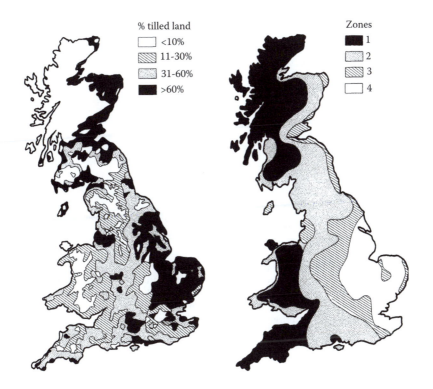

FIGURE 12.18

Changes in the status of sparrowhawks in relation to agricultural land use and organochlorine use. The agricultural map (left) indicates the proportion of tilled land, where almost all pesticide is used. The sparrowhawk map (right) shows the status of the species in different regions and time periods. Zone 1, sparrowhawks survived in greatest numbers through the height of the organochlorine era around 1960; population decline judged at less than 50% and recovery effectively complete before 1970. Zone 2, population decline more marked than in Zone 1, but recovered to more than 50% by 1970. Zone 3, population decline more marked than in Zone 2, but recovered to more than 50% by 1980. Zone 4, population almost extinct around 1960, and little or no recovery evident by 1980. In general, population decline was most marked, and recovery latest, in areas with the greatest proportion of tilled land (based on agricultural statistics for 1966). Reproduced from Newton, I. and Haas, M.B. (1984). The return of the sparrowhawk. *British Birds* 77, 47–70; reproduced in Newton, I. (1986). *The Sparrowhawk*. Calton, Poyser. With permission from Academic Press.

It should be added that the situation was different in North America. Peregrines were less exposed to cyclodienes but more exposed to p,p'-DDE (thus reflecting patterns of use of these insecticides; see also Chapter 15). Eggshell thinning was greater in North American peregrine populations than in British ones. In this case, there is strong evidence to suggest that p,p'-DDE was the principal cause of population decline, although the field data on populations is not so complete as that obtained in the UK.

The use of eggshell thinning as a biomarker is considered in Chapter 16.

12.7.3 The Boxworth Project (Experimental Analysis of the Effects of Pesticides on Farmland)

The Boxworth project was designed to assess experimentally the effects of three contrasting regimes of pesticide usage on the animals and plants, including crops, on a 300-hectare arable farm in eastern England. The aim was to investigate large-scale, long-term effects of

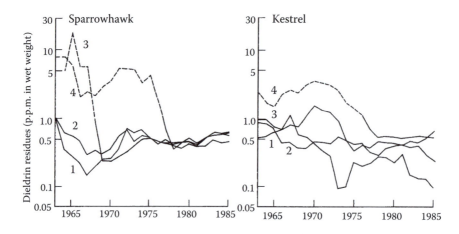

FIGURE 12.19

Dieldrin (HEOD) levels in the livers of sparrowhawks found dead in the four zones shown in Figure 12.18. HEOD is the active principle of the commercial insecticide dieldrin and accounts for some 80% of the technical product. Broken lines show periods when populations were depleted or decreasing; solid lines show periods when populations were normal or increasing. Population increase occurred when liver levels were less than about 1.0 ppm wet weight. Reproduced from Newton, I. (1988). Determination of critical pollutant levels in wild populations, with examples from organochlorine insecticides in birds of prey. *Environmental Pollution* 55, 29–40. With permission from Elsevier Science Ltd.

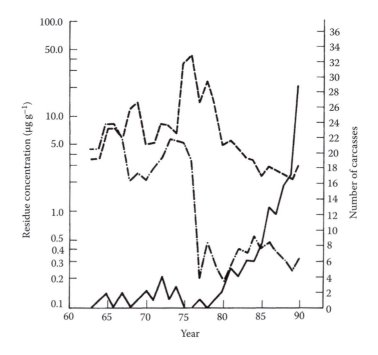

FIGURE 12.20

Number of sparrow hawk carcasses (————) received in a region of eastern Britain, together with concentrations of DDE (– – – –) and dieldrin (–·–·–·–) found in their livers. Reproduced from Newton, I. and Wyllie, I. (1992). Recovery of a sparrowhawk population in relation to declining pesticide contamination. *Journal of Applied Ecology* 29, 476–484. With permission from Blackwell Science Ltd.

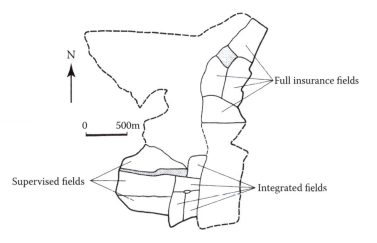

FIGURE 12.21

Map of the farm at which the Boxworth project was conducted, showing the locations of the fields treated with each pesticide regime. Reproduced from Greig-Smith, P.W. and Hardy, A.R. (1992). Design and management of the Boxworth project. In Greig-Smith, P.W., Frampton, G., and Hardy, T., Eds. *Pesticides, Cereal Farming and the Environment: The Boxworth Project.* London, HMSO. With permission.

pesticides under conditions as close as possible to farm conditions. It was a major project, lasting 7 years in its main phase, involving many scientific man-years. The account here is based on Greig-Smith et al. (1992b). Three regimes of pesticide usage were investigated. These were

1. Full insurance, in which relatively large amounts of pesticides were applied in advance of possible pest outbreaks, as "insurance"
2. Supervised, in which pesticides were applied only when needed, as assessed by monitoring pest, weed, and disease levels
3. Integrated, which was similar to supervised but incorporated some additional features of integrated pest management

A map of the farm is shown in Figure 12.21. Because large-scale effects were to be investigated, the farm was divided into three large areas, applying one treatment to each area. Because long-term effects were of interest, there was no swapping of treatments between areas.

These features of the experimental design mean that the experiment is unreplicated. For this reason, little assessment can be made of whether the variation between areas is such as might have occurred by chance in the absence of pesticide treatments. The lack of replication precludes the application of the statistical tests normally used in agricultural trials. This could have been remedied by allocating pesticide regimes to fields on, for example, a random basis. Other designs are possible that allow for factors known to cause variation between fields, such as soil type and the amount of noncrop habitat.

The lack of replication dissipates much of the power of the experimental approach to attribute pesticide cause to ecological effect. However, some idea of natural and baseline variation was obtained by monitoring the areas for 2 years before starting the trials. The pesticide regimes were applied for 5 years (1984–1988), and some residual monitoring has continued since.

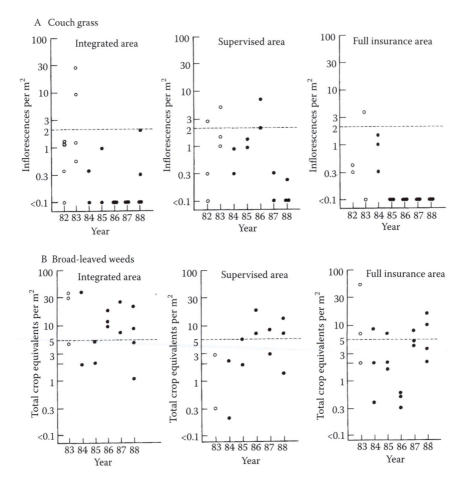

FIGURE 12.22

Efficacy of the Boxworth pesticide regimes on the densities per m² of (A) the weed grass couch (*Elymus repens*) in July and (B) the broad-leaved weeds in spring; 1982 and 1983 were baseline years before the pesticide regimes were applied. The dotted lines show spray decision thresholds used in deciding whether to apply pesticides. Reproduced from Marshall, J. (1992). Weeds. In Grieg-Smith, P., Frampton, G., and Hardy, T., eds. *Pesticides, Cereal Farming and the Environment. The Boxworth Project*. London, HMSO. With permission.

In the event, there was little difference between the supervised and integrated regimes (S + I henceforth). The analyses therefore generally consist in comparing the effects of full insurance with those of S + I.

Botanical monitoring showed that patterns of grass weed densities reflected the efficiency of the weed control regimes. An example is shown in Figure 12.22A. Full insurance was most effective: It held all grass weeds at low densities. By contrast, the densities of the broad-leaved weeds varied little between regimes (Figure 12.22B). Further botanical comparisons were hampered by high variation between years, including the two baseline years.

Turning to the invertebrates, it appears that, overall, the density of herbivorous invertebrates was about 50% less under full insurance than under S + I. Worst affected were certain nondispersing species. Carnivorous invertebrates (predators and parasites) showed a similar pattern, but detritivores were unaffected.

It proved particularly difficult to ascribe cause and effect in small mammals and birds, partly because of their mobility. The diet of common shrews (*Sorex araneus*) reflected the distribution of the invertebrates, described above, that form their diet. In particular, leatherjackets (crane-fly *Tipulidae* larvae) were less frequent in the stomachs of shrews caught in the full insurance area than in the supervised area. However, there was no evidence of long-term effects of the different pesticide regimes on any of the small mammal populations studied. Autumn application of slug pellets containing methiocarb had an immediate local impact on adult woodmice (*Apodemus sylvaticus*), but the population recovered quickly by immigration of juveniles from nearby woods, fields, and hedges.

The only bird species affected by the full insurance regime was the starling (*Sturnus vulgaris*). Part of the observed reduction in numbers of starlings nesting in the full insurance area relative to the S + I area may have been due to the effects of the pesticides on leatherjackets.

The economics of the pesticide regimes were also evaluated. Yields of wheat in the full insurance fields were 0.92 ton ha^{-1} higher than in the supervised fields and 1.35 ton ha^{-1} higher than in the integrated area, although there was considerable variation from year to year. Grain quality also appeared better under full insurance. However, the extra costs inherent in full insurance meant that the supervised approach was as profitable as full insurance. The integrated regime used in the study was less profitable. However, truly integrated low-input systems might do better.

The Boxworth project shows clearly that the experimental study of pesticide effects in the field is a very expensive business. A replicated experimental design would have allowed stronger conclusions to be drawn, but because the farm would have been divided into smaller experimental plots, mobility between areas would have increased and mobility was already a significant factor affecting the distributions of small mammals and birds. The lack of replication means that the Boxworth project was essentially a pilot study. It does nevertheless suggest which forms of wildlife were, and which were not, affected by high rates of application of pesticides.

12.8 Modeling the Effects of Insecticides on Skylarks for Risk Assessment Purposes

Assessments of the risks posed by chemicals to mammals and birds in specified geographical areas are routinely performed for the purposes of risk management. Such assessments typically start with results from laboratory tests on species such as rats or bobwhite quail. There is then the difficulty of extrapolating from effects in the laboratory to what is likely to happen in the field. Ideally this would be estimated from comprehensive large-scale field trials, but these are prohibitively expensive, as was seen in the previous section. A much cheaper alternative is computer simulation of the possible patterns of application of the chemicals, together with their likely effects on populations of the species of interest. Here we consider an example of this approach, published by Topping et al. (2005). The starting point was a very detailed agent-based model (ABM) of individuals moving around a particular mapped landscape (Figure 12.23; Topping et al., 2003). The resolution of the map was one meter, so that hedgerows and ditches were mapped, together with the crops and plants growing in individual fields and habitats. The time resolution was one

FIGURE 12.23
A 2.5 km section of the map used in simulations of the effects of insecticides on skylarks, showing the level of mapping detail. This section shows agricultural land containing individual farm buildings and fields, bordered by a small town (NE) and some wooded areas (SW). Reproduced from Topping, C., Sibly, R.M., Akcakaya, H.R. et al. (2005). Risk assessment of UK skylark populations using life history and individual-based landscape models. *Ecotoxicology* 14, 925–936. With permission from Springer-Verlag.

day. Farm ownership and the strategies of farm managers were modeled, so that planting and harvesting appeared within the model. The daily growth of plants was modeled, so that the height of plants—often of importance to mammals and birds—was known each day. Within these detailed and changing environments, the behavior, movement, reproduction, and death of individuals of key species were modeled using all that was known of the species' behavioral ecology.

This general model was used to investigate the likely effects of a fictitious insecticide on skylark populations in the UK. Two types of scenarios were investigated: winter wheat, in which winter wheat was planted on 42% of the arable land, and broad habitats, in which only half as much winter wheat was planted. Each scenario was investigated with and without the application of the insecticide by spraying. Forty simulation replicates were run for 40 years, but the first 4 years of each simulation were discarded for technical reasons that need not detain us here. The main results were that the skylark populations declined exponentially in the winter wheat scenarios and went extinct if insecticide was applied (Figure 12.24A, B). By contrast, the populations reached carrying capacity in the broad habitats scenarios because these contained more good habitat. However, application of the insecticide resulted in a marked reduction in carrying capacity (Figure 12.24C, D).

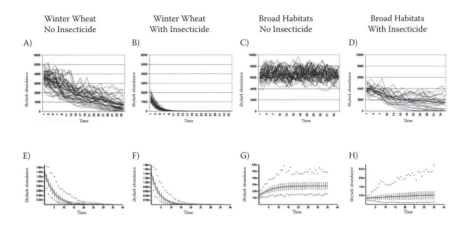

FIGURE 12.24

The results of two models of the effects of an insecticide on skylarks. Population abundance is plotted against time for four scenarios (columns), modeled by an agent-based landscape model (ABM, A–D, top row) and modeled by a life history model (E–H, bottom row). In A–D the individual runs are shown, and the thick line represents their mean. In E–H bars indicate standard deviations, and the range over 1,000 simulations is also indicated. Reproduced from Topping, C., Sibly, R.M., Akcakaya, H.R. et al. (2005). Risk assessment of UK skylark populations using life history and individual-based landscape models. *Ecotoxicology* 14, 925–936. With permission from Springer-Verlag.

This type of model incorporates autonomous individual agents, making independent decisions, in a specific landscape described in great detail. This is why it is described as an agent-based landscape model or ABM. A widely used alternative to an ABM is a life history model, and one such was investigated so that its results could be compared with those of the ABM. At first sight the life history model appears much simpler than the ABM. In the life history model, juvenile vital rates were distinguished from those of adults. The values of these vital rates, and their temporal variation, were taken from published studies, but there was no attempt to model spatial variation. Density dependence was included, and was assumed, for simplicity, to act according to a logistic equation (see Section 12.4) with carrying capacity set at 2.9 million skylarks, representing their abundance in the UK in the late 1960s. So far, so easy. The major stumbling block with the life history model was modeling the effects of the pesticide on the vital rates. The available data were laboratory data on reproduction in quail, and the problem lies in extrapolating this to what is likely to happen in the field. This problem was addressed in a separate modeling exercise. Once the effects of the chemicals on the vital rates were estimated, the model was run, and the results are shown in Figure 12.24E–H. The results are qualitatively similar to those obtained with the ABM.

What can we conclude from this exercise? Is one type of model better than the other? Can we rely on their results? No consensus has yet emerged on these questions. Points that can be made are that the life history models are very quick to set up and to run, but their apparent simplicity masks the difficulty of estimating the effects of chemicals on life histories in the field. When these difficulties are properly addressed, the scale of the simulation starts to resemble that of the ABM. Indeed, it is arguable that the quality of the simulation achieved by the ABM is superior.

Validation of any of these models is a thorny issue. However, there seems little doubt that complex detailed mechanistic models are here to stay. Because computing power is increasing and becoming cheaper, and because all the detailed mechanisms of the ABMs can in principle be separately checked and verified, it is likely that increasing reliance will be placed on them in the years ahead.

12.9 Summary

The effect of a pollutant on a population is assessed by the effect it has on the population's growth rate. Population growth rate has usually been calculated from experiments in which the life history of the organism is recorded at a number of concentrations of the pollutant. Such experiments generally show that increasing the concentration of the pollutant has the effect of decreasing birth rate and/or increasing mortality rate. Decreasing birth rate and increasing mortality rate both decrease population growth rate. In consequence, an increase in the concentration of a pollutant generally results in a decrease in population growth rate.

Experiments of this type have usually been conducted at low population density. In the absence of stressors, populations at low population density undergo exponential population growth if per capita population growth rate is constant. Populations cannot expand indefinitely, however, because they eventually exhaust the available resources. As this starts to happen, population growth rate starts to decrease, until at very high population densities population growth rate is negative. In between, there is an equilibrium population density for which population growth rate is zero. This population density is called the carrying capacity of the environment in which the population lives. To date, there has been insufficient attention paid to the joint effects of pollutants and population density on population growth. Future work should, among other things, examine the effects of pollutants on carrying capacity.

Establishing cause and effect in the field would ideally be achieved using large-scale replicated field experiments. Examination of three intensive case studies reveals some of the many difficulties of using this approach to assess the risks posed by new chemicals. Apart from ethical issues, the logistical problems and expense are generally prohibitive. How then should the judgments about risk be made? The only practicable solution would seem to be some form of computer simulation. An example is described, but validation of such models poses problems.

Further Reading

Begon, M., Mortimer, M., and Thompson, D.J. (1996). *Population Ecology: A Unified Study of Animals and Plants.* Provides an introduction to population ecology, including the analysis of density dependence and interactions between species.

Evans, P.R. (1990). Provides a useful discussion of the population effects of pesticides on birds and mammals.

Forbes, V.E. and Calow, P. (1999). Reviews the literature and concludes that population growth rate is a better measure of responses to toxicants than are individual-level effects.

Lammenga, J. and Laskowski, R., eds. (2000). A book about the effects of chemicals on life histories and populations.

Levin, L. et al. (1996). A classic recent example of an investigation of the effects of a pollutant on life history traits and population growth rate.

Linke-Gamenick, I. et al. (1999). Report of similar experiments that also examines the effects of density dependence.

Sibly, R.M. et al. (2005). Brief review of methods of population modeling with particular relevance to birds and mammals.

Stark, J.E. and Banken, J.A.O. (1999). Provides suggestions for the experimental design of population experiments.

13

Evolution of Resistance to Pollution

Evolutionary responses to pollution are referred to as resistance. It is implicit that resistance has a genetic basis, and what is known of its genetic inheritance is outlined in Section 13.5. This chapter considers how genetic changes in resistance come about. Resistance represents an evolutionary response to environmental changes resulting from (among other things) pollution, and the first sections of this chapter describe the general phenomenon of evolutionary response to environmental change. Pollution represents an environmental change for the worse, and resistance generally defends organisms against the deleterious consequences of pollution. Such defense may reduce an organism's mortality rate, but this is sometimes expensive, using energy and/or nutrients that could otherwise have been used for reproduction or somatic growth. Defense may, therefore, involve a trade-off between production and survival: Increased survival may be obtained only at a cost of reduced growth or reproduction. As a result, resistance may have a fitness cost in an unpolluted environment. The possible physiological basis of this trade-off has been considered in Chapter 8, and the evolutionary implications are considered here in Section 13.3. Five cases studies of evolutionary responses to pollution at the end of the chapter consider the evolution of pesticide resistance, of metal tolerance in plants, of industrial melanism, of tributyltin (TBT) resistance in dog whelks, and of resistance to pollution in estuaries. These include some of the best-documented studies of evolutionary responses to environmental changes. They show that the evolution of resistance generally consists of the selection of preexisting genes rather than the appearance of new genes. The first sections of this chapter are quite theoretical, and an attractive approach for those encountering this material for the first time may be to begin by reading the case studies.

13.1 Chronic Pollution Is Environmental Change

Transient pollution by definition has only passing effects and so is unlikely to change gene frequencies. Chronic pollution, however, can have lasting effects because it changes the environment in which organisms live. In considering evolutionary responses to chronic pollution, we are therefore dealing with a particular case of the general phenomenon of evolutionary responses to environmental change. Before considering the effects of environmental change, however, we need to know what happens in unchanging environments.

13.2 Evolutionary Processes in Constant Environments

Consider Figure 13.1, which provides a simple graphical account of the main ideas. The account here is based on Sibly and Antonovics (1992). The evolutionary process is

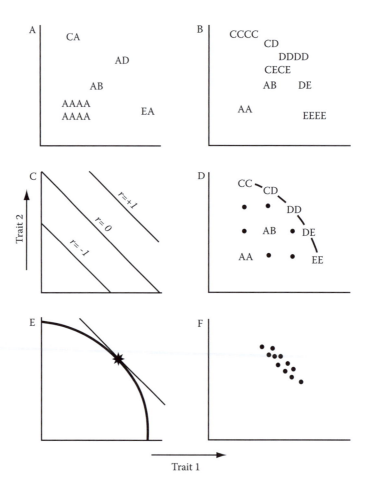

FIGURE 13.1
Simple example of an evolutionary process. Axes represent two life history traits. Note that the alleles far from
the origin (C–E) have increased in numbers between graph A and graph B, whereas those near the origin (A)
have decreased. Graph C shows the per copy rates of increase, i.e., fitnesses, here labeled *r*. Graph D shows
the genetic options set with genotypes obtainable by recombination represented as dots. The boundary of
the options set is the trade-off curve (thick line). Graph E shows the optimal strategy (evolutionary outcome),
starred. Graph F shows the genetic options that may persist in the population at the end of the evolutionary
process. See text for further details.

envisaged, put very simply, as consisting of the creation (by mutation) of new alleles,
which either displace or are displaced by their counterparts. In Figure 13.1, alleles A–E
affect two life history traits of carriers. These might, for example, be juvenile growth rate
and juvenile survival. If the constant environment assumed here is a polluted environment, the alleles C–E might represent resistant alleles that confer improved survival (e.g.,
allele C) or growth (e.g., allele E). Alleles A and B would then represent nonresistant
alleles, whose carriers on average have reduced survival or growth rate. Note that most
individuals carry the A allele in Figure 13.1A and so have small values of both traits, but
a few individuals have larger values, so that overall there is a small positive correlation
between individuals. If there is not much effect of environmental variables, this reflects
a positive genetic correlation (e.g., the two traits are correlated because they are determined by alleles that have an effect on the magnitude of both traits).

Figure 13.1B represents a hypothetical situation at some later time in the selection process. Now most of the small-trait A alleles has disappeared, but the numbers of the large-trait alleles (C–E) have increased. Furthermore, the genetic correlation between individuals is now negative, whereas earlier it was positive.

Clearly, all depends on whether or not an allele spreads in the population, i.e., on the rates of increase of the alleles. The per capita (or more correctly per copy) rate of increase of an allele is here called its fitness. Because the rates of increase of alleles depend on their effects on life histories, the rates of increase (fitnesses) can be plotted out in the space of Figure 13.1, as shown in Figure 13.1C. In our example, the small-trait alleles have negative rates of increase because they are declining in the population. For these alleles, fitness is negative (e.g., $r = -1$ in Figure 13.1C). On the other hand, the large-trait alleles have positive rates of increase because they are spreading. For them, fitness is positive (e.g., $r = +1$). Evolutionary change can also occur, however, if both small- and large-trait alleles increase but at different rates.

In general, as selection proceeds, the cloud of points in Figure 13.1 changes shape. In the absence of environmentally caused variation, the shape of the cloud is measured by genetic correlations and variances, and as the cloud changes shape the genetic correlations and variances change accordingly (cf. Figure 13.5). In the absence of further mutation, where would this process end up?

In considering the eventual outcome of this selection process, it must be remembered that we are here restricting attention to a constant environment. It is important to realize that the environment of an individual not only depends on physical features (e.g., temperature, rainfall, concentration of a pollutant) and biotic features determined by other species (e.g., food availability, predation) but also has characteristics determined by conspecifics, such as territory size, availability of mates, and competition for food.

In this environment, many alleles will affect life history components. Plotting all these genetically codable options (including all possible recombinants) in a space such as that of Figure 13.1A gives us a set of points that we shall call the genetic options set (Figure 13.1D). Although the exposition here is in terms of two-dimensional examples, the concepts all have natural generalizations to three or more dimensions (Sibly and Antonovics, 1992).

Of particular interest, because it limits selection, is the boundary of the options set. We shall call this the trade-off curve (Figure 13.1D). This curve represents the best that this organism can achieve genetically in the study environment.

Putting together the information about fitness (Figure 13.1C) with the information on options sets (Figure 13.1D), the optimal strategy is readily identified (Figure 13.1E) as that having the highest fitness in the study environment. This point represents the eventual outcome of selection in this environment.

In this section, a distinction has been made between the process and the outcome of selection. The process is modeled by quantitative genetics. This determines the short-term evolutionary trajectories within local populations from knowledge of the fitness surfaces together with the genetic options set, characterized by genetic correlations and variances. The outcome of selection can be identified by ecological optimality theory, given knowledge of the shapes of the trade-off curve and the fitness contours. Note that the outcome of selection may be a number of alleles having similar fitness, i.e., a thin oval set of genetic options on or near the trade-off curve in the neighborhood of the optimal strategy. In the example of Figure 13.1F, these genetic options would be characterized by a negative genetic correlation. In this way, genetic correlations can provide evidence about the shapes of trade-off curves.

If there are no trade-offs, so that life history traits can be genetically altered independently of each other, then there is always selection to increase fecundity, decrease mortality rate, and breed early. One way to achieve early breeding is through faster growth and development, so a corollary of selecting for early breeding is that there is always selection for faster growth and development—in the absence of trade-offs.

In this section, we have seen that the fitness of an allele can be measured by its rate of increase or, more specifically, by its per capita rate of increase. In population genetics, it is usual to make the definition of fitness relative to the rate of increase of the most successful allele or genotype, but this approach is not followed here. The advantage of the present approach is that the fitness of an allele can be related directly to the life cycle of its carriers. Fitness is increased by reductions in mortality rate, increases in fecundity, or breeding earlier. Thus alleles are selected that reduce their carriers' mortality rate, increase their fecundity, or make them breed earlier. It is important to note that the fitness of an allele depends on the environment in which its carriers live. If the environment changes, the fitness of the allele may change.

In these terms, an allele can be defined as resistant if it increases the fitness of its carriers in a polluted environment. Nonresistant alleles are also known as susceptible alleles. Thus resistant alleles are favored in polluted environments. What happens in unpolluted environments? If resistant alleles are then outperformed by susceptibles, the resistant alleles are said to have a fitness cost. Fitness costs are discussed further in Section 13.4.

13.3 The Evolution of Resistance When There Is a Mortality–Production Trade-Off

It follows from the above that if alleles exist that affect production rate but not mortality rate, selection acts to maximize production rate. Conversely, if alleles affect mortality rate but not production rate, selection acts to minimize mortality rate. However, it may not be possible to alter one life history trait without affecting the other if mortality and production are involved in a trade-off. A decrease in mortality rate can then be achieved only at the expense of a decrease in production rate. Possible physiological reasons for such a trade-off were described in Chapter 8 (Section 8.5).

The action of selection on such a trade-off has been analyzed by Sibly and Calow (1989). They focused on the juvenile phase, although the analysis can equally well be applied to adults. For the purposes of the analysis, they defined somatic growth rate as the reciprocal of development period. This definition makes most biological sense if egg size and adult size are not subject to selection, at least for the period of time analyzed.

For the reasons given in Chapter 8, it is likely that a decrease in mortality rate can be achieved only at the cost of a decrease in somatic growth rate. Protective mechanisms that decrease mortality rate are often thought of as defenses. Resources allocated to these defenses are not available for somatic growth, and it follows that increasing allocation to defense implies reduced somatic growth. Defensive structures and processes therefore have two key characteristics, from an evolutionary point of view. These characteristics are their effects reducing (i) mortality rate and (ii) growth rate. Plotting the different genetic possibilities reveals the form of the trade-off curve (Figure 13.2A, cf. Figure 8.14).

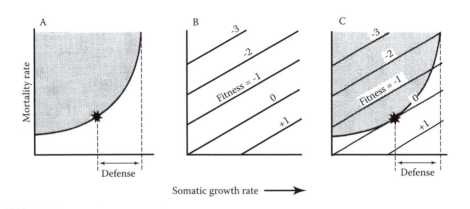

FIGURE 13.2
(A) Genetic options set (shaded) and trade-off curve. (B) Fitness contours. Note that the zero fitness contour goes through the origin. (C) Superimposing (A) and (B) allows identification of the evolutionary outcome as the allele achieving highest fitness.

It is important to remember that the form of the trade-off curve depends on the organism's environment. What is an effective defense in one environment may have no impact in another: Defenses against pollution are of no use in unpolluted environments. Thus trade-offs are environment dependent.

The fitness contours for this trade-off are straight lines, and the most interesting contour, that on which fitness is zero (stable population), goes through the origin (Figure 13.2B). Superimposing Figures 13.2A and 13.2B reveals the position of the optimal strategy (cf. Figure 13.1E) representing the evolutionary outcome in the studied environment (Figure 13.2C). This is the strategy toward which selection drives the population. It represents the optimal trade-off between mortality rate and growth. It indicates the optimal amount that the organism should spend on defense.

This analysis depends, as has been emphasized, on the constancy of the environment. What happens if the environment changes, as when it becomes polluted?

13.4 Evolutionary Responses to Environmental Changes

Pollutants may affect the shape and/or the position of the genetic options set. Changes in position will be discussed first. Pollution may damage organisms, either with lethal effect or with some detriment to production rate. Such processes increase mortality rate or reduce somatic growth rate, moving the options set either vertically upward or horizontally to the left.

The outcome of the evolutionary response to such changes in the position but not the shape of the genetic options set is shown in Figure 13.3. The analysis is straightforward. As in Section 13.3, attention can be restricted to steady-state outcomes, in which the final value of fitness is zero. As before, the zero-fitness contours are straight lines going through the origin. Inspection of Figure 13.3 shows that the evolutionary response is less defense if either mortality is increased, moving the options set vertically upwards, or production

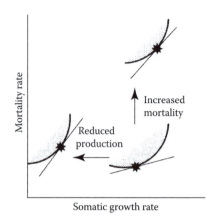

FIGURE 13.3
The evolutionary outcomes of long-term mortality and production stresses. The straight lines are zero fitness contours.

is reduced, moving the options set horizontally to the left. In other words, either mortality stress or production stress elicits the same evolutionary response—less defense.

This prediction—less defense in polluted environments—appears paradoxical, and it should be emphasized that it applies only if the shape of the trade-off curve remains unchanged when its position is shifted. If more defense evolves in polluted environments, then the inference must be that the shape of the trade-off curve has changed. A hypothetical example is shown in Figure 13.4. In this example, genes for optimal defense are selected in the polluted environment and genes for no defense are selected in the unpolluted environment, as shown by the stars in Figure 13.4.

In general, pollution does change the shape of the options set. This may happen because pollution elicits the expression of genes that would not otherwise have been expressed. For instance, enzymes may be induced or rates of pumping or molting may increase with consequent effects on life histories. Genetic variation that was of little consequence in the unpolluted environment may now distinguish survivors from nonsurvivors. Survival may depend on having alleles that increase relatively impermeable exterior membranes (e.g., Oppenoorth, 1985; Little et al., 1989), more frequent molts (and consequent removal of toxicant in shed skin, e.g., Bengtsson et al., 1985), a more comprehensive immune system,

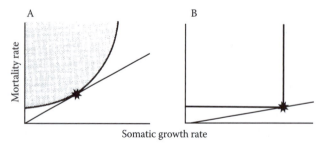

FIGURE 13.4
If the trade-off curves in (A) polluted and (B) unpolluted environments have different shapes, then the evolutionary outcomes (starred) may involve more defense in polluted environments, as shown.

TABLE 13.1

Fitness Advantages of Resistant Strains in Environments with or without Pesticides[a]

Species	Pesticide	Fitness Advantage with Pesticide	Fitness Advantage without Pesticide
Mosquito (*Anopheles culifacies*)	DDT	0.34	−0.23
Mosquito (*Anopheles culifacies*)	Dieldrin	1.50	−0.49
Rat (*Rattus norvegicus*)	Warfarin	>0	−0.62

Sources: Data from Bishop, J.A. (1981). A neoDarwinian approach to resistance: examples from mammals. In Bishop, J.A. and Cook, L.M., eds., *Genetic Consequences of Man-Made Change*, pp. 37–51. London, Academic Press; and Curtis, C.F., Cook, L.M., and Wood, R.J. (1978). Selection for and against insecticide resistance and possible methods of inhibiting the evolution of resistance in mosquitoes. *Ecological Entomology* 3, 273–287.

[a] Fitness advantage = fitness$_{resistants}$ − fitness$_{susceptibles}$. Fitness cost is the negative of fitness advantage.

or detoxication enzymes (Terriere, 1984; Oppenoorth, 1985), as described in Chapters 4, 5, and 7. Many examples are given in this book.

Figure 13.4 shows that if genes are expressed in polluted environments that before were silent or absent, then evolutionary outcomes may differ between polluted and unpolluted environments, and their populations may be genetically distinct. Such genetic differentiation can be documented and used experimentally in a transplant experiment. The method involves transplanting individuals from a number of environments to a common environment in which their life history components are then measured. Care must be taken that there are no maternal or other residual nongenetic effects carried over from the previous environment. For instance, mothers of measured individuals must be in equivalent condition, because maternal quality can affect offspring performance (maternal effect). Since in a transplant experiment all individuals are assessed in a common environment, effects of source population reflect genetic differences. If the transplanted populations are genetically distinct, it is unlikely that the transplanted populations, carrying alleles that evolved in other environments, will be superior in the study environment to the population that evolved there. Hence in each environment, the population that evolved there should outperform the others. This prediction has been confirmed by reciprocal transplant experiments (especially in plants) showing that resident populations outperform aliens (Sibly and Antonovics, 1992). Examples comparing resistant and susceptible strains in polluted and unpolluted environments are given in Table 13.1. The three studies shown in the table all indicate that, as predicted, resistant strains are fitter in polluted environments and that susceptible strains are fitter in unpolluted environments. When this happens, there is a fitness cost of resistance, meaning that resistant alleles, which are fitter in the polluted environment, are less fit than susceptibles in the unpolluted environment. Note, however, that whereas resistant strains are often much fitter than susceptibles in the polluted environments, the fitness of the resistants in the unpolluted environment is sometimes not much less than that of susceptibles (e.g., 0.23–0.62 in Table 13.1). In general, the fitness costs of resistance depend on the resistance mechanism involved. Examples are discussed in the case studies at the end of this chapter. One experimental problem that arises in practice is that if one is interested in the fitness benefits and costs of a single allele, then it is desirable to study the allele and its alternate on a common genetic background.

In general, then, chronic pollution results in a change in the shape of the genetic options set. Whereas before the environmental change the genetic options set may have been lined up on or alongside the trade-off curve and parallel to a fitness contour (Figure 13.1F), after the environmental change, with genes expressed that before were silent, the genetic

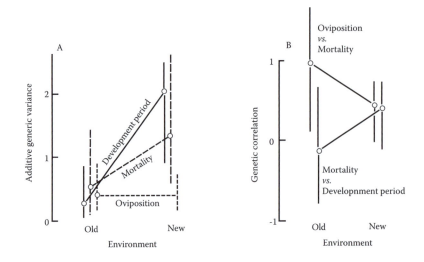

FIGURE 13.5

Additive genetic variance in development period and mortality rate increased as predicted after introduction of rice weevils *Sitophilus oryzae* to a new toxin-rich environment, although the changes were not significant. Units are eggs day–1 female–1 (oviposition rate), 4 days (development period) or 4×10^{-5} (mortality rate). (B) Changes in genetic correlations. Vertical bars represent 95% confidence intervals. Reproduced from Holloway, G.J., Sibly, R.M., and Povey, S.R. (1990). Evolution in toxin-stressed environments. Functional Ecology 4, 289–294. With permission from Blackwell Science Ltd.

options set is quite likely to bulge out and to lie some way from the new trade-off curve. These predictions can be tested by measuring quantitative genetics parameters, which give some idea of the shape of the genetic options set. The degree to which the points are spread out in Figure 13.1F is measured by additive genetic variance. The prediction here is that additive genetic variance will increase if new genes are expressed in response to environmental pollution. The correlation between the points in Figure 13.1F is measured by genetic correlation, and because in the unpolluted environment we argued that the genetic correlation would be tight (close to –1, Figure 13.1F), any loosening of the correlation will move it away from ±1.

Holloway et al. (1990) tested these predictions by transplanting a population of weevils from the environment in which it had evolved into a new toxin-rich environment. The new environment consisted of yellow split pea and the old environment of wheat, on which substrate the population had been maintained for 50 generations. Measurement of quantitative genetic parameters was achieved in a full-sib/half-sib breeding experiment, using over 50 males and 250 females. This illustrates the general point that quantitative genetic experiments are very demanding in terms of the numbers of animals needed. Because the population declined initially after transfer to the toxin-rich environment, genetic analysis could not be carried out until the fifth generation after transfer. The results are shown in Figure 13.5. Additive genetic variance in development period and mortality rate increased as predicted after introduction to the new toxin-rich environment. Genetic variance in oviposition rate did not change. Changes in genetic correlation were quite large in two cases (shown in Figure 13.5B). The genetic correlation between oviposition and juvenile mortality rate changed as predicted from a tight correlation close to +1 to a looser correlation of 0.4 in the toxin-rich environment (Figure 13.5B). A correlation of +1, not –1 as in Figure 13.1F, was predicted in the unpolluted environment, because although the association between fitness and oviposition rate is positive, that with mortality rate is negative. There was no change in the correlation of mortality and development period.

This example shows how the genetic options set, measured by quantitative genetic parameters, may change when the environment becomes polluted. If the environment remains chronically polluted, selection will then favor alleles close to the trade-off curve in the neighborhood of the new optimum. A new evolutionary process begins, but it is a process of the same type as that described in Section 13.2.

13.5 Monogenic Resistance

Alleles selected in polluted environments, with positive fitnesses, are said to be resistant, or, in some contexts, tolerant. Individuals that are homozygous resistant are referred to as resistant individuals. The relative resistance of heterozygotes measures the degree of dominance of the resistant allele. An example is shown in Figure 13.6. Pesticide resistance often shows simple inheritance, with resistant genes being semidominant (i.e., heterozygotes halfway between the homozygotes).

The degree of dominance affects the speed of spread of an allele but not the final outcome of the evolutionary process (except when the allele is overdominant, i.e., the heterozygotes outperform both homozygotes). Advantageous dominant alleles spread faster initially than recessive alleles.

Complications arise if more than one locus is involved in resistance. To tell how many loci are involved, it is generally necessary to carry out a breeding experiment lasting several generations. For such experiments, homozygous strains are required. These are usually obtained by mass selection in the laboratory of field-collected strains. A breeding experiment starts by crossing a homozygous resistant strain with a homozygous susceptible

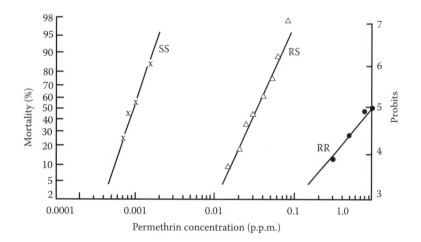

FIGURE 13.6
Dose–response curves for the mosquito *Culex quinquifasciatus* tested with permethrin (NRDC 167). Percentage mortality is plotted on a probit scale. SS shows the response of homozygous susceptible individuals, RS of heterozygotes, and RR of homozygous resistant individuals. Reproduced from Taylor, C.E. (1986). Genetics and evolution of resistance to insecticides. *Biological Journal of the Linnean Society* 27, 103–112.

strain. The offspring are necessarily heterozygous at all loci. These offspring are back-crossed with the parental strains. Half the offspring of these back-crosses are expected to be heterozygous if only one locus is involved. On the other hand, if more than one locus is involved, less than half these offspring show resistance. The exact prediction depends on the degree(s) of dominance of the resistant allele(s). Statistical techniques are available for estimating the number of genes involved, but these make assumptions whose validity has to be checked. When discrimination between genotypes is difficult, as it can be in resistance studies, further experiments may be necessary. Repeated back-crossing can be useful. Genetic markers that map the positions of the resistant genes on the chromosomes have also proved particularly useful in difficult cases.

The general conclusion of studies of this type is that major genes (i.e., genes with large effects) are found in most cases of resistance, including resistance to insecticides, acaricides, fungicides, herbicides, and heavy metals. Resistance is not always monogenic, but in most cases where resistance causes problems in pest control, resistance appears to be largely controlled by one, or occasionally two, loci (Roush and Daly, 1990). Minor genes and modifier genes may, however, still have some small effects.

13.6 Case Studies

Examples of evolutionary responses to pollution include some of the best-known studies of the evolutionary process in the field. The reason for this is not hard to find. To study the evolutionary process, it is usually necessary to find a population exposed to an environmental change, and in recent centuries most environmental changes have been effected by man. Many of these are cases of episodic or chronic pollution.

13.6.1 Evolution of Pesticide Resistance

The evolutionary response to pesticides has been staggeringly varied and successful. The number of species in which resistance is known rose from 30 in the 1950s to over 450 in the 1980s. The account here is based on Taylor (1986), Mallet (1989), Roush and Daly (1990), and Bass and Field (2011).

Most insecticides that have been widely used belong to one of five classes: (i) organochlorine insecticides, such as DDT, γ-HCH, and dieldrin; (ii) organophosphorous compounds (OPs), such as malathion and dimethoate; (iii) carbamates, such as carbaryl and temik; (iv) synthetic pyrethroids, such as permethrin and cypermethrin, and most recently (v) neonicotinoids such as imidacloprid and thiacloprid. Insects may be resistant to more than one insecticide, and often to insecticides in more than one class. When resistance to more than one insecticide is achieved by a single mechanism, this is true cross resistance, but when several resistance mechanisms are involved, this is called multiple resistance. In some cases, the situation may be uncertain, and these terms are then used more loosely.

All members of these five classes of insecticides act on the nervous system, although in different ways. The organochlorines and pyrethroids act upon sodium channels, thereby disturbing electrical conduction along the axons, whereas the OPs and carbamates act as inhibitors of acetylcholinesterase (Chapter 7) and the neonicotinoids act upon nicotinic receptors for acetylcholine. The last two classes disturb synaptic transmission across cholinergic synapses.

TABLE 13.2

Primary Mechanisms of Resistance

Mechanism	Insecticides Affected
1. Increased Detoxification	
DDT–dehydrochlorinase	DDT
Microsomal monooxygenase	Carbamates
	Pyrethroids
	Organophosphorous compounds
Glutathione transferase	Organophosphorous compounds
Hydrolases, esterases	Organophosphorous compounds
2. Decreased Sensitivity of Target Site	
Decreased sensitivity of acetylcholinesterase	Organophosphorous compounds
	Carbamates
Decreased nerve sensitivity (K_{dr})	DDT
	Pyrethroids
Decreased nerve sensitivity (GABA receptor)	Dieldrin, other cyclodienes, γ-HCH
3. Behavioral	
Increased sensitivity to insecticide	DDT
Avoidance of treated microhabitats	Many
Decreased cuticular penetration	Most insecticides

Source: Modified from Taylor, C.E. (1986). Genetics and evolution of resistance to insecticides. *Biological Journal of the Linnean Society* 27, 103–112.

The mechanisms by which insects have evolved resistance to these chemicals are summarized in Table 13.2. The main mechanisms are the first two: increased detoxication, which increases the rate at which the insecticide is broken down, and decreased sensitivity of the target site, as a result of mutation and selection of the target protein. The other mechanisms of resistance are less important. Behavioral mechanisms include avoidance of an insecticide after low-level exposure to it. Decreased cuticle penetration may result if the cuticle is relatively impermeable to insecticides in resistant strains.

Resistance mechanisms are the consequence of genetic differences between susceptible and resistant strains of the pest species. Most commonly, a resistant strain possesses a highly active form (or forms) of a detoxifying enzyme or one or more genes encoding for an insensitive form of the target site. Usually these are genes that do not occur in susceptible strains of the same species. Recently, however, a number of instances have been found where the strains differ in the number of copies of the same gene (Bass and Field, 2011). The resistant strain contains more copies of the gene than does the susceptible strain. In the case of increased detoxication, this allows resistant individuals to synthesize greater quantities of the detoxifying enzyme in question. This phenomenon has been termed gene amplification and was the subject of a classical study on the resistance of the peach potato aphid (*Myzus persicae*) to organophosphorous insecticides (Figure 8.16, Devonshire and Sawicki, 1979). A role for gene duplication or amplification in resistance has now been demonstrated in at least 10 insect and mite species, either in increased detoxication or in decreased sensitivity of the target site (Bass and Field, 2011). Resistance due to gene amplification has been particularly associated with the operation of detoxifying mechanisms. However, there are examples of this phenomenon in the case of Category 2 as well—i.e., resistance due to an insensitive site of action. One example is the resistance of certain

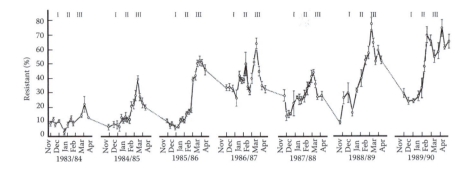

FIGURE 13.7

The evolution of pyrethroid resistance in cotton budworm, *Helicoverpa armigera*, in the Namoi–Gwydir cotton-growing region of New South Wales, Australia. The stages of the resistance management program are indicated by roman numerals at the top of the graph. The period of pyrethroid use each year (stage II) is shaded. Percentage resistant refers to the percentage surviving a dose of pyrethroid fenvalerate that killed 99% of susceptibles. Vertical bars are standard errors. From Forrester, N.W., Cahill, M., Bird, L.J., and Layland, J.K. (1993). Management of pyrethroid and endosulfan resistance in *Helicoverpa armigera* (Lepidoptera: Noctuidae) in Australia. *Bulletin of Entomological Research* 1, 1–132. With permission of CAB International.

clones of *M. persicae* to the cyclodiene endosulfan, which involves duplication of a subunit of a GABA receptor (Anthony et al., 1998).

All of these resistance mechanisms could involve fitness costs in areas where there is no insecticide. Avoidance of certain microhabitats could result in reduced nutrient uptake, increased detoxication may be energetically expensive, and reduced sensitivity or penetrance of insecticides may involve a costly reduction in sensitivity or penetrance of other substances. Although the fitness costs of resistant alleles in untreated areas have rarely been studied, it is known that they can be large (Table 13.1). Roush and Daly (1990) conclude that the most serious and consistent fitness costs are those associated with general esterases (e.g., carboxylesterases that hydrolyze naphthyl acetate). By contrast, in some studies, little or no reproductive disadvantage has been found associated with malathion-specific carboxylesterases, increased oxidative detoxication, mutant acetylcholinesterase, or knockdown resistance-like mechanisms.

Insecticide resistance is characterized by rapid evolution under strong selection. However, resistant alleles do not generally spread to fixation, completely displacing competitor alleles, in the field. This is probably because pesticide treatments are stopped when they become ineffective. However, inability to spread to fixation would also result if resistant alleles have a fitness cost in untreated environments. Untreated populations mix with treated populations by dispersal. Immigration from untreated populations can considerably delay the increase of resistance, depending on the relative sizes of the treated and untreated populations and on the degree of migration.

One of the most spectacular examples of the evolution of insecticide resistance is that shown in Figure 13.7. As part of a resistance management strategy, pyrethroid use was restricted to a short period (stage II) each year in the middle of the cotton-growing season. Endosulfan use was permitted in stages I and II; other insecticides could be used at any time. Figure 13.7 shows that the proportion of individuals that were resistant rose each year during and immediately after pyrethroid application. The apparent rise immediately after pyrethroid application ceased because of a time lag in measurement; affected individuals were not tested themselves, instead the test was carried out on their field-collected eggs. There was thus a delay while individuals matured and reproduced before the test was performed. After pyrethroid application, the proportion of resistant individuals declined. This could be the result of a fitness cost of pyrethroid resistance, but it could also be because of immigration of susceptibles from adjoining untreated areas. Note that overall the frequency of pyrethroid resistance rose during the course of the 7-year program.

13.6.2 Evolution of Metal Tolerance in Plants

The mining of metals inevitably leads to pollution of the soils around mines, and spoil heaps are often rich in copper, zinc, lead, or arsenic. The spoil heaps of Devon Great Consols, for example, which was the richest copper mine in Europe in the late nineteenth th century, contain more than 1% copper and 5% arsenic in places. Such concentrations are highly toxic to most plants. Yet plants can often be found growing on spoil heaps, as a result of the evolution of metal tolerance (Macnair, 1981; Macnair, 1987; Shaw, 1989; Schat and Bookum, 1992; Kruckeberg and Wu, 1992). These plants are genetically more tolerant than plants from neighboring populations. Methods of quantifying metal tolerance in plants are described in Section 6.3.4.

As an example, consider the wind-pollinated grass *Agrostis tenuis*, copper-tolerant genotypes of which grow on the spoil heaps of copper mines. Taking a transect through such a mine, as in Figure 13.8, one finds copper tolerance on the copper-rich soils of the mine but no tolerance outside it in the upwind direction. Some tolerance is, however, found immediately downwind as a result of pollen or seeds of tolerant parents being blown off the mine. Because tolerance is found on the mine but for the most part not off it, it is clear that tolerant alleles have increased in numbers on the mine. This is what is meant by saying that tolerant alleles are fitter in the polluted environment. Conversely, tolerant alleles seem to be outcompeted off the mine, indicating that there they are less fit. In other words, there appears in this case to be a fitness cost of tolerance.

Figure 13.8 also contrasts the copper tolerance of individuals grown from seeds (white bars) with that of adults (black bars). Comparison of the two shows how selection works at different places on the transect. Thus adults on the mine are more tolerant than seeds, indicating selection for tolerance on the mine. Conversely, downwind of the mine adults are less tolerant than seeds, indicating selection against tolerance off the mine.

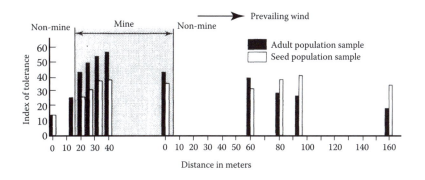

FIGURE 13.8
Copper tolerance in the grass *Agrostis tenuis* along a transect on the surface of a copper mine. The copper-impregnated part of the mine is shaded. (Data of McNeilly, T. (1968). Evolution in closely adjacent plant populations. III. *Agrostis tenuis* on a small copper mine. *Heredity* 23, 99–108.; redrawn by Macnair, M.R. (1981). Tolerance of higher plants to toxic materials. In Bishop, J.A. and Cook, L.S., Eds. *Genetic Consequences of Man-Made Changes*. London, Academic Press. With permission.

TABLE 13.3

Percentage of Copper-Tolerant Individuals in Normal Populations of Grass Species
That Are Commonly Found near Mines in Britain in Relation to Whether Copper-
Tolerant Populations of These Species Have Been Found on Copper Mines

Species	Occurrence of Tolerant Individuals in Normal Populations (%)	Presence (+) or Absence (–) of Tolerant Populations on Copper Mines
Holcus lanatus	0.16	+
Agrostis capillaries	0.13	+
Festuca ovina	0.07	–
Dactylis glomerata	0.05	+
Deschampsia flexuosa	0.03	+
Anthoxanthum odoratum	0.02	–
Festuca rubra	0.01	+
Lolium perenne	0.005	–
Poa pratensis	0.0	–
Poa trivialis	0.0	–
Phleum pretense	0.0	–
Cynosurus cristatus	0.0	–
Alopecurus pratensis	0.0	–
Bromus mollis	0.0	–
Arrhenatherum elatius	0.0	–

Source: Data of C. Ingram reported in Macnair, M.R. (1987). Heavy metal tolerance in plants: A model evolutionary system. *Trends in Evolution and Ecology* 2, 354–359.

It is generally believed that there is a fitness cost of tolerance, i.e., that tolerant genes are disadvantageous under normal conditions. If this were not the case, then tolerance genes would occur widely, at least in some species, given their competitive edge in polluted environments. However, there is disagreement as to the severity of the cost of tolerance. Some authors consider that these costs may be great (see Baker, 1987). Others suggest that they may be small on the basis of studies that carefully distinguish metal tolerance from other evolved attributes of mine populations (Macnair, 1987). For example, tolerant strains may be those that grow more successfully in the nutrient-poor conditions often found in mine waste (Ernst et al., 1992).

Ernst (1976) suggested that the fitness costs of tolerance are a result of the energy needed by the mechanisms of tolerance. Energy spent on detoxication is not available for growth, and Ernst suggested that this is why many tolerant plants grow more slowly and produce less biomass than nontolerant conspecifics. This is the situation illustrated in Figure 13.2A.

Why do some species evolve metal tolerance and others not? Presumably, similar selection pressures apply to all. Selection alone, however, is not enough; there must also be present some suitable tolerance genes on which selection can act. Table 13.3 shows that normal populations of species able to evolve metal tolerance contain tolerance genes at low frequencies. Theory shows that whether tolerance genes exist in normal populations depends on mutation rate, i.e., the rate of creation of tolerance genes by mutation, on the fitness cost of tolerance, and on the population size. Tolerance is more likely to evolve if the mutation rate is high, the fitness cost of tolerance is low, and the species is common.

13.6.3 Evolution of Industrial Melanism

The blackening of industrialized parts of the countryside was a result of the Industrial Revolution that began in Britain in the eigteenth century. One result was that an area west of Birmingham became known as the Black Country. Blacker (melanistic) forms of animals are better camouflaged in such environments and may thus be better protected against predators. This may give melanistic forms a survival advantage. Melanism is found in many species of arthropod, but most examples are in moths. The account here is based on Brakefield (1987), Majerus (1998), and Moriarty (1999).

The incidence of melanism has risen steadily since about 1850. Over 100 of the 780 species of larger moths in Britain now commonly include melanic forms. Similar changes have occurred in Europe and North America, usually in industrial areas. In many cases, the melanic forms have spread very rapidly and have become predominant.

The first and best-studied example is the peppered moth, *Biston betularia* (Kettlewell, 1973). J. W. Tutt suggested in 1896 that the nonmelanic form is well concealed (cryptic) when at rest on the pale bark of trees in rural areas, where it resembles thalli of foliose lichens. By contrast, it was conspicuous on the completely black surfaces of trees and walls in industrial Britain. The blackening was mostly due to soot, but lichens were also killed, mainly by sulfur dioxide. Tutt suggested that in this environment the nonmelanic form was conspicuous to avian predators; conversely, melanic forms were cryptic.

The first melanic specimen was caught in Manchester in 1848, and by 1895 98% of the moths in that area were melanic. This corresponds to a 50% increase in the chances of survival of the melanic forms, according to calculations by J. B. S. Haldane (see Brakefield, 1987). Increases in melanic forms were recorded during this period in a number of industrialized zones. The phenomenon has been of major interest to evolutionary biologists since the 1920s.

Kettlewell organized surveys between 1952 and 1970 of the geographical distribution of melanism in *B. betularia* in Britain (Figure 13.9). Three forms of the moth were recognized. The nonmelanistic form, *typica*, shows a range of coloration from heavily speckled individuals to others that are almost white with fine, granular black markings. The common melanic form, *carbonaria*, produced by a dominant allele at a single locus, is usually all black but sometimes has some light-colored spots or patches. A third intermediate form, *insularia*, is also recognized. Figure 13.9 shows that in general the melanic forms occurred in the industrialized regions of Britain and in the areas downwind (the predominant winds in Britain are from the southwest). Kettlewell suggested that the downwind occurrence of the melanistic forms might be because soot had blackened the downwind areas, but it could also be the result of passive wind dispersal of larvae. When eggs hatch, the very small larvae suspend themselves on silk threads and are dispersed by air currents. It is possible that the high incidence of melanism in East Anglia is the result of long-distance dispersal of larvae from industrial zones in London and the English Midlands.

Interestingly, there has been a further environmental change in recent years consequent to clean-air legislation and the creation of smokeless zones in the late 1950s. This led to a rapid fall in emissions of smoke and sulfur dioxide. Tree surfaces became lighter, and the relative frequency of melanics fell (Figure 13.10). This phenomenon has been described as evolution in reverse! The time lag between the reduction in pollution and the evolutionary response is probably a result of slow change in the composition of the lichens covering the tree branches on which the moths rest (Cook et al., 1999).

Although it is generally accepted that melanism in *Biston betularia* is a result of atmospheric pollution, the detailed working of the selective mechanisms is still not fully

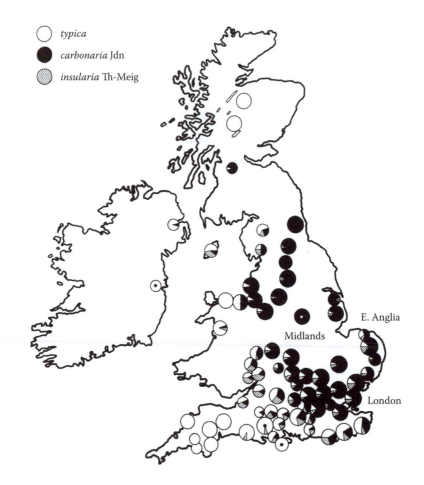

FIGURE 13.9
The relative frequencies of the normal and two melanic forms of the peppered moth *Biston betularia* in Britain. The results are based on more than 30,000 records collected from 1952 to 1970 at 83 sites. Reproduced from Kettlewell, B. (1973). *The Evolution of Melanism.* Oxford, Clarendon Press. With permission from Oxford University Press.

understood. The fitness advantages of the melanic form in industrial areas and of the non-melanic forms in rural areas were demonstrated by Kettlewell in mark-release-recapture experiments. Marked individuals of melanic and normal forms were released in rural and industrial areas (Table 13.4). More melanic individuals were recaptured in the industrial area and more nonmelanic in the rural areas.

To go further and to estimate the predation rate by birds, it is necessary to have a detailed knowledge of the time budgets of adult moths of different ages and both sexes during the reproductive period. Peppered moths emerge shortly before dusk. After a dispersal flight, females rapidly attract mates and pair shortly after dusk. Thereafter, they do not fly and walk only short distances. They remain in copula for nearly 24 hours. Many moths rest during the day underneath or on the side of small branches in the tree canopy. Studies of visual predation on the melanic and nonmelanic forms have, however, generally been carried out on tree trunks. Although the findings are qualitatively in keeping with the camouflage protection hypothesis, the experiments need repeating on more realistic substrates.

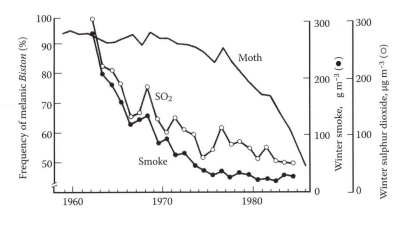

FIGURE 13.10

The decline in the frequency of the melanic form of the peppered moth near Manchester after clean-air legislation reduced emissions of smoke and sulfur dioxide. Reproduced from Brakefield, P.M. (1987). Industrial melanism: do we have the answers? *Trends in Ecology and Evolution* 2, 117–122. With permission from Elsevier Trends Journals.

TABLE 13.4

Relative Recoveries of Marked Individuals of Two Forms of the Peppered Moth, *Biston betularia* (*typica* and *carbonaria*) from Two Sites, One Rural and One Industrial

Site	Form	No. Released	No. Recaptured	Recapture (%)
Rural Dorset	*Typica*	496	62	12.5
	Carbonaria	473	30	6.3
Industrial Birmingham	*Typica*	201	32	15.9
	Carbonaria	601	205	34.1

Source: Moriarty, F. (1999). *Ecotoxicology*, 3rd ed. London, Academic Press, presenting data from Kettlewell (1955, 1956). With permission.

13.6.4 Evolutionary Response of Dog Whelks, *Nucella lapillus*, to TBT Contamination

The use of dog whelks as biological monitors of tributyl tin (TBT) pollution is described in Chapter 16. Briefly, TBT pollution sometimes causes imposex in female dog whelks. Imposex is a condition in females involving the growth of male organs, and this may prevent successful reproduction in areas polluted by TBT. Because migration is very limited in this species, population declines attributed to TBT pollution have been reported in many areas adjacent to harbors. One such area is the north coast of Kent, where dog whelks are known to have been abundant in pre-TBT times. However, Gibbs (1993) discovered that a population in this area at Dumpton Gap had evolved modified genitalia that allowed it to persist. The account here is based on Gibbs's paper.

The Dumpton Gap population is characterized by the absence of a penis or an undersized penis in about 10% of the males (absence of a penis in males is unknown elsewhere). The vas deferens and prostate are incompletely developed in affected males. Gibbs has labeled these abnormalities the Dumpton syndrome. Laboratory-bred animals of Dumpton Gap stock display the same characteristics, suggesting that the character is genetically determined. It cannot be due to some aberrant feature of the Dumpton environment such as predators or parasites because neither predators nor trematode parasites were present in the laboratory. About 75% of the Dumpton Gap females showed little or no imposex, whereas usually all females are expected to show imposex at the TBT levels experienced at Dumpton Gap. This phenomenon has also been reported in France (Huet et al., 1996).

The evidence therefore suggests that the Dumpton syndrome is the result of a genetic mutation that reduces imposex in females, allowing them to breed successfully even though it prevents breeding in some 10% of males. Overall, the gene must confer a marked fecundity advantage in environments contaminated with TBT. In TBT-free environments, however, the gene would be disadvantageous because a significant proportion of males are infertile (cf. Table 13.1). It seems, therefore, that the Dumpton syndrome carries a fitness cost and so would be selected against in TBT-free environments.

13.6.5 Evolution of Resistance to Pollution in Estuaries

Further striking examples of the evolution of resistance to pollutants have been reported from polluted harbors and estuaries in the American Northeast. Over the period 1953–1979, a battery factory on the Hudson River in New York released over 50 tons of cadmium and nickel hydride wastes into Foundry Cove, a tidal estuarine cove that became in consequence one of the most metal-polluted areas in the world. Resistance to cadmium evolved in the most common aquatic benthic species in the cove, an oligochaete worm *Limnodrilus hoffmeisteri*, and was subsequently lost after cadmium was removed from the cove in a cleanup program. The following account is based on Levinton et al. (2003). Resistance in the Foundry Cove worms was the result of a very marked increase in synthesis of a metal-binding protein, produced by a genetic change at a single locus. Resistance spread very rapidly, perhaps in as little as four generations, through the Foundry Cove population in the mid-1950s. In 1994 and 1995, however, a major cleanup program removed most of the cadmium from the cove, and the subsequent loss of genetic resistance was monitored by comparing the Foundry Cove worms with worms from an unpolluted cove 1 kilometer away. The study showed that genetic resistance was rapidly lost from the population, and by 2002 it could no longer be detected. This was 7 years, corresponding to something between 9 and 18 generations, after the cleanup.

The loss of resistance after removal of the cadmium suggests there was a fitness cost of resistance to cadmium, so that worms lacking the resistant genes were able to outcompete resistant worms (see Section 13.3). An alternative explanation for the loss of resistance is that there was an influx of susceptible genes carried by immigrant worms from surrounding unpolluted areas. This explanation can be discounted, however, because measurements suggest that immigration occurred only at a low rate in Foundry Cove.

The evolution of resistance in *L. hoffmeisteri* was important to other species in the food chain too, because *L. hoffmeisteri* is important in sediment turnover and is the most abundant benthic animal food source for higher trophic levels. Subsequent to the cleanup program, body burdens in *L. hoffmeisteri* returned to normal, so ecological restoration was able to reduce the trophic transfer of cadmium into the local ecosystem.

Further up the North American coast, genetic resistance to PCBs has evolved in a nonmigratory estuarine fish, *Fundulus heteroclitus*, resident in New Bedford Harbor in Massachusetts, which has been heavily polluted with PCBs from industrial discharges since the 1940s. The evolution of resistance to PCBs was inferred from a comparison of the fish of New Bedford Harbor that had been exposed to high levels of PCBs with fish from a nearby area where contamination was much less. The following account is drawn from a paper by Nacci et al. (2002). The second-generation progeny of fish (embryos and larvae) collected from New Bedford Harbor survived levels of 3,3′,4,4′,5

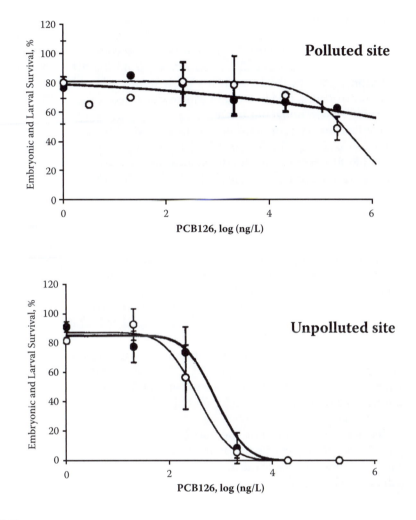

FIGURE 13.11

The survival of progeny of the mummichog *Fundulus heteroclitus* in aquaria containing water contaminated with various concentrations of PCB 126. The fish in the upper graph came from a PCB-contaminated site, New Bedford Harbor; those in the lower graph came from a nearby unpolluted site. Unfilled symbols and light lines represent data from first-generation progeny; second-generation progeny are represented by filled symbols and heavy lines. Bars represent standard errors; each treatment was replicated either two or three times. Reproduced from Nacci, D.E., Champlin, D., Coiro, L. et al. (2002). Predicting the occurrence of genetic adaptation to dioxin-like compounds in populations of the estuarine fish *Fundulus heteroclitus*. *Environmental Toxicology and Chemistry* 21, 1525–1532. With permission from the Alliance Communications Group.

pentachlorobiphenyl (PCB 126) three orders of magnitude greater than the progeny of fish from a control area (Figure 13.11; compare the positions of the closed symbols in the upper and lower panels). Coplanar PCBs like PCB 126 (see Section 1.2.2) are referred to as dioxin-like compounds because they produce toxic effects similar to dioxin (see Section 1.2.3). The forebears of the tested fish had been kept under standardized unpolluted conditions for two generations, which carried two benefits. First, it allowed enough time to eliminate residues of highly persistent PCBs, which might have caused physiological differences between the groups; second, it minimized effects due to differences in the condition of the mothers (i.e., maternal effects) from the two contrasting environments. In fact, any such effects must have been small or nonexistent, because the F1 and F2 generations of fish from the polluted site performed similarly (compare filled and unfilled symbols in Figure 13.11), and the same was true for the control site. In summary, genetic resistance to PCBs was inferred from the differences in survival from PCB pollution shown by the two strains (New Bedford Harbor and control), after the forebears of both strains had been kept in standardized unpolluted conditions for two generations to remove nongenetic effects.

Fish from New Bedford Harbor also showed lower induction of cytochrome $P_{450}1A$ in response to dioxin-like compounds than did fish from control areas (Nacci et al. 1999; Bello et al., 2001). EROD (ethoxyresorufin deethylase) activity provides a measure of levels of cytochrome $P_{450}1A$ (see Section 5.1.5), and both are sensitive indicators of a process that follows from binding of coplanar PCBs and other organic pollutants to the aryl hydrocarbon receptor (Ah receptor) of the cytosol. When pollutants bind to the Ah receptor, there follows a complex pattern of responses that, while known for enhancing the detoxication of some pollutants (see, for example, Section 5.1.5), actually increases the toxicity expressed by dioxin-like compounds. Thus it is understandable that alterations minimizing Ah receptor responses can enhance fitness during chronic exposures to dioxin-like compounds and provide a mechanism for resistance. However, the specific biochemical and genetic mechanisms of PCB tolerance in *F. heteroclitus* are not yet understood.

Interestingly, *F. hetereoclitus* populations also reside in other highly polluted estuarine sites. Has genetic resistance evolved there, and are the evolved mechanisms the same? Certainly tolerance to local contaminants is found (e.g., Wirgin and Waldman, 2004), and it is often accompanied by down-regulation of P_{450} 1A (e.g., Hahn, 1998). However, in one case this tolerance was persistent through one generation but declined over the second or third generations of laboratory rearing (Meyer et al., 2002). Here, therefore, tolerance does not appear to have a genetic basis.

For further discussion of Ah receptor–mediated toxicity, see Ahlborg et al. (1994), Walker (2001), and Box 5.1 of the present text. It is interesting to compare the present example, where resistance is a consequence of the insensitivity of the mechanism responsible for Ah receptor–mediated toxicity, to the examples of mechanisms of resistance shown by insects to insecticides in Table 13.2. Three of these examples identify an insensitive target as the factor conferring resistance.

Another example of the dioxin-like toxicity of coplanar PCBs being expressed in the field is given in Section 16.2 of the present text

13.7 Summary

The case studies presented above are good examples of the evolutionary process because they report evolutionary responses to known environmental changes. In each case, the change is in response to chronic man-made pollution that started at most a few centuries ago. For evolutionary responses to occur, genetic variation has to be present on which selection can act. Evolutionary responses occur when the change to the environment alters the relative fitnesses of different alleles. In each case, there was strong selection for resistant alleles in polluted environments, and evolutionary responses to pollution occurred within tens of generations. Where the genetic mechanisms have been investigated, they consist for the most part of one or occasionally two major genes. Although it is extraordinarily hard to study, it seems likely that resistance generally entails a fitness cost. In other words, resistant alleles are favored at polluted sites but selected against at unpolluted sites. Last, it should be emphasized that although our outline understanding seems secure, in each case there is still some uncertainty as to the detailed working of the evolutionary process.

Further Reading

The account of the evolutionary process given here is based on Sibly and Antonovics (1992). This approach was selected because it is accessible and gives insight into how selection acts on life history characteristics such as juvenile growth rate, fecundity, and survival. Those wanting a population-genetic text should consult Hartl and Clark (1989); Bishop and Cook (1981) remains a useful review of the genetic consequences of man-made environmental changes.

Forbes, V.E., ed. (1999). *Genetics and Ecotoxicology*. Contains a number of case studies of adaptations to polluted environments.

Macnair, M.R. (1987). Gives a short but useful introduction to the evolution of heavy metal tolerance in plants.

Majerus, M. (1998). *Melanism: Evolution in Action*. Contains an extended discussion of the peppered moth story.

Mallet, J. (1989). Gives a brief but accessible introduction to the evolution of pesticide resistance.

Roush, R.T. and Daly, D.C. (1990). Provides a heavyweight, authoritative view of resistance research, invaluable to anyone starting work on any aspect of resistance.

Shaw, A.J. (1999). Reviews what is known about the evolution of heavy metal tolerance in plants.

14

Changes in Communities and Ecosystems

14.1 Introduction

In the two preceding chapters, effects upon populations of individual species were given particular attention. This is the approach usually followed when dealing with larger species. By contrast, microbiologists are particularly concerned with effects on communities and ecosystems, a subject area sometimes termed synecology. The measurement of effects of pollutants upon ecosystems has certain strengths and limitations. It has the advantage of being a holistic approach, which can take into account the overall functional state of an ecosystem (Freedman, 1989).

There are two distinct approaches to studying changes in communities or ecosystems. One is structural, the other functional. Structural changes relate to changes in composition. In the extreme case, species may disappear altogether from communities and ecosystems in which they are usually found. Examples include the disappearance of the dog whelk from many coastal areas of southern England because of the effects of tributyltin (TBT) (Section 16.3) and the disappearance of the sparrowhawk from large areas of eastern England because of the effects of organochlorine insecticides (see Chapter 12). Changes as severe as these indicate a reduction in biological diversity. More commonly, however, changes are only shifts in the balance of species in defined habitats. Changes on this scale may be evident from the use of biotic indices such as RIVPACS (Section 11.2.2). Monitoring such changes provides an early warning system; it may give an early indication that pollution is causing changes that will eventually bring about a serious reduction in biological diversity if remedial action is not taken.

In contrast to these indications of structural changes in community composition, a functional approach can provide a simple measure of the state of a community or ecosystem. It is relatively straightforward to measure the operation of the carbon cycle or the nitrogen cycle in soils and to determine whether it is adversely affected by pollutants. Such a holistic approach gives no information on the status of individuals in a community or ecosystem, and there are advantages in combining it where possible with a structural approach. It makes sense to relate structure to function in communities as well as in individual organisms.

An ecosystem approach has been used with some success to study the effects of pollutants upon soils. The features of soil communities that make them amenable to scientific study are that (i) the boundaries of communities can be easily and clearly defined, and (ii) similar communities exist in vast numbers in the field. These features make it possible to investigate the factors determining their field distribution and to carry out field experiments.

These useful attributes of soil communities are not shared with most larger-scale communities and ecosystems. Exceptions are the aquatic communities inhabiting lakes and rivers. Lakes exist in enormous numbers in Canada, Siberia, and Scandinavia. The major types of pollution are acidification and metal pollution, and their effects can be readily

observed in loss of species and diversity (see Section 14.3). Field experiments are in principle a possibility, although because of their scale they are expensive.

Other types of terrestrial and aquatic communities are not so amenable to study. This
may be because their constituent species are highly mobile, so that their boundaries are
hard to define, or because no two pollution events are the same. For instance, severe oil pollution rarely affects the same type of community twice, and the boundaries of the affected
marine communities are ill defined. Similar consideration applies to air pollution and to
radioactive pollution as a result of nuclear warfare or accidents at nuclear power stations.

In the extreme case of an ecosystem analysis, the composition of the atmosphere can
be taken as an index of the state of health of the entire planet. The recent increase in CO_2
levels in the atmosphere gives evidence of pollution on the global scale. This presents the
threat of global warming because of the so-called glasshouse effect. Likewise, the disappearance of the ozone layer above Antarctica has been attributed to the appearance of
chlorofluorocarbon (CFC) gases in the stratosphere. Reduction of the ozone layer brings
an increase in ultraviolet radiation reaching the Earth's surface and a consequent threat of
cellular damage to living organisms (e.g., skin cancer in humans). In both these examples,
there is a suggestion that ecological damage on the global scale may result from changes
in natural processes caused by pollutants—namely, changes in the carbon cycle and the
oxygen–ozone cycle. This view of pollution in terms of global ecology is discussed by
Lovelock in his elaboration of the Gaia theory (see Section 14.4).

14.2 Changes in Soil Processes: The Functional Approach

Soil communities are complex associations between a variety of micro- and macro-organ-
isms, minerals, and nonliving organic materials (refer to Chapter 5). The carbon cycle and
the nitrogen cycle operate in soils, and both provide indices of the function of soil communities. Stages in the cycles—e.g., CO_2 production (carbon cycle) and nitrification (nitrogen
cycle)—can be measured in the presence and absence of pollutants.

The basic carbon cycle, shown in Figure 14.1, is an example of a nutrient cycle. Organisms
that obtain their carbon from organic compounds are termed heterotrophs. Animals and
many microorganisms fall into this category. Organisms that obtain their carbon from
carbon dioxide are termed autotrophs. Autotrophs include green plants, green algae, and
certain bacteria and play a vital role in fixing atmospheric carbon dioxide into organic
compounds. The heterotrophs then complete the cycle by converting organic compounds

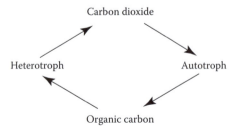

FIGURE 14.1
Carbon cycle.

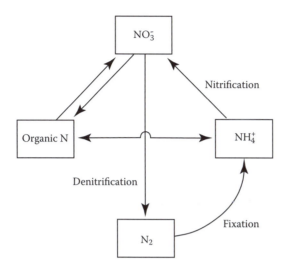

FIGURE 14.2
Nitrogen cycle.

to carbon dioxide. Thus the operation of the cycle may be affected by the action of pollutants upon autotrophs, heterotrophs, or both.

A simplified version of the nitrogen cycle in soil is shown in Figure 14.2. Atmospheric nitrogen can be fixed by certain bacteria, including both free living species (e.g., *Azotobacter* spp.) and symbiotic bacteria (e.g., *Rhizobium* spp., which are found in the root nodules of leguminous plants). Fixation involves conversion of molecular nitrogen to ammonia, which then forms ammonium ions (NH_4^+) in soil water. NH_4^+ can be oxidized sequentially to nitrite ions (NO_2^-) and nitrate ions (NO_3^-) by soil bacteria (nitrification). Plants and bacteria can take up NH_4^+ and/or NO_3^- and use them for the biosynthesis of organic nitrogen compounds (e.g., amino acids and purines). Heterotrophs can then utilize organic nitrogen compounds as sources of nitrogen and can release NH_4^+ and NO_3^- from them. Finally, some microorganisms can convert NO_3^- to nitric oxide (NO), nitrous oxide (N_2O), and nitrogen, a process termed denitrification.

Microorganisms have an important role in the operation of the carbon and nitrogen cycles. Under normal environmental conditions, these cycles are subject to the influence of environmental factors such as temperature, pH, and soil water content. The degree to which these environmental factors vary in time and space is dependent upon geographical location and the inherent properties of the soil (e.g., clay content, organic matter, and $CaCO_3$ content). In evaluating the effects that pollutants may have on soil processes, it is important to see them in relation to the fluctuations that occur in unpolluted sites. Effects of pollutants can be regarded as serious and significant if they go beyond the normal variations in operation of the cycles.

A widely used method of estimating hazards of chemicals to microorganisms involves determination of the rate of carbon dioxide formation in soil samples. Usually, a source of carbon (e.g., plant or horn meal) is added to soil to stimulate CO_2 formation, and a comparison is made between soils with and without the test chemical (Figure 14.3).

The rate of production of nitrite and nitrate from soil ammonium (nitrification) can also be readily measured (Table 14.1). This is regarded as a valuable method of testing the effects of pollutants because of the importance of nitrification in relation to soil fertility. A

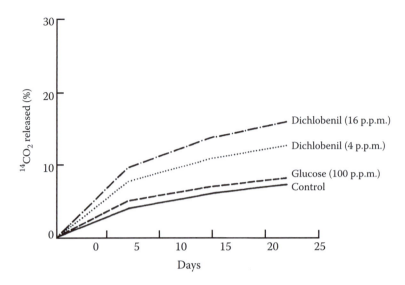

FIGURE 14.3

Effect of an herbicide (dichlobenil) on the rate of CO_2 production in soil. {^{14}C}-Glucose was added to soil as a carbon source for microorganisms. The addition of the herbicide dichlobenil caused an increase in the rate of release of $^{14}CO_2$ (derived from {^{14}C}-glucose) over a 22-day period. The rate of $^{14}CO_2$ release from soil is expressed as the percentage added ^{14}C, which appears in this form. From Somerville, L. and Greaves, M.P. (1987). *Pesticide Effects on Soil Microflora*. London, Taylor & Francis. With permission.

further example is measurement of the rate of nitrogen fixation by root nodule bacteria in a leguminous crop, which may be affected by soil pollutants.

Another factor to be considered is the duration of any effect produced by a chemical. Pollutants can change the composition of communities of soil microorganisms (Chapter 5). When a chemical has an immediate effect on soil processes as a result of chemical toxicity, there may be subsequent population growth in the species/strains of microorganisms that can metabolize it and use it as a nutrient source. Thus the effects of an organic chemical will be relatively short-lived, because the increase in numbers of these microorganisms will lead to a more rapid breakdown of the compound in soil. It has been suggested that effects of chemicals lasting up to 30 days be regarded as normal (1), up to 60 days as tolerable (2), and beyond 60 days as critical (3). In reviewing effects of pesticide on soil microflora, some 90% of all cases fell into the first category.

TABLE 14.1

Effects of Pollutants on Nitrification in Soil[a]

Treatment	mg NO_3^+ N/day	% of Control
Control	0.53	100
5 × Normal application rate of pesticide	0.54	102
1 ppm Nitrapyrin	0.14	26

Source: Somerville, L. and Greaves, M.P. (1987). *Pesticide Effects on Soil Microflora*. Taylor & Francis. With permission.

[a] Horn meal was added to soil as a nitrogen source. The rate of production of nitrate was unaffected by a high application of pesticide. It was, however, strongly inhibited by nitrapyrin, a bactericide used as a positive control.

A general problem when performing soil tests to evaluate the effects of chemicals is choosing the type of soil to use and the appropriate operating conditions. Attempts have been made to define standard soils, but the trouble here is that these are only representative of particular geographical and climatic areas. Also, a standard soil may not represent a worst-case scenario—it may not be the soil most likely to show the effect of a particular chemical.

Soil tests of the type described are of particular interest with regard to the testing of pesticides. Herbicides, fungicides, and insecticides are chemicals with high biological activity that may be expected to have effects on soil organisms and consequently upon soil fertility.

14.3 Changes in Compositions of Communities: The Structural Approach

Responses of communities to pollution in terrestrial and aquatic ecosystems have been covered briefly in Chapter 11 as part of the discussion on type 1 in situ biological monitoring (see Section 11.2 for an introduction to some of the principles involved). In fresh waters, methods such as the river invertebrate prediction and classification scheme (RIVPACS) are well established. In marine systems, studies such as those by Clarke and Warwick (1998) and Warwick and Clarke (1998) have shown how a variety of indices (in their case a "taxonomic distinctiveness index" utilizing nematodes) can be used to assess the health of sediments affected by pollution. The marine planktonic community is more difficult to monitor, but some success has been achieved by maintaining organisms within outdoor enclosures. Jak et al. (1998) were able to show that tributyltin (TBT) reduced grazing activities of zooplankton (principally copepods), resulting in an algal "bloom."

Before and after studies are rare in ecotoxicology because it is often impossible to predict where a pollution event is likely to occur. Furthermore, it is difficult to get long-term funding for basic ecological monitoring. Environmental impact assessment of, for example, the site of a new industrial development may have to rely on a brief survey rather than a more extensive data set gathered over several seasons. Many of the most successful pollution and recovery examples rely on data collected by natural historians that give a baseline with which the impact can be compared (e.g., the status of dog whelks before TBT, eggshells before DDT, lichens before air pollution).

Fortunately, enough such data existed to enable accurate assessment of the environmental impact of the Sea Empress, a tanker that spilled 70,000 tons of crude oil into Milford Haven, Wales, in February 1996 (Crump et al., 1999, 2003). Permanent quadrats had been established before the spill on rocky shores, which became heavily smothered with oil. These were not sprayed with detergents and were allowed to clean up naturally. Normally, percentage cover of the rocks with seaweed is low because of the grazing activities of limpets (mainly *Patella vulgata*). After the spill, limpet mortalities were very high and resulted in a dramatic green phase of the seaweeds *Enteromorpha* followed by a flush of *Porphyra* and ultimately a brown phase of *Fucus*. Limpets subsequently recolonized the rocks (probably recruited from small individuals that survived the pollution in deep crevices) and eventually restored the natural ecological balance by renewed grazing.

In a rare example of the experimental approach, the effects of an oral contraceptive on the aquatic communities that live in the water bodies into which municipal wastewater is discharged were studied by Kidd et al. (2007, 2010). After two years of baseline research, EE2 (ethinylestradiol), an estrogen used as an oral contraceptive, was added three times a week during the summers of 2001–2003 to a lake in the experimental lakes Area of Ontario in Canada, and concentrations in the surface waters reached 5–6 ng/L⁻¹. Nearby lakes containing similar communities were used as controls and all lakes were monitored subsequently. Exposure to estrogens was assessed by measuring the presence of a bio-marker, vitellogenin, an egg-yolk protein (see Section 10.4.5). Production of vitellogenin in the fishes increased up to 15,000 times. There were no effects on the abundances of plank-ton or benthic macroinvertebrates, but more than 40% of male fathead minnow developed intersex in 2001; their population bred poorly in 2002 and 2003, and the population became almost extinct. One third of male pearl dace developed early stage oocytes in 2001 and their population decreased. The top predator, lake trout, showed no gonadal abnormali-ties. Their population decreased but probably due to loss of their prey species. This whole ecosystem study shows that discharge of an estrogen into wastewater can damage fish populations if concentrations in the surface waters reach 5–6 ng/L⁻¹. For further informa-tion about EE2, see Section 1.2.15.

14.3.1 Changes in Soil Ecosystems

Examples similar to the above are much harder to come by with soil ecosystems. One of the reasons for this discrepancy is the lack of user-friendly identification keys for some of the most speciose groups (e.g., mites, enchytraeid worms). Nevertheless, progress is being made toward a soil prediction and classification scheme (SOILPACS), although it will be some years before this can be considered as routine as RIVPACS and the like. In the remainder of this section, examples will be given of a few of the methods that have been used to assess the impact of aerial deposition of metals on soil and leaf litter com-munities. It does not pretend to be comprehensive; further details can be found in the excellent reviews of Posthuma (1997), Van Straalen (1997, 1998, 2004), and Van Straalen and Kammenga (1998).

One of the characteristic features of terrestrial ecosystems contaminated by aerial deposition of metals is an accumulation of undecomposed leaf litter. This is attributed to a reduction in the populations of invertebrates that feed on dead plant material. Their feces provide a more favorable substrate for microbial decay than intact leaves. Oligochaetes are particularly susceptible to the pollution because the moist body sur-face allows metals dissolved in the soil pore water to diffuse into their body tissues. In coniferous forests, enchytraeids are usually the dominant macroinvertebrates. However, in the vicinity of a copper-nickel smelter in Finland, numbers of enchytraeids were sub-stantially lower than in more distant sites, and there was a much thicker layer of unde-composed pine needles on the soil surface (Haimi and Siira-Pietikainen, 1996).

In temperate grassland and deciduous woodland, lumbricid earthworms dominate the invertebrate decomposer community in terms of biomass and consumption of leaf litter. At Avonmouth, UK, deposition of aerial emissions of cadmium, copper, lead, and zinc from a smelting works have heavily contaminated soils in the vicinity of the fac-tory (for further details, see Hopkin, 1989). About 600 m from the plant (sites 1 and 2 in Figures 14.4 and 14.5; Table 14.2), dead grass lying on the soil surface contains more than 2% lead and nearly 4% zinc on a dry weight basis. Detailed sampling by Spurgeon and Hopkin (1999) during the spring, summer, autumn, and winter revealed a complete lack of earthworms from the two closest sites (< 0.6 km) and significantly reduced numbers compared with controls at a further five sites (< 3 km) (Figure 14.4). Species richness was lowest at sites near the factory (Table 14.2). Worms such as *Aporrectodea caliginosa* and *Allolobophora chlorotica* that were dominant at relatively clean sites farther from the

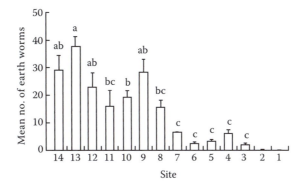

FIGURE 14.4

Mean abundance of earthworms collected from six 25 × 25 cm quadrants taken at 13 sites along a gradient of contamination in the Avonmouth area and a control (site 14) 100 km from the smelting works (error bars indicate SE values) in April 1996. Worms are absent from the most heavily contaminated sites (1 and 2) and are present in reduced numbers at sites with a medium level of contamination (sites 3–7) in comparison with relatively uncontaminated localities (sites 8–14). Sites sharing the same letter indicate no significant differences at P > 0.05 as given by Tukey's test for the multiple comparison of means. Reproduced from Spurgeon, D.J. and Hopkin, S.P. (1999). Seasonal variation in the abundance, biomass and biodiversity of earthworms in soils contaminated with metal emissions from a primary smelting works. *Journal of Applied Ecology* 36, 173–183. With permission from the British Ecological Society.

smelter were absent from the most contaminated soils. Multivariate cluster analysis (Figure 14.5) indicated that sites could be split into three groups based on relative species composition. Studies such as these provide strong circumstantial evidence that the accumulation of organic material (and the concomitant disruption of nutrient cycling) is related to the absence of key invertebrates that act as decomposition "catalysts."

According to Van Straalen (1997), a combination of ecophysiology (i.e., understanding the biology of the organisms involved) and multivariate statistics holds greatest promise for the future development of a SOILPACS technique. At Avonmouth, this approach has already been adopted in a study of ground-running invertebrates (Read et al., 1998). Principal components analysis and canonical correspondence analysis were used to

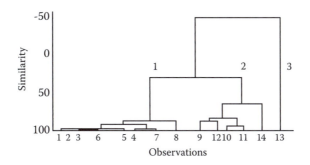

FIGURE 14.5

Dendrograms of earthworm communities at Avonmouth sampled in April 1996 from the sites described in Figure 14.4 ordered by cluster analysis using Euclidean distance and Ward's minimum variance method. Reproduced from Spurgeon, D.J. and Hopkin, S.P. (1999). Seasonal variation in the abundance, biomass and biodiversity of earthworms in soils contaminated with metal emissions from a primary smelting works. *Journal of Applied Ecology* 36, 173–183. With permission from the British Ecological Society.

TABLE 14.2

Shannon-Weiner Diversity Indices of Earthworm Communities at
Avonmouth Sampled during the Period from Spring 1996 to Winter 1997
from the Sites Described in Figure 14.4

Site No.	Spring	Summer	Autumn	Winter
14	1.83	1.55	1.37	1.56
13	1.1	1.46	1.22	1.76
12	1.65	0	1.38	1.69
11	1.46	1.3	1.56	1.68
10	1.45	0.9	1.13	1.56
9	1.66	1.39	1.41	1.53
8	1.95	1.36	1.44	1.49
7	0.31	0	0.67	0.93
6	1.03	1	0.83	0.24
5	0.80	0	0	0.87
4	0.88	0	0.68	0.85
3	0.64	0	0	0.64

Source: Reproduced from Spurgeon, D.J. and Hopkin, S.P. (1999). Seasonal variation in
the abundance, biomass and biodiversity of earthworms in soils contaminated
with metal emissions from a primary smelting works. *Journal of Applied Ecology*
36, 173–183. With permission of the British Ecological Society.

Note: Note that the diversity is invariably lower in sites closest to the smelting works
(sites 3–7) in all seasons than in more distant sites (sites 8–14).

show that the dominant factor accounting for differences in the species composition of
woodlands was the degree of contamination with metals.

14.3.2 Acidification of Lakes and Rivers

An example of changes in a community caused by pollution is given by the consequences
of the acidification of certain lakes as a result of acid rain. The pH of surface waters is
determined by both abiotic and biotic factors. Such factors as the composition of the base
rock and the precipitation of acids in rain or snow have a strong influence upon pH. Also
important are the release of acids when organic residues are decomposed by microorganisms and the influence of neighboring forest land. pH, in turn, influences the composition
of aquatic communities. In particular, reduction of pH below 6 can be harmful to many
species. Thus the pH of surface waters is affected both directly and indirectly by pollutants
(pollutants may affect pH of surface waters indirectly through action upon microorganisms or plants). In summary, the pH of surface waters is dependent upon the operation
of natural processes as well as pollution, and provides an indication of the diversity and
health of aquatic ecosystems (Barth, 1987).

The deposition of acid rain has resulted in the acidification of weakly buffered surface waters in many areas, including Scandinavia, eastern Canada, and the northeastern
United States. For example, the pH of 21 water bodies in central Norway decreased from
an average of 7.5 to 5.4–6.3 between 1941 and the early 1970s; the pH of 14 surface waters
in southwestern Sweden decreased from 6.5–6.6 before 1950 to 5.4–5.6 in 1971; the average
pH of seven rivers in Nova Scotia decreased from 5.7 in 1954–1955 to 4.9 in 1973. Although
doubts have been expressed about the accuracy of older pH data, there is a broad scientific

consensus that there has been recent acidification of many surface waters. The material in this section is based in part on Freedman (1995).

The effects of acidification on aquatic communities are sometimes dramatic, as in the loss of fish populations from a number of highly acidified waters. For instance, commercially valuable salmonids have been lost from many surface waters in Scandinavia (example in Figure 14.6). A survey of more than 2,000 lakes in southern Norway showed that one third have lost their fish populations since the 1940s. The fishless lakes generally had a pH < 5.0. The local salmonids vary in their sensitivity to pH; requirements range from pH > 4.5 for nonmigratory brown trout (*Salmo trutta*) to 5.0–5.5 for Atlantic salmon (*Salmo salar*). Juveniles are generally more sensitive than adults, but adult deaths are sometimes recorded in the spring when water pH is lowest. Similar results involving more species have been recorded in many studies in Canada and the northeastern United States.

Field experiments are possible with lakes, just as they are with soil communities, but the costs are greater. The best example is the experimental acidification of lake 223, a 27-ha oligotrophic lake in the experimental lakes area of northwestern Ontario. The lake was extensively studied in 1974 and 1975 to provide baseline data, and then the lake was progressively acidified so that pH fell from 6.5–6.8 to 5.0–5.1 by 1981, after which pH was maintained in the range 5.0–5.1 until 1983. This was achieved by adding sulfuric acid; 27,400 l had been added by 1983. Experimental acidification has subsequently been carried out at two other lakes, one in the experimental lakes area, the other in northern Wisconsin. These lakes have different physical settings and originally had different assemblages of plants and animals, including fish.

Despite their initial differences, the responses of these three lakes to acidification were in many respects remarkably similar, particularly with regard to biogeochemical processes and effects on the lower trophic levels (Schindler et al., 1991; Schindler, 1996). Although acidification disrupted nitrogen cycling in all three lakes, each generated some buffering capacity internally.

The phytoplankton communities of all three lakes had originally been dominated by chrysophyceans and cryptophyceans. Acidification changed the dominant species and decreased diversity. Phytoplankton production and standing crop were somewhat increased, probably because light penetration increased. The littoral zones of the lakes, however, became dominated by a few species of filamentous green algae. These formed mats or clouds of algae that changed the entire character of the littoral zones when the pH fell below 5.6.

Acidification also changed the zooplankton communities, with cladocerans becoming increasingly dominant. As acidity increased, *Daphnia catawba* took over from other *Daphnia* species; the identity of the dominant rotifer species changed, and several sensitive zooplankton species disappeared.

Acidification also caused the loss of several species of large benthic crustaceans. The responses of fish species, however, were varied and appeared to depend on the sensitivity of key organisms lower down the food chains.

When acidification was reversed in lake 223 between 1984 and 1989, there was rapid partial recovery of the lacustrine communities, although the assemblages of phytoplankton and chironomids retained an acidophilic character.

Overall, acidification consistently reduced diversity, through loss of species and through the increased dominance of a small number of acidophilic taxa. These responses are similar to those found in atmospherically acidified lakes.

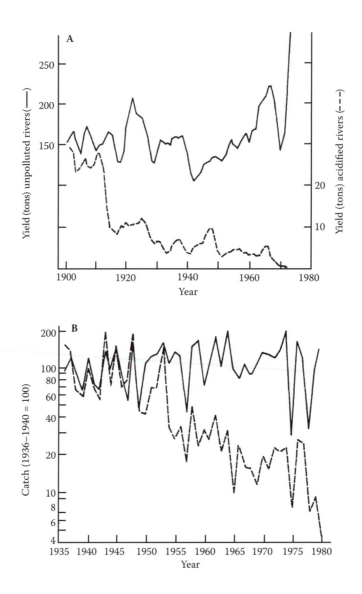

FIGURE 14.6

The declining catch of Atlantic salmon in acidified rivers (– – –) compared with less polluted rivers (——). (A) In southernmost Norway: seven acidified rivers compared with the rest of the country. Modified from Leivestad, H., Hendry, G., Muniz, I.P., and Snekvik, E. (1976). Effects of acid precipitation on freshwater organisms. In Brakke, F.H., Ed., *Impact of Acid Precipitation on Forest and Freshwater Ecosystems in Norway*, pp. 87–111. Research Report FR 6/76. Oslo, SNSF Project. (B) In Nova Scotia: 12 rivers with pH > 5.0 compared with 10 with pH ≤ 5.0 in 1980. Reproduced from Watt, W.D., Scott, C.D., and White, W.J. (1983). Evidence of acidification of some Novia Scotia rivers and its impact on Atlantic salmon, *Salmo salar. Canadian Journal of Fisheries and Aquatic Sciences* 40, 462–473. With permission from the National Research Council of Canada.

14.3.3 Mesocosms

As discussed in Section 6.6, "mesocosms" are sometimes used for testing the effects of environmental chemicals (see Ramade, 1992; Crossland, 1994; Caquet et al., 2000). These systems can be replicated, thus allowing the inclusion of reliable controls in the experimental design. Changes in the composition of communities or ecosystems in response to treatment with chemicals can be assessed with the aid of advanced statistical techniques. Mesocosms are usually aquatic, and examples include experimental ponds, simulated streams, and enclosures within ponds, lakes, estuaries, or the marine environment; enclosures may take the form of limnocorrals, pelagic bags, or littoral enclosures (Caquet et al., 2000). With the controlled conditions of most aquatic mesocosms, it is possible to study the question of bioavailability: how this can be reduced by adsorption to suspended solids or sediments, or by increased bioturbation or biomagnification.

The experimental pond is perhaps the most widely used type of mesocosm. In one study, eight steel-reinforced concrete experimental ponds were used to assess the environmental impact of the pyrethroids cypermethrin and lambda cyhalothrin (Kedwards et al., 1999). The ponds were stocked with natural hydrosoil and seeded with aquatic organisms other than fish collected from natural sources. Prior to application, a baseline year study was carried out to document the development of populations. When designing the experiment, the mesocosms were first split into two blocks of four to allow for a known environmental gradient in ambient light and temperature. Then, within each block, four treatments were assigned at random. Thus a stratified randomized design was achieved, capable of properly accounting for the effects of the known environmental gradient.

BOX 14.1 STATISTICAL METHODS IN ECOTOXICOLOGY

Up to now, standard methods of statistical analysis have been adequate for most of the studies described in this book. Thus ecotoxicologists want to know the effects of chemicals on some specified aspect of an organism—for instance, body weight. They design replicated experiments exposing organisms to each of a number of concentrations of the chemical. They weigh the organisms and use statistical analyses such as ANOVA to see whether the body weights vary more between concentrations than between replicates. The response variable is body weight, and the concentrations are entered into the analysis as factors. This method works well with parts of animals, aspects of whole animals, and populations, provided we abide by the basic principles of experimental design and include independent replicates of each treatment. In practice it is hardest at the population level, because the replicates have to be populations, and the experiments become expensive unless the organisms are small.

When moving beyond the population level to consider communities, a new set of problems is encountered. Communities are complex and contain within them many different types of interacting organisms such as predators, parasites, and competitors. As a result of this complexity, communities are difficult to characterize precisely. Simple characterizations of communities are possible in terms of total community biomass and in terms of diversity, measured as number of species or using the Shannon–Weiner diversity index. In these cases, ANOVAs like those described above can be used to analyze the effects of chemicals on these specified aspects of the community, if the variances are normally distributed. Although these analyses are informative, they are clearly simplistic. To go beyond them, however, requires a means of characterizing communities. So far, community ecologists have not found this easy.

In some ways, ecotoxicologists are well placed to advance the science of community ecology, because they address straightforward questions (e.g., what are the effects of chemicals?) and they are able to conduct well-designed experiments including replicate communities (e.g., in mesocosms). Some progress is being made by applying multivariate statistics. The basic idea is that many variables will be measured—for instance, the abundance of each species present. Advanced statistical methods are then employed to construct indicator variables that represent the major responses of the communities to the chemical stressors. Methods deployed include MANOVA, principal components analysis, and principal response curves. An introduction to current methods is provided by Maund et al. (1999).

The four treatments were a control, the application of cypermethrin, and the application of lamda cyhalothrin at two different rates. Organisms were sampled at appropriate intervals before and after treatment with the insecticides, which were formulated as commercial emulsifiable concentrates. The response of the macroinvertebrate community to the insecticides was evaluated using the multivariate statistical techniques redundancy analysis (RDA) and principal response curve (PRC) analysis. A significant change from the control community structure was observed after treatments with both cypermethrin (Figure 14.7) and the highest level of lamda cyhalothrin. This was due principally to a decline in both Gammaridae and Asellidae. In the case of cypermethrin treatment, the Asellidae population appeared to be recovering by the end of the experiment.

A wide-ranging study of the environmental impact of pyrethroids used to control pests of cotton involved the use of three different pond systems located in Great Britain and the United States in seven separate experiments (Giddings et al., 2001). The detailed description of the different experimental designs used in this work lies beyond the scope of the present text but can be found in works listed at the end of this chapter. Results from the mesocosm studies were compared with those from related field studies as part of a risk

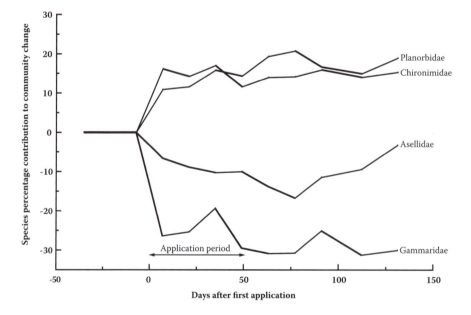

FIGURE 14.7
Change in the community structure of an aquatic mesocosm following application of cypermethrin at a rate of 0.7 g a.i./ha. Figure supplied by T. Kedwards. With permission.

assessment exercise that also utilized toxicity data for the insecticides (Solomon et al., 2001). The different taxa showed the following range of sensitivities to cypermethrin and esfenvalerate, measured in terms of abundance: amphipods, isopods, midges, mayflies, copepods, and cladocerans (most sensitive) ranging to fish, snails, oligochaetes, and rotifers (least sensitive). Values for lowest-observed adverse effect concentrations for cypermethrin and esfenvalerate were derived from this investigation.

Outdoor stream mesocosms have also been used for the purposes of risk assessment. In one study, an anionic alkylsulfate surfactant (AES) was tested in experimental stream mesocosms located in Lafayette County, Mississippi (Lizotte et al., 2002). Twelve experimental stream mesocosms were used. Each consisted of a 180-liter mixing chamber, a riffle section, and an 870-liter pool section. Five duplicate nominal exposure concentrations were randomly assigned to the stream mesocosms; two of them were kept as controls. Samples for analysis of invertebrates were taken before treatment and then at 7, 14, 21, and 30 days after treatment. A level of AES of 10.18 mg/L significantly decreased population densities of members of the following invertebrate groups: Annelida, Amphipoda, Copepoda, Trichoptera, Cladocera, and Diptera. Multivariate cluster analysis and nonmetric multidimensional scaling ordination confirmed distinct structural effects of AES on the invertebrate communities at this concentration in comparison to the lower levels of treatment (0.7, 1.27, 2.2, and 4.31 mg/L) and the control. A no-observable-effect concentration (NOEC) of > 2.0 mg/L for this ecosystem was deduced from the study.

A somewhat different approach to mesocosms involves the establishment of enclosures within a relatively large area of surface water such as a lake, estuary, or offshore marine area. One such study was conducted in Lake Rotomanuka, New Zealand, to assess the seasonal variations in response of plankton communities to pentachlorophenol (Willis et al., 2004). These mesocosms took the form of 10 immersed polyethylene bags of ca. 860-liter capacity, covered to exclude rainwater and fouling by birds. In this case, seven unreplicated treatments were used with three controls in winter and spring, but only two controls in summer. Samples were taken for plankton analysis just before treatment, and then 4, 10, 24, 36, and 54 days after treatment. Experiments were conducted in winter, spring, summer, and autumn, and results were analyzed using the multivariate statistical techniques of PRC analysis and principal components analysis (PCA). The copepod *Calamoecia lucasi* dominated the zooplankton community and was most affected by pentachlorophenol, showing a reduction in numbers at the relatively low concentration of 24 micrograms/L. Overall, the response of the plankton was complex and dependent upon season. For this plankton community, an NOEC value of 24–36 micrograms/L was estimated.

Mesocosms are attractive as experimental systems, but the critical question is how comparable they are to the real world. The interpretation of the data that they generate and the extrapolation of the results from them to the living environment is a continuing problem. There is an ongoing debate about their value in the process of risk assessment.

14.4 Global Processes

The atmosphere enshrouds the whole planet and provides a link between oceans and continents. Changes in its composition and character can have far-reaching effects upon ecosystems across the globe. The air masses that constitute the lower atmosphere are in a constant state of circulation, individual gaseous components moving by diffusion processes through the air masses (Chapter 3). The composition of the atmosphere is dependent not only on contamination by anthropogenic pollutants but also on the operation of living processes at the Earth's surface. Levels of carbon dioxide, oxygen, and nitrogen are dependent upon living processes. Thus the effects of pollutants on ecosystems can, at least in theory, influence the operation of these processes globally and alter the composition of the atmosphere. Some scientists have a conception that the entire Earth is a single living entity (as, for instance, in the Gaia theory of Lovelock).

Looking at things in this global way, the effects of pollutants on living organisms may be direct or indirect. Inhibitors of photosynthesis, such as urea and triazine herbicides, are examples of pollutants that act directly on plants. By contrast, CFC gases can affect plants indirectly by their action on the ozone layer. A reduction of the ozone layer will lead to more ultraviolet radiation reaching the Earth, causing damage to both plants and animals. In turn, damage caused to plants, whether direct or indirect, can affect the rate of photosynthesis globally, which means a change in the rates of assimilation of carbon dioxide and the release of oxygen—events that can, in due course, alter the composition of the atmosphere. This illustrates the close reciprocal relationship between the functioning of ecosystems and the state of the global environment.

Over the last 40+ years, the level of CO_2 in the atmosphere has increased by ca. 10%, and there is continuing debate about the causes and the environmental effects of this increase. It is thought that most of the increase is due to the combustion of fossil fuels (coal, oil, etc.). It has also been argued that a substantial reduction in the area of forest land has contributed to it, because trees are responsible for locking up considerable quantities of the gas. The concern over the rise of atmospheric CO_2 is that it is causing a greenhouse effect, retaining more heat at the Earth's surface and thereby causing global warming. A very serious long-term consequence of global warming would be melting of the polar ice caps and the rising of the sea level. To complicate matters further, there is now evidence of global dimming, also caused by pollutants. Particulate aerial pollutants arising, for example, from the combustion of fuels, interact with water droplets. The cloud formations resulting from this type of interaction can strongly reflect solar radiation, with a consequent reduction in the solar energy reaching the Earth. It is now argued that global dimming caused by aerial pollution has to some extent counteracted global warming due to greenhouse gases. So ongoing measures to reduce air pollution caused by burning certain fuels may actually accelerate global warming. However, the main point is that these discoveries strengthen the case for reducing carbon dioxide emissions globally, an issue touched upon in Section 2.4.

The issues raised here are long-term ones that are unlikely to go away. The effects of pollutants on the global scale can influence the state of ecosystems over the entire planet and are likely to be of increasing concern throughout the twenty-first century.

14.5 Summary

Earlier chapters of this book have considered the biochemical, physiological, and behavioral effects of pollutants on individual organisms. In the Introduction, we defined ecotoxicology as the study of harmful effects of chemicals on ecosystems. Chapter 14 examines the most important aspect of ecotoxicology, the damaging effects of pollution on communities of organisms, and wider ecological processes. Absence of sensitive species may not just result in lower overall biodiversity but also in disruption of essential ecological processes. For example, the reduction in populations of earthworms in soils contaminated with metals may lead to an accumulation of undecomposed leaf litter. This would normally be converted rapidly to feces for more rapid microbial decay. Nutrients are locked up in the dead leaves that would otherwise be returned to the soil for new plant growth. In the aquatic environment, acidification of lakes results in lower overall diversity, although there are a few species that can take advantage of the reduced competition to reach higher population densities than in unacidified waters. One of the themes that runs throughout such studies is the need for long-term monitoring of natural uncontaminated ecosystems, and a good understanding of the biology of organisms, if we are to be able to recognize when ecosystems depart from their normal state.

Mesocosms are enclosed systems, usually aquatic ones, that simulate communities in the natural environment. They are more under the control of experimenters than natural communities can be and are carefully designed to permit the inclusion of proper controls and the use of advanced multivariate statistical techniques to aid the interpretation of results. There is an ongoing debate about their potential value in environmental risk assessment.

In the final analysis, communities and ecosystems need to be seen within the wider context of pollution of the global environment. Issues such as global warming caused by greenhouse gases, and the reduction of the ozone layer by the action of CFCs, are of long-term importance in ecotoxicology.

Further Reading

Caquet, T. et al. (2000). A review of aquatic ecosystems—the principles involved and their design and operation.

Freedman, B. (1995). *Environmental Ecology: The Environmental Effects of Pollution Disturbance and Other Stresses*, 2nd ed. Excellent coverage of the effects of pollution on ecosystems. In particular, it is useful for the effects of acidification on aquatic ecosystems.

Lovelock, J. (1989). *The Ages of Gaia: A Biography of Our Living Planet*. An exposition of the Gaia hypothesis.

Maund, S.J. et al. (1999). An editorial article discussing the use of multivariate statistics in ecotoxicological field studies.

Newman, M.C. (1998). *Fundamentals of Ecotoxicology*. Useful text on ecological effects of pollutants.

Pankhurst, C.E. et al., eds. (1997). *Biological Indicators of Soil Health*, and Van Straalen, N. and Løkke, H., eds. (1997). *Ecological Risk Assessment of Contaminants in Soil*. These include several useful chapters concerning effects of pollutants on soil communities.

Schindler, D.W. et al. (1991): *The Environmental Effects of Pollution Disturbance and Other Stresses*, 2nd ed. A review of the experimental acidification of lakes.

Solomon, K.R. et al. (2001) and Giddings, J.M. et al. (2001): A series of five papers bearing the general title *Probabilistic Risk Assessment of Cotton Pyrethroids*. Taken together they present an interesting account of the employment of different techniques to carry out risk assessments at the level of community/ecosystem.

Wood, M. (1995). *Environmental Soil Biology*, 2nd ed. Gives an account of the soil processes that may be affected by pollutants.

15

Extrapolating from Molecular Interactions to Consequent Effects at Population Level

15.1 Introduction

As explained in the Introduction, a major objective of this book has been to argue from cause to effect. Thus it has been shown how certain molecular interactions between pollutants and their sites of action in living organisms can lead to sequential changes starting at the biochemical and physiological levels, progressing to effects on the whole organism, and eventually to effects at the levels of population, community, and ecosystem. Of course, many effects at the biochemical level are not translated into consequent changes at higher levels. A fundamental problem in ecotoxicology is to distinguish between situations in which such molecular interactions cause harmful effects at the population level from situations that do not.

The text has been structured to underline this approach. The first part covered pollutants, their properties, and their fates in the environment—biotic and abiotic. The processes described here determine how much of a pollutant (or its metabolite) reaches sites of action in living organisms. This is a critical factor in determining environmental risk. The second part of the text deals with the effects of pollutants on individual organisms at biochemical, physiological and whole organism levels, and measuring the effects. An important issue has been the development of biomarker assays in the laboratory that may be used to measure harmful effects in the field. Now, in the third and final part of the book, attention turns to effects at the levels of population, community, and ecosystem.

The next two chapters will address the question of how we can establish relationships between events at the molecular level (described in Parts 1 and 2) and consequent changes at higher levels of biological organization. The importance of doing this can hardly be overstated. It is one thing to establish a relationship between the level of a chemical in the environment and a population decline—it is another to prove that the chemical has actually caused the population to decline. Proof is required of causality. How can this be provided?

In Chapters 12 and 14 we suggest that the evidence may come from experiments of appropriate design. A chemical is applied to some randomly chosen populations and not to others that serve as controls. Statistical tests then establish whether the treated populations declined relative to the controls. This approach is often difficult, time-consuming, and expensive, and not feasible for many risk assessments. This is where the biomarker approach holds great potential.

The biomarker approach depends on the availability of biomarkers that provide measures of toxic effect, as described in Chapter 10. Many such biomarkers actually measure the operations of toxic mechanisms and will be called mechanistic biomarkers for the purposes of the present discussion.

How can these biomarkers provide evidence of causality? The answer is developed in this chapter. To look ahead briefly, the method depends on a comparison of a dose-related biomarker response with a dose-related population decline in both laboratory and field. Under controlled laboratory conditions, a population of a test species can be exposed to a range of doses of a pollutant. For this population, changes in numbers and biomarker responses can be measured in relation to the dose received and the resulting data can be used to construct (i) a dose–response curve relating population numbers to the dose given; and (ii) a second dose–response curve recording the magnitude of one or more biomarker responses in relation to dose over the same dose range.

These two types of dose–response curves can then be compared with similar dose response curves for the same species based on field data. In this way, field biomarker measurements can demonstrate that a chemical in the environment actually operates toxic mechanisms in free-living organisms and that the operation of a toxic mechanism is related to a population decline. Such an approach provides powerful evidence of causality.

As will be discussed in Section 15.4 , this approach may make an important contribution to risk assessments of environmental chemicals. The number of environmental chemicals is dauntingly large, but the number of toxic mechanisms relevant to ecotoxicology appears to be relatively small (see Section 15.2). Testing large numbers of chemicals according to a statutory protocol is extremely expensive. In the longer term, the development of effective biomarker strategies could be very cost effective, focusing the science on the operation of critical mechanisms rather than automatically applying a range of statutory toxicity tests to a large number of compounds. Testing protocols may be carefully selected according to the nature and properties of the compound, and maximum use could be made of comparisons with existing compounds of the same type, where ecotoxicological data already exist by using QSAR techniques.

This chapter gives an overview of these questions and serves as a forerunner to the final chapter that will present examples of the use of biomarkers in population studies.

15.2 Translation of Toxic Effects across Organizational Boundaries

In concept, toxicity begins when a toxic molecule interacts with a site of action in a living organism (Figure 15.1). This interaction typically leads to a localized cellular response that in turn may cause physiological, biochemical, or behavioral changes at the whole organism level. The effects at the whole organism level may manifest as characteristic symptoms of poisoning, for example, neurophysiological effects with associated disturbances of behavior, disruptions of endocrine function, or changes in blood pressure. The following account will consider the relationship between the primary interaction of a chemical with its site of action and consequent localized effects before discussing subsequent effects on the whole organism. After that, effects occurring later at the levels of population and above will receive attention.

15.2.1 From Effects at Site of Action to Localized Cellular Disturbances

Chapter 7 reviewed biochemical effects of pollutants; Chapter 8 covered physiological effect. In most cases, the primary toxic interaction of a chemical occurs at the biochemical level, frequently with a macromolecule. Thus organophosphates bind to the acetylcholinesterase

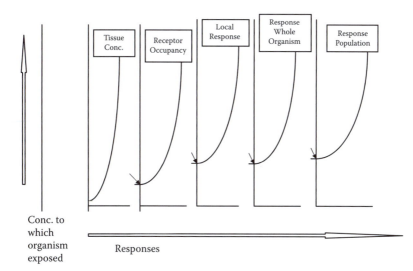

FIGURE 15.1
Responses to pollutants at different levels of biological organization. The threshold tissue concentrations of biological responses for which no response is measurable are indicated by arrows. These thresholds tend to increase with movement toward higher levels of biological organization.

of postsynaptic membranes; pyrethroids bind to sodium channels of nerve membranes; warfarin and other anticoagulant rodenticides compete for vitamin K binding sites; certain endocrine disruptors bind to the estrogen receptor; and active metabolites of benzo(a) pyrene bind to DNA. Many other examples could be cited. Occasionally the primary mode is better described as physiological, for example, where physical poisons such as ethers or esters change the properties of a biological membrane when they dissolve in it.

A vital question is the quantitative relationship of the initial interaction of a chemical at its site of action (usually biochemical) and consequent localized cellular disturbances (often physiological). There is a degree of compensation in the response to the initial interactions of toxic molecules (Figure 15.1). Thus knock-on effects at the cellular level are unlikely to be shown if a chemical binds to only a small percentage of the target sites. It is important to understand (i) how high the concentration of the chemical must be at the site of action before consequent effects occur at the next level and (ii) after effects start to be expressed, the relationship between the concentration of chemical that is present and the degree of adverse change at the cellular level.

Such relationships are represented by dose–response curves. In fact, two sequential processes are implicated in this scenario: (i) the binding of the chemical to its site of action and (ii) the subsequent response. Considering binding first, individual toxic chemicals sharing the same mode of action differ widely in their affinities for target sites. Where affinity is strong, binding to most or all the available sites may occur at relatively low concentrations. Conversely, where affinity is low, high concentrations cause only limited binding.

However, strength of binding does not simply relate to strength of response. In pharmacological terms, a molecule that elicits a normal physiological response when it interacts with a receptor (site of action for a natural transmitter or drug) is called an agonist (Main, 1980). Other molecules may bind just as strongly to the same receptor and elicit only a partial response (partial agonists) or no response. In the latter case, where they oppose a typical pharmacological response to a normal chemical messenger, they are known as antagonists.

Anticoagulant rodenticides such as warfarin function as antagonists at vitamin K binding sites located in a carboxylase enzyme complex of the hepatic endoplasmic reticulum (Thijssen, 1995). Similarly, some hydroxy metabolites of planar PCBs act as thyroxine antagonists at binding sites on the complex transthyretin protein, thereby disrupting the transport of thyroxine and retinol (Figure 7.6; Brouwer et al., 1990).

Not surprisingly, based on the complications noted above, the relationship between the concentration of a chemical at its site of action and the consequent cellular response is not simple. However, similarities in toxic effect may be found among members of homologous series of compounds sharing the same mode of action. Thus a detailed study of the toxic effects of one member of a homologous series may yield data that can be extrapolated to other members of the same group through QSAR modeling (Bawden et al., 1991).

Four examples demonstrate the relationships of levels of pollutants to consequent toxic effects at the cellular level. These same examples will be followed across the different organizational boundaries described in Section 15.1 to illustrate how interactions at the molecular level may eventually be translated into adverse effects at the population level and above (Figure 15.2).

15.2.1.1 Example A: Action of Organophosphates on Acetylcholinesterase of Nervous System

A very important site of action in ecotoxicology is the enzyme acetylcholinesterase (AChE) on the postsynaptic membranes of cholinergic synapses of both the central and peripheral nervous systems of animals (Ballantyne and Marrs, 1992). This is the site of action for organophosphorous and carbamate insecticides and chemical warfare agents such as sarin and soman (also organophosphorous compounds). Active oxon forms of these insecticides and chemical warfare agents bind very strongly to AChE, thereby causing strong and lasting inhibition (Figure 7.3). Indeed, if aging of the phosphorylated enzyme occurs, reactivation becomes impossible.

Many in vitro and in vivo studies have shown the relationship between the concentration of an organophosphate and the percentage inhibition of AChE. The percentage inhibition in this case provides a measure of the proportion of active sites (serine residues) that are bound to organophosphate moieties. When the AChE studied exists largely in one form, simple first-order inhibition plots are found when plotting log% inhibition against time (see Main, 1980B). If more than one form is present, a more complex response is to be expected.

Considering now the events in the vicinities of nerve synapses of living animals continuously exposed to an organophosphate, inhibition of synaptic AChE will progressively increase until it reaches a point at which the rate of hydrolysis of acetylcholine begins to fall. This will lead to increased concentration and longer persistence of the transmitter molecule in the vicinity of the acetylcholine receptor following its release from nerve endings. This will be followed in the short term by a prolongation of transmission of action potential from the postsynaptic membrane and along the nerve. If strong inhibition of the enzyme persists, depolarization of the postsynaptic membrane and associated synaptic block (loss of transmission across the synapse) will follow.

15.2.1.2 Example B: Action of p,p′-DDT on Voltage-Dependent Sodium Channels of Axonal Membranes

Both *p,p′*-DDT and pyrethroid insecticides such as permethrin and cypermethrin bind reversibly to voltage-dependent sodium channels of nerve axons (Eldefrawi and Eldefrawi,

EXAMPLE	A	B	C	D
SITE OF ACTION AND POLLUTANT	ACh-ase + OP	Na CHANNEL + +DDT	TARGET IN EGGSHELL GLAND MEMBRANE + DDE	OESTROGEN RECEPTORS /VARIOUS TISSUES + EE2
LOCALIZED TOXIC EFFECTS	Build up of acetylcholine in synaptic cleft.	Prolongation of sodium current across axonal membrane following stimulation.	Retardation of Ca++ transport across the eggshell gland membrane.	In liver, vitellogenin synthesis. At other sites, responses leading to female secondary sexual characteristics.
TOXIC EFFECTS AT THE WHOLE ORGANISM LEVEL	Various neurophysiological and behavioral disturbances which, if severe enough, can lead to death.	Prolonged neurophysiological and behavioral disturbances including tremors and convulsions.	Production of eggs with thin eggshells. When thinning exceeds 17-18% eggs begin to break with consequent breeding failure.	Oestrogenic effects lead to male fish developing female sexual characteristics [intersex] and consequent sterility.
EFFECTS AT THE POPULATION LEVEL	Increased mortality due to lethal toxicity. Sublethal effects may also be important.	Increased mortality due to lethal toxicity. Prolonged sub lethal neurotoxic effects may also be important.	Reduced birth rate can lead to reduced breeding success and consequent population decline in some predatory birds.	Development of sterility in male fish may lead to population decline.

FIGURE 15.2
The translation of toxic effects across organizational boundaries after exposure to pollutants. Examples A through D are described in Section 15.2 of the text. The use of biomarker assays to monitor the operation of these pathways is discussed in Section 15.2.1. Examples A and C are discussed further in Chapter 16, where they are used to illustrate the use of biomarker strategies in field studies.

1990). When the sodium channels function normally, they are involved in the formation and propagation of action potentials. They respond to changes in membrane potential with increases in permeability to sodium ions for a few milliseconds.

After stimulation of a nerve, a Na+ current flows briefly through these sodium channels and is quickly terminated when permeability is lost from closure of the sodium channels. If *p,p´*-DDT or pyrethroids are bound to the channels, their closure is delayed, and the Na+

current flows too long. There follows a prolongation of the action potential that may lead to repetitive discharges from a single stimulation of the nerve—a characteristic of DDT poisoning. The p,p′-DDT disturbs the transmission of messages along nerve fibers.

15.2.1.3 Example C: Action of p,p′-DDE on Transport of Calcium into Eggshell Glands of Birds

As noted earlier, *p,p′*-DDE is a persistent and stable metabolite of *p,p′*-DDT, the principal insecticidal component of commercial DDT. At very low concentrations (greater than 4 ppm by weight in the diets of some species of predatory birds) it can cause thinning of eggshells.

This pollutant can restrict the transport of Ca^{++} into the eggshell glands of birds (Lundholm, 1987, 1997), thereby retarding the calcification of developing eggshells. Controversy still surrounds the exact molecular site of action of *p,p′*-DDE (Lundholm, 1997). Calcium ATPase is involved in the active transport of Ca^{++} ions across the walls of shell glands and is known to be sensitive to the action of *p,p′*-DDE. Strong evidence indicates that eggshell thinning may be caused by inhibition of membrane-bound calcium ATPase by the DDT metabolite. However, other studies suggest that eggshell thinning may also be an effect of *p,p′*-DDE on prostaglandin metabolism. Whatever mechanisms are involved, the critical effect is a failure of Ca^{++} transport into eggshell glands after exposure to *p,p′*-DDE.

15.2.1.4 Example D: Action of 17A-Ethinylestradiol (EE2) on Estrogenic Receptors of Fish

The following account concerns the effects of the potent synthetic estrogen 17A-ethinylestradiol (EE2) that is widely used in oral contraceptives. EE2 has been found at low concentrations in some sewage effluents (e.g., 0.2 to 7 ng/l in samples from a study in the UK; Desbrow et al., 1998; Routledge et al., 1998).

Although some species of teleost fish are functionally hermaphroditic, most species have separated sexes (Tyler et al., 1998). EE2 has been shown to cause endocrine disruption in male fish of a number of species including the rainbow trout (*Onchorhynchus mykiss*) and roach (*Rutilus rutilus*). In female fish, the natural estradiol-17B and estrone estrogens are secreted by follicle cells surrounding the oocytes of the ovary (Tyler et al., 1998). After secretion, they are transported to other body tissues where they may interact with estrogen receptors located there.

Three different types of receptors are known to be present in most fish, ER-α, ER-β, and ER-βII (sometimes called ER-γ). Interactions of estrogens with these receptors at different locations in the body are responsible for the development of feminine secondary sexual characteristics, the operation of female reproductive cycles, and the maintenance of fertility. In most fish, the interactions of estrogens with receptors in the liver leads to the production of the vitellogenin protein. This is a general characteristic of oviparous (egg-laying) vertebrates; vitellogenin is the precursor of the major constituents of egg yolk. In some fish, e.g., zebra fish (*Danio rerio*), this interaction also occurs in the ovaries.

The exposure of male rainbow trout to concentrations of EE2 comparable to those reported in sewage effluent produced rapid synthesis of vitellogenin and retardation of the growth and development of testes in maturing males (Tyler et al., 1998). These effects are attributed to action of the synthetic estrogen on receptors in the liver and testes, respectively. Note that similar effects may be produced by other natural or synthetic estrogens (Section 7.4.7) and that effects are likely to be approximately additive when fish are exposed to mixtures of estrogens (Section 9.2).

Exposure of zebra fish to EE2 during early life stages (encompassing the period of sexual differentiation) results in retarded differentiation of testes, associated with elevated expression of genes cyp.19b and sox 9 (Ball et al., 2004). Similarly, in roaches, exposure to EE2 during early life results in retarded testicular development, enhanced expression of cyp.19a mRNA in the gonads, and enhanced expression of cyp.19b mRNA in the brain (Lange et al., 2001).

15.2.2 From Cellular Disturbances to Effects at Whole Organism Level

Localized effects can be transmitted around the organism and exert knock-on effects elsewhere in a number of ways. Effects arising in neurons can quickly cause wider disturbances of the nervous system and the disturbances are likely to occur rapidly if the initial disruption is in the central nervous system. Disturbances of the nervous system can affect the functions of organs and tissues and produce typical symptoms of poisoning such as muscular twitches, spasms, and convulsions. Persistent effects on the nervous system may cause prolonged behavioral disturbances, as shown in studies of the organochlorine insecticide dieldrin in experimental animals and occupationally exposed humans (Jaeger, 1970; Walker, 2001).

Blood can also play a part in disseminating toxic effects around the body. This may occur when metabolic products from the local action of a toxic molecule move into the blood. One example is the release of precursors of clotting proteins from liver into blood after poisoning by warfarin and other anticoagulant rodenticides (Section 7.4.4). Over time, the precursors replace the functional clotting protein and the blood loses its capacity to coagulate throughout the entire organism. A further example is the disturbance of steroid metabolism by inhibitors of aromatase, with consequent changes in circulating levels of steroid hormones (Tyler et al., 1998). More generally, living animals and plants function in an integrated way, and the dysfunction of one organ or tissue is likely to produce knock-on effects elsewhere.

If we consider any particular site of action, only a limited number of structurally related chemicals can interact with it to cause a toxic response. In other words, most molecular interactions leading to toxicity are relatively specific to certain types of compounds that have the necessary structural features to enable interaction with a specific target. However, with movement across organizational boundaries, along the chain of causality, observed changes become less specific for any one type of compound.

Progressing to higher levels of organization, effects become less and less readily attributable to any particular type of compound. For example, when studying the effects of a neurotoxic chemical at the level of the whole organism, the neurophysiological or behavioral disturbances it causes may be similar to those produced in response to other neurotoxins working through different mechanisms. Although the disturbances originate at different sites of action, the consequent effects may become manifest through the same tracts of the nervous system and produce similar symptoms of intoxication. The implications of this for biomarker strategies will be addressed in Section 15.3. We continue with specific examples.

15.2.2.1 *Example A: Actions of Organophosphates on Acetylcholinesterase of Nervous System*

The effects of organophosphates at the whole organism level are diverse and vary among species. In some studies, effects were related to the percentage of inhibition of brain acetylcholinesterase. Typically for birds, at around 40% to 50% inhibition, mild physiological and behavioral symptoms are seen (Grue et al., 1991). The birds become listless and inactive. At 70% inhibition and above, severe symptoms of neurotoxicity begin to appear, leading to death. In severe cases, synaptic block may cause respiratory failure.

Studies with fish have revealed behavioral effects caused by organophosphates starting at around 20% inhibition of brain acetylcholinesterase (Beauvais et al., 2000). In experimental animals such as rats, early stages of organophosphate poisoning were detected using electroencephalogram measurements and testing motor reflexes.

A variety of effects may result from cholinesterase poisoning. Some effects taken alone are also symptomatic of the actions of other neurotoxins. However, the relationships between percentage inhibition of acetylcholinesterase and certain characteristic neurotoxic and/or behavioral effects can be represented as a number of dose–response curves that collectively provide clear mechanistic evidence for the operation of this particular toxic mechanism.

15.2.2.2 Example B: Action of p,p′-DDT on Voltage-Dependent Sodium Channels of Axonal Membranes

Studies with animals and birds have shown that DDT can affect all major parts of the central and peripheral nervous system (Hayes and Laws, 1991; Environmental Health Criteria 9, 1979). At an early stage in DDT poisoning, motor unrest, abnormal spontaneous movement, and abnormal susceptibility to fear occur. As poisoning increases, hyperirritability develops and tremors are seen. Later, severe tremors and convulsions will lead to death.

As noted earlier, DDT poisoning can cause repeated firing of neurons following a single stimulus, thus giving rise to multiple action potentials. Disturbances in electroencephalogram traces can be seen from an early stage in DDT poisoning. Thus a number of neurophysiological disturbances with associated behavioral changes are characteristic of DDT poisoning in vertebrates.

15.2.2.3 Example C: Action of p,p′-DDE on Transport of Calcium into Eggshell Glands of Birds

The effect of very low concentrations of *p,p′*-DDE on females of sensitive species such as the peregrine, sparrowhawk, and gannet is to reduce Ca^{++} transport from blood into the eggshell glands, resulting in a dose-dependent reduction in eggshell thickness (Section 16.1; Figure 16.2; Peakall, 1993). A small degree of thinning has little effect on the stability of eggshells, but 17% to 18 % thinning produces a marked effect and can lead to egg breakage and breeding failure (Section 16.1).

15.2.2.4 Example D: Action of 17A-Ethinylestradiol on Estrogenic Receptors of Fish

The estrogenic action of this compound causes male fish to develop female sexual characteristics. Juvenile zebrafish exhibited retarded differentiation of testes after exposure to EE2 (Ball et al., 2005). Such estrogenic effects can lead to the development of intersex, a condition in which ova begin to appear in the testes (ova testes). A major concern in ecotoxicology is infertility in male fish exposed to estrogens such as EE2.

The degree of sexual disruption (intersex) in roaches is correlated with reduced reproductive capability; fish with gonads severely affected in this way produce lower quantities of sperm of reduced quality, an effect that leads to reduced fertility (Jobling et al., 2002a and b). Fish severely affected by intersex are unable to reproduce (Jobling et al., 2002b). Population effects of controlled discharge of EE2 into water bodies documented in an experiment by Kidd et al. (2007, 2010) are described in Section 14.3.

15.2.3 From Effects on Whole Organism to Population Effects

At this stage of the narrative, we move on from effects of chemicals upon individual organisms, which can be studied in the laboratory, to the more complex situation that exists in the field. Here, the exposure of organisms to pollutants, often in the long term, is a critical issue; but it can be very difficult to measure or estimate. It is not just a question of levels of chemicals present in water, soil, or air. How available are they to animals and plants, and what internal exposure occurs? In determining exposure in the field, chemical analysis has a prominent role, but evidence may also come from biomarker measurements and bioassays. Mesocosm systems provide something of a halfway house between the field and the laboratory and permit the study of effects of chemicals at the population level under relatively controlled conditions (see Section 14.3.3).

The importance of population growth rate and the factors influencing it were discussed in Chapter 12. In general terms, pollutants reduce population growth rates by decreasing birth rates or increasing mortality rates. When a natural population is at or near the environment's carrying capacity, the growth rate approximates to 0. Small effects of pollutants on birth rate or mortality are unlikely to change population growth rate, because they will be compensated for by the operation of density-dependent factors (Figure 15.1). For example, a small increase in the annual mortality of wood pigeons (*Columba palumbus*) caused by dieldrin seed dressings had no long-term effect on population numbers. Dieldrin merely replaced starvation as the cause of death for a certain percentage of the population. Some 90% of the UK woodpigeons die from starvation during autumn and winter (Murton, 1965).

The following discussion addresses specific questions. Have environmental levels of pollutants A through D been high enough to operate the toxic mechanisms described above? If so, to what extent? What have been the ecological consequences?

15.2.3.1 Example A: Actions of Organophosphates on Acetylcholinesterase of Nervous System

Organophosphates are usually subject to rapid metabolism by vertebrates, so effects are likely to be of shorter duration than those due to persistent organochlorine compounds (see Examples B and C). However, certain organophosphates are known to cause delayed neuropathy (Walker, 2009; Section 12.2.4.) in which the effect is present long after the chemical causing it has disappeared from the tissues.

When organophosphates were more widely used in agriculture than they are today, many reports of lethal toxic effects on vertebrates in the field led to studies of local, short-term effects on population numbers. For example, large numbers of migrating grey lag geese (*Anser anser*) and pink-footed geese (*Anser brachyrhynchus*) died as a result of poisoning by the carbophenothion seed dressing in east-central Scotland in 1971 and 1972 (Hamilton et al., 1976). The hazard to the relatively small population of migrating geese was considered sufficiently high to lead to the withdrawal of this organophosphate as a seed dressing for winter wheat and barley in the area in question. In a study in New Brunswick, Canada, to be discussed in Chapter 16, considerable mortality of forest birds was attributed to the wide-scale use of organophosphates to control an insect pest (Ernst et al., 1989).

Clearly, organophosphates have increased mortality rates of wild vertebrate populations. It is not clear, however, whether they exerted effects on birth rates. Because they are neurotoxic, it is not surprising that behavioral and other sublethal effects were noted in field and laboratory studies (Grue et al., 1991). Whether such effects, that are likely to be of short duration, produced serious impacts at the population level is still an open question.

15.2.3.2 Example B: Action of p,p´-DDT on Voltage-Dependent Sodium Channels of Axonal Membranes

The neurotoxic action of DDT has caused mortality of vertebrates in the field. For example, American robins (*Turdus migratorius*) were lethally poisoned by this chemical on the East Lansing campus of Michigan State University in the early 1960s (Bernard, 1966). This and other studies established that wild vertebrates sometimes acquired lethal levels of $p,p´$-DDT when it was widely used. Residue studies from that period also established that many animals and birds found dead or sampled in the field in the 1960s and early 1970s carried sublethal levels of this insecticide (Edwards, 1976), raising questions about the importance of sublethal effects in the field.

Neurotoxic pesticides can cause physiological and behavioral disturbances in vertebrate animals at sublethal levels (Hayes and Laws, 1991; Walker, 2003). Thus the possibility should be considered that persistent neurotoxic compounds such as $p,p´$-DDT and certain other organochlorine insecticides may produce effects on populations not only through lethal toxic actions but also through sublethal effects.

Sublethal effects may increase mortality rate and/or decrease birth rate. A reduction in birth rate may result from effects on adults (including effects on breeding behavior) or on embryos or young. Thus persistent lipophilic neurotoxins such as $p,p´$-DDT pass to the fetuses of mammals and the eggs of birds, fish, and amphibians where they are only slowly degraded and can cause toxic effects during the development of embryos.

The possible importance of sublethal effects at the population level of other persistent neurotoxic organochlorine compounds, such as dieldrin, when they are widely used are discussed in Chapter 12 (see also Walker, 2001, 2003).

15.2.3.3 Example C: Action of p,p´-DDE on Transport of Calcium into Eggshell Glands

When DDT was widely used, levels of $p,p´$-DDE high enough to cause marked eggshell thinning and egg breakage were found in a number of species of predatory birds worldwide. The toxic action was likely to affect birth rate but not adult mortality. The impact of eggshell thinning caused by $p,p´$-DDE and consequent egg breakage on populations of certain predatory birds is discussed in Chapter 16.

15.2.3.4 Example D: Action of 17A-Ethinylestradiol on Estrogenic Receptors in Fish

EE2 is relatively persistent and can undergo strong bioconcentration in fish (Larsson et al., 1999; Gibson et al., 2005). The tendency for its glucuronide conjugate to undergo enterohepatic circulation extends its residence time in fish. Detectable levels of EE2 were reported in sewage effluents and surface waters (Desbrow et al., 1998; Routledge et al., 1998; Roefer et al., 2000). Levels as high as 50 ng/l have been recorded in extreme cases.

EE2 concentrations below 0.05 to 9 ng/l have been reported in surveys of North American and European surface waters (Larsson et al., 1999; Ternes et al., 1999).

The recent discoveries that exposure of naturally breeding zebra fish to 5 ng/l over a 3-week period can reduce fecundity and breeding success and that lifelong exposure to the same concentration can cause complete population failure with no fertilization (Nash et al., 2004), are matters of serious concern. Infertile males still showed normal male reproductive behavior, competing with healthy males and thereby reducing the fertilization success of the latter. The retention of normal behavior patterns by infertile males may thus have an adverse population effect. It is worth recalling that it has been possible to control insect pests by releasing sterile males.

These observations were made at environmentally realistic concentrations of EE2 and raise the possibility that EE2, acting in concert with other naturally occurring estrogens, may affect birth rates of fish inhabiting polluted waters. If severe enough, such an effect could cause fish populations to decline (Jobling and Tyler, 2003).

15.2.4 Complete Causal Chain

We now consider the complete sequence of events from the initial interaction of a pollutant with one or more sites of action to the appearance of adverse effects at the level of population. In a natural environment, the situation is complex. Living organisms are usually exposed to mixtures of pollutants, albeit in most cases at levels too low to produce effects at the population level. Furthermore, chemicals are not the only stressors that exert adverse effects on populations. Factors such as disease, nutritional deficiency, temperature, and drought can create stress and work in concert with toxic chemicals to produce harmful effects in free-living organisms. Thus we sometimes need to consider the overall effects of stressors of different kinds including chemicals on natural populations (Van Straalen, 2003).

We chose examples of pollutants that, acting individually or in combination, have sometimes reached sufficiently high environmental concentrations to cause adverse effects at the population level on their own, regardless of possible adverse effects of other stressors. They have been chosen to illustrate how it is possible even in complex natural ecosystems to follow the sequence of adverse changes produced by specific chemicals or groups of chemicals. In other words, the response to a chemical may follow a characteristic sequence of interlinked events, and this sequence can be distinct from others produced in response to different chemicals that may be present in the same ecosystem at the same time.

The complete causal sequences for the four examples used here are summarized in Figure 15.2. The use of biomarker assays to monitor the operation of these pathways will be discussed in Section 15.3.1.

15.3 Biomarker Strategies

The foregoing discussion presented a mechanistic view of ecotoxicology, explaining how pollutants can cause adverse physiological and biochemical effects that may be damaging at the population level. The question now is how knowledge of these mechanisms can be used in practice to study the effects of pollutants now in the environment and assess the risks presented by new chemicals that may be released into the environment in the future. What technology exists or must be developed to achieve these goals?

15.3.1 Establishing Causality Where Pollution Already Exists

In the field, chemical pollution is sometimes associated with population decline and reduction of biological diversity in communities and ecosystems. Correlations may be established between environmental concentrations of chemicals in water, soil, or biota and adverse effects at the population level. While such correlations may suggest adverse effects caused by chemicals, the question remains whether the correlations represent causal relationships.

Environmental factors such as disease, limited food supply, and chemicals other than those determined in a pollution study may all cause population declines and may happen to correlate with the concentrations of pollutants determined in a field study. However, establishing that chemicals are present in the environment at concentrations known to produce adverse effects on key species under laboratory conditions can provide strong evidence of causality. Comparisons can be made between dose–response curves obtained in the laboratory and dose–response data from the field.

Adopting such a simple approach is not without problems. It can be difficult to establish accurately the actual exposure in the field, especially with long-term chronic exposure, where levels present in water or food may fluctuate a great deal. Episodic pollution is a frequent complication. Also internal exposure is a critical factor — how available are chemicals present in water, sediments, or food for uptake by free-living organisms? A further complication is that many organic pollutants are rapidly metabolized, so their concentrations within organisms cannot be determined accurately. On the other hand, as we have seen, pollutants may cause characteristic time-dependent sequences of changes in living organisms that provide evidence for the operation of particular toxic mechanisms. These changes can be measured by the use of mechanistic biomarker assays. Moreover, such biomarker responses can be quantified and related to dose both in the laboratory and in the field. Thus the employment of mechanistic biomarkers in field studies can be used to produce evidence of causality in field studies. Such studies can show, for example, that a population decline is not related only to the environmental concentration of a chemical but also to typical harmful physiological and biochemical effects known to be caused by that chemical at the concentrations in question.

Figure 15.2 summarizes the sequence of toxic changes for the examples discussed above. In all cases it is possible, at least in concept, to follow the course of toxic change using biomarker assays in combination with microarray technology. Thus in Example A, the percentage of brain cholinesterase inhibition has been related to behavioral disturbances in fish (Beauvais et al., 2000; Sandahl et al., 2005). It should also be possible to use chemical and electrophysiological assays to follow changes in synaptic levels of acetylcholine and neural transmission, respectively, to obtain a more complete picture. The development of microarray technology (Box 15.1) can give supporting evidence, establishing the points at which certain genes are switched on during the same time scale.

In the case of Example B, p,p'-DDT is a very persistent chemical, and residues found in tissues can be related to toxic effects in laboratory studies. Comparisons can then be made between laboratory data and residue levels in tissues of animals sampled or found dead in the field. Thus predictions of toxic effect can be made on the basis of residue levels, as in the investigation of the effects of dieldrin on predatory birds (Section 12.6.2). In theory, it would be useful to measure the extent of binding of p,p'-DDT to sodium channels, but no technology is currently available to do this.

In surviving animals, neurophysiological disturbances caused by p,p'-DDT can be measured using an electroencephalogram or by inserting electrodes into isolated nerves. Also behavioral disturbances caused by this chemical can be measured. Thus it should be

---

**BOX 15.1   USE OF MICROARRAYS TO MEASURE EXPRESSION
LEVELS OF GENES INVOLVED IN STRESS RESPONSES**

All organisms respond to environmental stressors by regulating the expression of genes. Some genes are turned on and some off, while others modify their levels of expression. Microarrays provide a powerful new technology that allows us to measure these changes. The products of gene expression (mRNA converted to cDNA) are measured by hybridizing them to complementary sequences of DNA printed onto a glass slide or microarray. After processing, genes that have increased their expression appear red; those that decreased expression appear green. The color intensity is proportional to the level of expression.

This new technology can be used to identify genes that respond to specific environmental stressors. Microarrays for sequenced organisms (*Drosophila melanogaster*, *Caenorhabditis elegans*) are readily available. For other organisms, it may be necessary to make a microarray containing relevant genes. The technique of subtractive suppression hybridization is used to isolate the DNA of genes whose expression levels change when stressors are applied. Although, at the time of writing, this new technology is in the early stages of development, it is already clear that it will have many applications in the field of ecotoxicology (Snell et al., 2003; Snape et al., 2004). Pioneering studies identifying the genes involved in stress responses in yeasts have already been completed (Causton et al., 2001; Chen et al., 2003).

---

possible with appropriate assays to follow the sequence starting with tissue levels of *p,p′*-DDT → degree of binding of the chemical to sodium channels → neurophysiological disturbances of progressing severity → behavioral disturbances by using appropriate assays. This may be supported, as mentioned above, by microarray technology.

The case of C is discussed in chapter. Example D is the subject of ongoing research that is described briefly in Section 15.2. All these examples illustrate the central point: biomarker assays and associated microarray technology (Box 15.1) have been or can be used to provide evidence of causality when attempting to relate population effects to levels of pollutants in the field. The following sections deal with certain aspects of this issue in more detail.

15.3.2 Biomarker Strategies in the Field

When biomarkers are used in the field, two contrasting situations exist. During field trials (Somerville and Walker, 1990), known quantities of chemical are released into the environment and samples are taken from free-living organisms so that biomarker assays can be performed. Such trials are performed on agricultural land, usually to assess the effects of pesticides. They are also sometimes carried out in lakes or watercourses. The point is that the exposure to a chemical (or chemicals) is under the control of the experimenter in this situation.

By contrast, many field studies are carried out to investigate an existing pollution problem so it is not possible to control the release of the chemical. However, this shortcoming may be partially overcome by the deployment of organisms. Species that exist in the environment in question or appropriate surrogates for them can be maintained under clean conditions so that biomarker assays can be performed to obtain control values. Representatives of the same stock can then be deployed into polluted environments. Biomarker assays may again be carried out on the deployed organisms to establish whether there is a response that may be attributed to the pollutant(s) under investigation. In this way, dose–response curves can be constructed under both laboratory and field conditions and then compared. If dose–response relationships in the field are similar to those found in the laboratory under comparable conditions, this provides powerful evidence of an effect caused by a chemical (or chemicals) under field conditions.

Deployment of organisms can help overcome the difficult question of controls when studying existing pollution problems. Comparisons are often made in the field between polluted and clean areas. The fundamental problem here is the difficulty—or impossibility—of finding a comparable area that is completely clean, i.e., free from the pollutant(s) in question. Pollutants become widely disseminated in the natural environment, and it is likely that no area of the globe is completely free of them. However, it may be possible to identify a pollution gradient, e.g., in a river downstream from a sewage outfall. Biomarker assays can be performed on organisms sampled at different points along the pollution gradient and a dose–response curve constructed for comparison with laboratory data.

15.3.3 Control Problems

When employing biomarker assays in free-living organisms, the selection of suitable controls with which valid comparisons can be made is a fundamental problem. In the definition of biomarkers in Chapter 10, reference was made to "any biological response to an environmental chemical … demonstrating a departure from normal status." What is normal status? In the case of deployed organisms, the answer may be relatively straightforward—an assay is carried out (e.g., of an enzyme activity) in an indicator species under clean laboratory conditions and the same measurement is performed on deployed individuals in a polluted environment.

The difference between the two measurements may be regarded as the biomarker response as long as the assay conditions are strictly comparable in the two cases. Such studies have, for example, been carried out with deployed fish to find evidence of estrogenic activity at sewage outfalls (Section 15.2, Example D). The issue of comparability deserves further consideration

Many endpoints used in biomarker assays (e.g., enzyme activities) show temporal variations. In the short term, diurnal variations may occur; in the long term, seasonal variation may depend, for example, on the operation of sexual cycles. Further, the magnitude of a response may depend on environmental factors such as temperature and pH. Temperature is clearly an important factor in the physiology of aquatic poikilotherms and pH is important for both aquatic and soil-dwelling organisms. Again differences between sexes and developmental stages are very marked for some assays.

In the end, comparisons should be made only between individuals of the same species (and where appropriate, strain), sex, and developmental stage; they should be made at the same temperature and at approximately the same time. The ambient medium should be approximately the same, e.g., treated and control aquatic organisms should be maintained in water of the same pH level and ionic composition.

While many of these objectives can be realized in the case of deployed organisms when maintained under closely controlled laboratory conditions, this ideal is unlikely to be fully achieved when comparisons are made between polluted and relatively clean areas in the field because of the inherent variabilities of sites that lie outside the control of investigators. Thus, the interpretation of results from field studies can be very difficult unless a striking and clearly marked dose-related biomarker response to a particular chemical is apparent at a number of different sites in a field study (see Sections 16.1 and 16.3).

15.3.4 Selection of Biomarkers for Field Studies

Field studies may be initiated when evidence suggests that environmental chemicals are causing adverse ecological effects. Commonly, an association is found between a population decline or a reduction in biological diversity and the presence of pollutants. The pollutants may have been determined by chemical analysis or circumstances may suggest that they are present (e.g., a sewage outfall or intensive use of pesticides on agricultural land). As explained in Chapters 6, 11, and 14, adverse ecological effects may become apparent when ecological surveys are carried out in the course of biological profiling using systems such as RIVPACS.

Harmful effects of chemicals can also be detected by simple rapid assays for toxic effects in the environment (Section 6.7.1; Persoone et al., 2000). Such tests may simply show toxicity to a microorganism or cell suspension or, as with the CALUX system, may measure a particular toxic mechanism (Section 6.7.1). Cellular systems may be used to detect estrogenic activity. An advantage of such assays is that they allow rapid and inexpensive screening of polluted ecosystems to identify hot spots where toxic effects are clearly caused by environmental chemicals, after which hot spots can be subjected to more intensive (and expensive) investigations. Detailed chemical analysis and the use of appropriate biomarker assays can establish the extent to which certain toxic mechanisms operate based on chemicals present in a specific ecosystem. In turn, the toxic effects measured using biomarker assays can then be related to adverse effects at the population level. Where effects are found, further evidence can be sought to establish causality (see Section 15.2.1).

In following this strategy, the judicious selection of appropriate biomarker assays and careful consideration of how they are employed are important for demonstrating that pollutants actually cause population changes. Consideration must be given to which species or life stage is likely to be critically affected. This is not simply a question of which species or life stage is thought to be the most vulnerable. Practical considerations of availability and suitability for laboratory study are also important. Thus in Example C, *p,p′*-DDE did not affect the survival of adult birds; it reduced eggshell thicknesses of sensitive species (certain raptors and piscivorous birds, but not the domestic fowl) and chick survival. Example D related to the fertility of male fish.

The timing of the assays may also be critical. Some effects (e.g., certain behavioral effects of neurotoxic pesticides) may be more important during the breeding season than at other times. Some lipophilic compounds such as dieldrin tend to exert their effects when fat is mobilized (during migration, egg laying, or starvation of birds) rather than in times of plenty. Assays should be conducted during vulnerable stages in the life cycle and at critical times of year when toxic effects are most likely to occur.

Most importantly, assays should be selected to yield maximum information about the toxic mechanisms known or believed to operate in the ecosystem in question. Ideally, the biomarkers selected should quantify the different stages of a specific toxic process in the indicator species over time. In other words, an integrated, quantitative, in-depth picture of the effects of a pollutant should emerge, and the question can then be asked whether these changes are causally related to population change. For example, can the toxic changes measured this way be causally related to changes in the terms used in population models such as that described in Box 12.1? Conceptually this seems reasonable, but the real problem is a shortage of the appropriate technology to achieve this end. At the time of writing, the number of rapid, inexpensive, user-friendly mechanistic biomarker assays is very small. More research and development are necessary before this approach can be more widely implemented.

15.4 Biomarkers and Environmental Risk Assessment

The concept of environmental risk assessment and the possible role of biomarkers in same have been briefly described in Sections 6.5. and 10.5, respectively. Very large numbers of new industrial chemicals are subject to environmental risk assessment to find what risks they may pose before they are actually marketed. Particularly stringent testing procedures are required for biologically active compounds such as pesticides and other biocides, which are deliberately released into the environment with the intention of causing toxic effects. There has been a tendency to carry out testing according to a strict protocol employing standard toxicity tests, which has been extremely expensive as well as time-consuming. Also, as we have seen, the value of some of the testing (e.g., values for median lethal toxicity) has been widely questioned (see Section 6).

As our understanding of mechanistic aspects of ecotoxicology grows, a rather different approach to ecotoxicity testing suggests itself. Whereas the number of chemicals that have to be tested is very large, the number of toxic mechanisms that have been shown to lead to adverse effects upon populations is, relatively speaking, very small. Testing every chemical according to a rigid protocol is likely to be of very limited scientific value as well as prohibitively expensive. Often, standard lethal toxicity tests would be of little value in predicting environmental risks.

On the other hand, improved knowledge of ecotoxicology supported by better mechanistic biomarker assays should generate more sophisticated screening methods for ecotoxicity, methods to establish whether realistic levels of exposure to novel test compounds are likely to have critical toxic effects at the population level. Such an approach would facilitate greater use of QSAR methods (Section 6. 6.5; Box 6.8) because such data can be used in QSAR models for predicting ecotoxicity. As explained earlier, the value of this approach in identifying neurotoxic compounds and endocrine disruptors is already clear. The challenge is to identify more precisely and in greater depth and detail the toxic changes that are particularly hazardous for ecosystems.

Pursuing such an approach would be a long-term and costly enterprise but should ultimately provide a more scientific and more economical approach to environmental risk assessment than the unsatisfactory and bureaucratic system in use now. Moreover, it should allay some of the concerns of animal welfare campaigners. Better science should mean a movement toward more fundamental assay systems involving in vitro testing and wider use of QSAR models and movement away from the old toxicity testing procedures that cause suffering to laboratory animals. Such issues surfaced during a lengthy debate about the pros and cons of the REACH proposals for assessing the risks of chemicals to humans, wildlife, and the environment adopted by the European Union (Combes et al., 2003; Walker 2006).

15.5 Summary

Molecular interactions between certain environmental chemicals and their sites of action within organisms may initiate a sequence of events, passing through different organizational levels and culminating in changes at the levels of population, community, and ecosystem. Thus initial interaction with a site of action leads to local physiological and biochemical disturbances that in turn become manifest at the whole organism level. This leads, in some cases, to adverse effects at the population level and above. Mechanistic biomarker assays developed in the laboratory can be used to identify and quantify such causal sequences in the field. Four contrasting examples illustrate this principle.

A. The action of organophosphates upon acetylcholinesterase of the nervous system
B. The action of *p,p'*-DDT on voltage-dependent sodium channels of the axonal membranes
C. The action of *p,p'*-DDE on the transport of calcium into eggshell glands
D. The action of 17A-ethinylestradiol on estrogenic receptors in fish

The appropriate use of biomarker assays can establish causality when investigating relationships between levels of pollutants and population declines in the field. However, there are many difficulties in using this procedure in an uncontrolled field situation. Of critical importance are the selection of appropriate combinations of biomarker assays and performing them at the relevant time, the use of appropriate controls, identifying appropriate species and life stages, and judicious use of deployed indicator species.

Only rarely are field trials carried out in cases of controlled releases of a chemical using biomarkers to measure effects. Mechanistic biomarkers have a potential role in the development of improved risk assessment strategies for environmental chemicals.

Further Reading

Nash, J.P. and Kime, D.E. (2004). A study of endocrine disruption in fish employing biomarker techniques with a view to relating them to ecological consequences.

Vasseur, P. and Leguille-Cossu, C. (2005). A review discussing the molecules-to-ecosystems approach.

Walker, C.H. (2009). *Organic Pollutants: An Ecotoxicological Perspective*. Mechanistic approach to ecotoxicology using biomarker strategies applied to major groups of organic pollutants including organochlorine, organophosphorous, carbamate, and pyrethroid insecticides, polycyclic aromatic hydrocarbons, PCBs and dioxins, organometallic compounds, and anticoagulant rodenticides.

16

Biomarkers in Population Studies

The causal chain linking initial biochemical and physiological changes caused by pollutants to consequent effects at the level of population and above was described in the introduction to this book. Chapter 15 discussed this question in more detail and provided examples. Emphasis was given to the potential role of biomarker assays for sequentially measuring the magnitudes of such changes at different levels of biological organization, culminating in an assessment of harmful effects at the level of population and above. This description was illustrated by examples A through D.

This chapter describes four case studies conducted in the field. Population effects were measured and biomarker assays were conducted to support the measurements and give evidence of causality. The examples are

1. DDE-induced eggshell thinning in raptorial and fish-eating birds
2. Reproductive failure of fish-eating birds on the Great Lakes of North America
3. Reproductive failure of mollusks caused by tributyl tin
4. Forest spray programs of eastern Canada

The first example expands on example C from Chapter 15. The fourth example deals with the effects of large-scale operational use of the fenitrothion organophosphorous insecticide on birds and illustrates the mechanism described in example A from Chapter 15. Examples 1 through 3 describe studies of unexpected adverse side effects of the use of pesticides and/or industrial chemicals.

16.1 DDE-Induced Eggshell Thinning in Raptorial and Fish-Eating Birds

In the early 1950s, Derek Ratcliffe of the British Nature Conservancy found what he considered to be an abnormal number of broken eggs in the eyries of the peregrine falcons in several regions of the British Isles. Later in that decade, racing pigeon enthusiasts made presentations to the British Home Office concerning the losses caused by peregrines preying on their birds. They lobbied for a change in the protected status of the peregrine because they claimed that the peregrine population had greatly increased. As a result of this pressure, the Nature Conservancy was asked to undertake a nationwide survey. It revealed that the peregrine population declined greatly and that fewer than one-fifth of the birds successfully raised young in 1961 and 1962 (Figure 12.15).

The report had an immediate impact in the U.S. A team surveyed peregrines in eastern North America, traveling 22,500 km and visiting 133 known eyries in 1964. They found every eyrie deserted. This finding spurred a conference to examine the population changes of the peregrine and other birds of prey. Interested people from many parts of the world attended the conference in Madison, Wisconsin, in 1965 (Hickey, 1969). Data presented showed that the species was extirpated in eastern North America and decreased markedly in many countries in Europe. In Scandinavia, the population had decreased to only 5% of its prewar numbers by the early 1970s. In Germany, the number of breeding peregrines decreased from 400 pairs to 40 pairs by 1975.

Based on the finding that egg breakage was common, it was likely that eggshells had become thinner. Ratcliffe examined the temporal variations in peregrine and sparrow-hawk eggshells collected in the UK after 1900. His findings of pronounced decreases in eggshell thickness in both species in the mid-1940s were published in *Nature* in 1967. A replotting of Ratcliffe's data for the peregrine over the critical period of change is shown in Figure 12.15. The first sign of change was in 1946, although the mean was not statistically significant. In 1947, the change was clear and statistically significant. In the decade that followed eggshells as thick as the prewar norm were rare.

The most dramatic case of eggshell thinning was the brown pelican (*Pelecanus occidentalis*) off the coast of California in 1969. The colony on Anacapa Island showed almost complete reproductive failure, with only four chicks raised from some 750 nests. Most nests were abandoned and remains of crushed eggshells were found throughout the colony (Figure 16.1A). On average, the thickness of the broken and crushed eggshells was only half the normal value. The reproductive failure of the colony continued for the next few years.

The productivity of double-crested cormorant (*Phalacrocorax auritus*) colonies was also close to zero and, again the main cause was egg breakage. The most detailed studies of eggshell breakage were made by the Canadian Wildlife Service on colonies of cormorants on Lake Huron in 1972 and 1973 (Weseloh et al., 1983). These workers visited all the known colonies of cormorants in Lake Huron. They found that 79% of the eggs were lost within 8 days of laying; by the end of the normal incubation period only 5% of the eggs remained in the nests. In about half the cases of lost eggs, eggshell fragments were found in or around the nest. The eggshells averaged 24% thinner than prewar values. Although this thinning was not as severe as that found in California, it was enough to cause almost complete reproductive failure. The subsequent recovery of this population is detailed in Section 16.2.

The importance of the North American findings was the clear linkage between reproductive failure and eggshell thinning (Risebrough and Peakall, 1988). Analytical work revealed that DDT (and its metabolites) and PCBs were the only contaminants present in appreciable amounts. This situation is different from that in the UK where dieldrin (Chapter 12) was the major factor in population declines of peregrines and sparrowhawks.

A pesticide was the suspected cause of eggshell thinning because the effect occurred when synthetic pesticides were introduced on a large scale and the declines in population were greatest in areas where pesticides were most heavily used. The analysis of residues of organochlorine pesticides was becoming more widespread and by the early 1970s it was possible to show a correlation between the levels of DDE in egg contents and shell thickness (Figure 16.2). This dose–response curve shows that a low concentration of DDE causes a large initial response and that tends to flatten with increasing doses.

These findings established a correlation between the concentration of DDE and the degree of eggshell thinning. Studies of the American kestrel established a cause-and-effect relationship. Furthermore, the relationship between the degree of eggshell thinning and DDE residue levels in the egg based on laboratory studies was the same as that found in the field. These studies also showed that PCBs did not cause eggshell thinning.

DDT triggered a major legal battle in 1971. It was the first pesticide subjected to environmental regulation and the producers feared, with some justification, that if they lost this battle they would lose others. The hearing that led to the ban of DDT in the U.S. was the

FIGURE 16.1
(A) Crushed eggs in the nest of a brown pelican, Anacapa Island, California, 1970. (B) Double-crested cormorant with deformed bill from a colony on Lake Michigan, U.S.

most extended and bitterly fought proceeding involving an environmental contaminant. It continued from August 1971 to March 1972 and a total of 125 expert witnesses produced over 9,000 pages of testimony in the presence of four batteries of lawyers. One side represented the chemical industry (twenty-seven companies acting as a group of petitioners) and the lawyers for the U.S. Department of Agriculture. The other side consisted of the U.S. Environmental Protection Agency and the Environmental Defense Fund.

Observations of eggshell thinning linked to DDT served as important evidence. Despite the verdict against DDT (and similar verdicts in other countries), the finding was criticized. Because eggshell thinning occurred so soon after the introduction of the insecticide (before it was widely used) DDT proponents argued that the chemical could not have caused the effect.

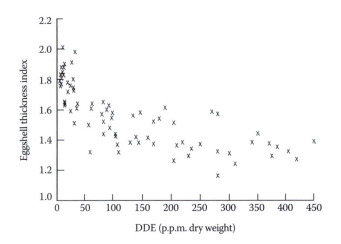

FIGURE 16.2
Relationship of eggshell thickness index to DDE residue levels in peregrine eggs collected from Alaska and northern Canada. (*Source:* Peakall, D.B. (1993). *Environmental Reviews* 1, 13–20. With permission.)

All the investigative work was conducted some ten to twenty years after thinning was first observed. The first measurements were made on DDE levels in peregrine eggs in 1962, whereas eggshell thinning started in 1946. However, DDE is so stable and analytical techniques are so sensitive that it proved possible to extract enough DDE from the dried membranes of eggs collected over the critical period (1946 and 1947) to demonstrate that enough DDE was present to have caused eggshell thinning.

Eggshell thinning in peregrines was found to be a global phenomenon and a close correlation was found between the degree of thinning and the health of the peregrine population. Studies from Australia, Europe, and North America showed that when eggshell thinning exceeded 17% to 18%, population declines followed. The relationship between eggshell thinning and population is shown in Figure 16.3. Four extirpated populations showed eggshell thinning of 18% to 25%; thinning in declining populations exceeded 17% and stable populations showed less than 17% thinning.

Considerable interest surrounded the mechanism by which DDE caused eggshell thinning. A number of mechanisms were proposed. However, the finding of thinning up to 50% in the brown pelican indicated that the site of impairment was in the shell gland and a reduction of 50% of calcium levels throughout the organism would cause profound physiological changes. This hypothesis was supported by the finding that the circulating level of calcium in the blood is normal in birds laying eggs with thin shells.

It is now generally agreed that transport of calcium across the eggshell gland mucosa is affected by DDE. Decreased activity of Ca-ATPase and effects on prostaglandins have been proposed as the key mechanisms. We have no reason to suppose that these mechanisms are mutually exclusive but no comprehensive explanation of the mechanistic basis of species variation of DDE-induced eggshell thinning has been put forward.

The evidence that DDT and dieldrin caused widespread decreases in several populations of peregrines and many other birds of prey eventually led to bans and restrictions. DDT and dieldrin were banned in the Scandinavian countries starting in 1969. In Holland, dieldrin was restricted in 1968 and DDT was banned in 1973. In Great Britain, voluntary restrictions on both DDT and dieldrin use started in 1962, although some small-scale uses

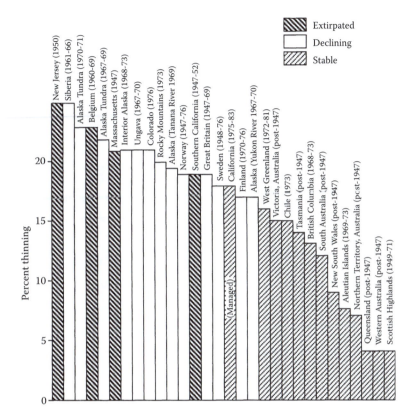

FIGURE 16.3

Relationship between degree of eggshell thinning and status of populations of peregrines. (*Source:* Peakall, D.B. (1993). *Environmental Reviews* 1, 13–20. With permission.)

were not banned until 1986. The approach in Great Britain was unlike the situation in the U.S. where the ban on DDT imposed in 1972 followed a long court hearing.

In the eastern U.S. and southern Canada, the peregrine disappeared completely by the early 1960s and reintroduction was the only feasible means of restoring the population. Large-scale breeding programs intended to raise birds for reintroduction, were started by the Canadian Wildlife Service at Wainwright, Alberta, in 1972 and by the Peregrine Fund at Cornell University in New York State around the same time. The first few captive-raised young were released in 1975 and the numbers soon increased. The first breeding attempt of released birds in the wild was in 1979 and progress has been rapid since then. We have now reached the point where releases are needed in only a few specific areas.

In other parts of the world, some breeding stock remained and reintroduction programs were not essential. Recovery started in Alaska, western Canada, the continental U.S., and Mexico in the later 1970s. In the British Isles, where the most detailed data are available, recovery was described as "virtually complete in overall numbers, though not in precise distribution." In other parts of Europe, the populations increased but are still low compared with prewar numbers, notably in the Czech Republic, Slovakia, Germany, and Poland, which formerly housed large populations. These and other recoveries of the peregrine are detailed in Cade et al. (1988).

An interesting aspect of eggshell thinning is the wide difference exhibited in the sensitivities of different avian species. Most sensitive are the raptors and the fish-eating birds; this is unfortunate because these species are exposed to the highest doses as a result of the bioaccumulation of DDE up the food chain. Raptors other than the peregrine have shown marked effects—the osprey (*Pandion haliaetus*) and the bald eagle (*Haliaetus leucocephalus*) in North America and the sparrowhawk in Europe. In contrast, the species most commonly used in in vivo toxicity tests (quail, pheasant, chicken) are almost completely insensitive, and others (ducks) are only moderately sensitive. Thus it is unlikely that even the studies now required before the registration of a new pesticide would have detected this specific adverse effect of DDT. However, other negative aspects of DDT, such as its strong tendency to biomagnify up the food chain and its toxicity to fish, should prevent registration.

No story is quite as simple as one would like. Although in many parts of the world it seems clear that DDT-induced eggshell thinning was the main cause of population declines of raptorial birds, in some areas other pesticides were certainly also involved. In the UK, direct mortality caused by dieldrin, another organochlorine that travels up the food chain, was considered the most important factor (Chapter 12). In North America, the levels of dieldrin recorded were well below the critical value.

The eggshell thinning story is one of the most comprehensive environmental investigations in the short history of environmental toxicology. An important feature of eggshell thinning and the subsequent collapse of the populations of peregrines and other birds of prey is that the effects occurred over wide areas far distant from the places of application of the pesticide.

16.2 Reproductive Failure of Fish-Eating Birds on Great Lakes of North America

The North America Great Lakes collectively constitute the largest body of fresh water in the world. The very vastness of the lakes became a factor in their pollution. They are so large that "dilution seemed the solution to pollution." The waterway through the lakes led to the development of the surrounding states. Towns and industries sprang up on their shores. The harnessing of the Niagara Falls to generate electricity led to major industrial development along the Niagara River. Now some 36 million people live within the Great Lakes basin, and over 13,000 manufacturing and industrial plants operate there. By the 1960s, serious wildlife problems had arisen.

Investigations of the problems with wildlife of the Great Lakes had two distinct beginnings—one was proactive and the other reactive. The proactive approach was led by Joe Hickey, a professor of wildlife ecology at the University of Wisconsin. Concern about DDT led to a study "to determine what pesticide residues, if any, are present in different trophic levels in a Lake Michigan ecosystem" and "to understand the biological significance, if any, of pesticide residues encountered in various layers of the Lake Michigan animal pyramid." The bioaccumulation of DDT residues up the animal pyramid was clearly demonstrated, and the findings are illustrated in Figure 16.4. Although bioaccumulation of DDT was known, Lake Michigan study was important because it demonstrated that these problems could occur in a large lake system.

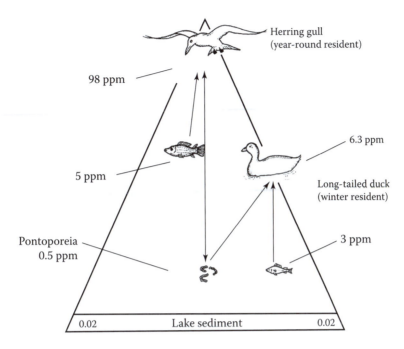

FIGURE 16.4
Accumulation of DDT up the Lake Michigan food chain. Values represent total DDT on wet weight basis. (*Source:* Hickey, J.J. et al. (1966). *Journal of Applied Ecology* 3, 141–154. With permission.)

The second starting point was more dramatic. Michael Gilbertson of the Canadian Wildlife Service visited tern colonies on Lake Ontario in 1970. He described his visit as follows.

As I walked about one of these islands, many birds whirled around my head and swooped down upon me again and again to prevent me from approaching their nests. Their shrill piercing cries rang in my ears. As I wandered about, I soon noticed that something was fundamentally wrong with the colony. While some young of varying age were found in the nests, the eggs in most had failed to hatch. On examining one of these eggs, I found that the young chick had died before it could completely crack open the shell. Several other eggs contained dead embryos. At the edge of a grass tussock, I also noticed an abnormal 2-week old chick, its upper and lower bill crossing over without meeting—a deformity which would result in certain starvation.

The herring gull became the key indicator species in the studies of pollution on the North American Great Lakes. The reasons for this choice are given in Box 16.1.

Although the first studies that demonstrated high embryonic mortality were carried out in the mid-1960s, not until the 1970s were systematic studies conducted by the Canadian Wildlife Service. The seriousness of the problem was shown by the reproductive failure of herring gulls on Lake Ontario—only one young produced by ten pairs of birds (normal production would have been more than one young per pair). The hypothesis that these severe effects over a wide area were caused by pollutants, notably the organochlorines, was immediately proposed. A program to monitor residue levels in herring gulls was set up in 1974, and the birds were collected from thirteen colonies throughout the Great Lakes. This program is continuing; some of the data generated are shown in Figure 16.5.

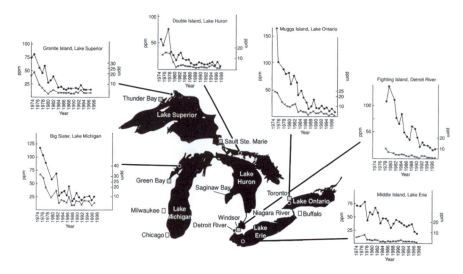

FIGURE 16.5
Residue levels of PCBs and DDE (ppm, wet weight) in herring gull eggs from six Great Lake colonies, 1974–1997. Left-hand axis (•) = PCBs. Right-hand axis (×) = DDE. (Data courtesy of Canadian Wildlife Service.)

BOX 16.1   HERRING GULL: KEY INDICATOR SPECIES ON GREAT LAKES

Herring gulls serve as key indicator species for the Great Lakes because:

1. They feed at the top of the food chain of the Great Lakes.
2. The adult is a year-round resident of the Great Lakes; comparatively little movement from lake to lake occurs.
3. They nest colonially. Colonial birds are probably the only groups of organisms for which it is possible to count entire breeding populations. Collection of eggs and assessment of reproduction are also easier in a colonial species.
4. The herring gull breeds throughout much of Europe and North America. This allows comparisons of contaminant levels, reproductive success, and other parameters of Great Lakes and coastal and European populations.

One interesting study aimed to establish the relative importance of effects of pollutants in adult birds that could cause behavioral changes (extrinsic effects) compared with effects of the pollutants in the eggs of developing embryos (intrinsic effects). To distinguish between these two causes of embryo mortality, an egg exchange experiment was devised. Eggs were moved from a highly contaminated (dirty) colony and placed under adults in a relatively uncontaminated (clean) colony and vice versa. It was also possible to incubate eggs from both clean and dirty colonies artificially to examine the effects of residues on embryos. The outline of this experiment is shown in Table 16.1.

In theory, the experiment appeared simple; in practice, it was not.

Transportation did not affect the viability of the eggs, but finding colonies containing fresh eggs in different parts of the country at the same time and moving them rapidly from place to place was a logistical nightmare. Nevertheless, the results obtained in 1975 were quite clear. Pollutants caused major effects in the eggs and also significantly affected

TABLE 16.1

Egg Exchange Experiments

Adult	Egg	Information Obtained
Clean	Clean	Normal reproduction in clean environment
Clean	Dirty	Effects of pollutants in eggs on embryos
Dirty	Clean	Effects of pollutants on reproduction via behavioral changes of adult
Dirty	Dirty	Impaired reproduction in dirty environment

the behavior of adults. We were foolish enough to repeat the experiment in 1976 when the results were less clear-cut, and again in 1977, when no differences were found. In fact, we noted a sudden marked improvement in the reproductive success of the herring gull in Lake Ontario.

Although the increased breeding success of birds on the Great Lakes was good news, it complicated the scientific investigation of the impacts of pollutants. The change arose from decreased inputs of some organochlorines into the environment. In 1972, DDT use was banned in the U.S. and curtailed in Canada. In the same year, Monsanto introduced its voluntary ban on the open circuit uses of PCBs. Thus the intensive phases of the studies on the herring gull were carried out against a background of decreasing inputs.

The effect of these restrictions on the residue levels of PCBs and DDE can be seen in Figure 16.5. Although the patterns vary somewhat, they are characterized by a steep decline in the 1970s (especially in areas where the initial levels were highest), followed by slower declines in the 1980s, and comparatively little change in the 1990s. It is clear that we now have chronic low-level pollution.

The adverse effects on fish-eating birds were not confined to gulls. Severe declines were noted in other species such as the cormorant and the bald eagle. The best estimate of the population of cormorants on Lake Ontario in the 1940s and 1950s was 200 pairs—and only 3 pairs by 1973. The main cause of failure was the breaking of eggshells: 95% of the eggs broke before hatching.

The recovery of the cormorant after the banning of DDT was rapid. By 1980, 375 pairs were found. The numbers increased to 6,700 pairs by 1990, 13,100 by 1995, and 18,800 by 1998. Obviously, density-dependent factors such as food supply and nesting habitat (Chapter 12) come into play at some stage, but this has not been reached yet.

An interesting question is why the population increased so much, far beyond the size of the initial population, after the pressure was released. The best explanation is a change in the fish stocks. Marked decreases in the population of the top predatory fish—lake trout and salmon—allowed increases in the populations of smaller fish such as alewife and smelt. The causes are complex and include overfishing and the effects of pollutants on the reproduction of the larger, long-lived fish. However, when reproduction of the cormorants improved as DDT levels decreased, the species could exploit the increased food base of small fish.

Abnormal young have been found in several species throughout the Great Lakes. The best data is on cormorants. Bird watchers who ring these birds were asked to record the numbers of young with abnormalities such as crossed beaks (Figure 16.1B) when they banded the young. The highest incidence of abnormalities was found in certain areas of Lake Michigan where 1 young in 100 was defective. The rate of occurrence was considerably lower in other parts of the Great Lakes. Thus although the occurrence of abnormalities was a potent reason for starting studies, the actual impacts on populations of fish-eating birds have been slight.

The situation on the Great Lakes changed by the early 1980s. The reproductive problems of fish-eating birds were confined to a few specific areas rather than lake-wide as they were a decade earlier. Other changes occurred after the early 1970s. With hindsight, one problem that existed then was environmental chemistry and experimental toxicology had not advanced enough. The confirmation that chlorinated dioxins (PCDDs) and chlorinated dibenzofurans (PCDFs) were present in the Great Lakes was not made until 1980. The chemistry of these compounds is discussed in Section 1.2.2.

Analytical chemists now routinely report levels of as many as 100 different organochlorines compared with a dozen a decade or so ago. Studies with specific congeners of PCBs, PCDFs, and PCDDs revealed that the toxicity of specific compounds varies by several orders of magnitude. Toxicologists demonstrated that 2,3,7,8-tetrachlorodibenzo-p-dioxin is one of the most toxic compounds known; however, molecular biologists have found it useful for their work on the isolation and characterization of receptors. The identification of the Ah receptor in the mid-1970s by means of its strong specific binding to dioxin was a key finding in bringing molecular biology into the realm of toxicology. The Ah receptor is responsible for the control of certain monooxygenase enzyme systems that metabolize against foreign compounds (Chapters 5, 7, and 9).

The ability of individual compounds to induce the monooxygenase system is greatly influenced by the degree of chlorination and the chlorine substitution pattern. The most toxic PCBs have no chlorine atoms next to the central bonds, allowing the molecules to assume a coplanar configuration (Chapter 1). Studies on the structure–activity relationships of a number of organochlorines show that the toxic effects and binding strengths to the Ah receptor are related.

The application of this fundamental biochemistry to field investigations involved expressing the complex mixtures of organochlorines as dioxin equivalents. The technique is based on the relationship between the concentration required to induce a specific monooxygenase activity (aryl hydrocarbon hydroxylase) and the concentration required for toxic effects for a large number of PCBs, PCDFs, and PCDDs. When calculating dioxin equivalents, the activity of the most powerful compound, 2,3,7,8-TCDD, is set at 1 and the potencies of the other compounds are calculated from their ability to induce the monooxygenase relative to the ability of 2,3,7,8-TCDD to do the same (Section 9.2; Safe, 2001). The results are called the toxic equivalent factors of the compounds.

For each compound, this number can then be multiplied by the concentration to calculate the toxic equivalence (TEQ) in terms of dioxin. Although the potencies of other compounds such as individual PCBs are much lower than the potency of dioxin, the concentrations are often much higher and therefore they frequently contribute more to the total dioxin equivalent than dioxin alone. For example, recent studies in Lake Michigan suggest that 90% of the dioxin equivalents in the eggs of fish-eating birds come from two specific PCBs. Table 16.2 shows calculations of dioxin equivalents. This approach is now

TABLE 16.2

Toxic Equivalency Factors and Dioxin Equivalents of Three Compounds

Compound	Concentration (pg/g)	Toxic Equivalency Factor	Dioxin Equivalent
2,3,7,8-TCDD	2	1	2
3,3′,4,4′,5-PCB	3,000	2×10^{-2}	60
3,3′,4,4′,5,5′-PCB	25,000	5×10^{-4}	12.5

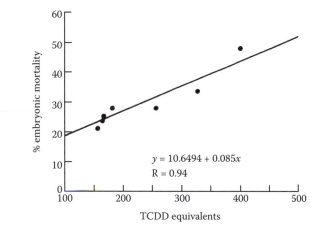

FIGURE 16.6

Relationship between dioxin equivalents and reproductive success in Caspian terns on the North American Great Lakes. (*Source:* Peakall, D.B. (1992). *Animal Biomarkers as Pollution Indicators.* Chapman & Hall. With permission.)

used in reverse; the degree of induction of the enzyme is measured and then converted into dioxin equivalents. This bioassay approach is rapid and inexpensive compared with the conventional chemical analysis by gas chromatography–mass spectrometry.

The dioxin equivalents of egg samples from fish-eating birds collected from colonies of cormorants and terns in Michigan and Ontario were determined. When the reproductive success of double-crested cormorant and Caspian tern (*Hydroprogne caspia*) colonies was plotted against dioxin equivalents of eggs from each colony, a high degree of correlation was found (Figure 16.6).

The reproductive failure of fish-eating birds in the North American Great Lakes was determined by what were initially two entirely different lines of research—molecular biology and investigations by field biologists. Their combined efforts eventually provided an answer to the problem.

16.3 Reproductive Failures of Mollusks Caused by Tributyl Tin

The general formula for tributyl tin (TBT) compounds is $(n-C_4H_9)3-Sn-X$, where X represents an anion such as a chloride or carbonate. TBTs have been used as molluscicides on boats, quays, and other marine structures and as biocides for cooling systems in pulp and paper mills. Their use as active ingredients in marine antifouling paints began in the mid-1960s, and for the next decade their popularity increased because these paint formulations were extremely effective. The worldwide production of organotins was estimated in 1980 to be about 30,000 tons annually; about a tenth was used in antifouling paints and a similar amount in wood preservatives.

Problems surfaced in the late 1960s with population declines of oysters and whelks in France and southern England and marine snails in Long Island Sound in the US Investigations of the declines of populations of dog whelk (*Nucella lapillus*) mollusks in

Plymouth Sound revealed a high degree of imposex. Imposex is the development by females of male characteristics, typically a small penis close to the right tentacle. A superficial vas deferens grows between the genital papilla and the penis. In the most extreme cases, it occludes the papilla, thus preventing egg liberation and reproduction.

Detailed laboratory and field studies including transplant experiments were conducted by the Plymouth Marine Laboratory after the first finding of imposex in dog whelks in 1969. These studies, reviewed in Matthiessen and Gibbs (1998), show abundant proof of the linkage of TBT with this irreversible sexual abnormality in female gastropods. A broader survey around the southwest peninsula of England revealed that imposex was widespread, with the most marked effects along the Channel coast.

One of the first findings was a clear relationship between the degree of imposex and the proximity of affected populations to harbors and marinas. This suggested a pollutant associated with the boating industry. Another suggestive finding was the marked increase of imposex after its discovery in 1969. These findings suggested that the causative agent was TBT. The correlations were confirmed by laboratory experiments that revealed a cause-and-effect link. The studies showed that a few nanograms per liter were enough to sterilize young whelks (Bryan et al., 1988).

Studies have shown that imposex is caused by elevated levels of testosterone that masculinize TBT-exposed females. The precise mechanism is not clear, but the evidence suggests that TBT acts as a competitive inhibitor of cytochrome P_{450}-mediated aromatase (Matthiessen and Gibbs, 1998).

Imposex has been widely reported in marine gastropods associated with marinas and harbors around the world (Champ, 1999). In addition to studies in the UK, France, and the eastern U.S., further studies were conducted in the western U.S., Canada, Alaska, southeast Asia, and New Zealand. Initially, the problem appeared restricted to areas near marinas and harbors. However, in the North Sea, imposex was found in the common whelk (*Buccinum undatum*), and the frequency correlated well with shipping traffic intensities. It is now clear that imposex is a global phenomenon, with seventy-two species of forty-nine genera of gastropods affected.

Laboratory experiments and in situ transfer experiments indicate that imposex may be initiated in dog whelks with concentrations of TBT as low as 1 ng/l. The no-observed-effect concentration (NOEC) for development of imposex cited by the International Programme on Chemical Safety Environmental Health Criteria Document in 1990 is less than 1.5 ng/l. Sterilization of some females occurs at localities containing 1 to 2 ng/l and is total in areas averaging 6 to 8 ng/l. Affected populations suffer reproductive failure and local extinctions have occurred around marinas (Gibbs and Bryan, 1986).

In France, restrictions on the use of TBT initiated in 1982 were later extended to a complete ban of the use of organotin compounds in antifouling paints. In the UK, the use of TBT-containing antifouling paints on small boats and aquaculture cages was banned in 1987, and the environmental quality standard was set at 2 ng/l TBT. Vessels longer than 25 m can still use TBT paints, based on the premise that they are in open waters much of the time and released TBT will be effectively diluted. Many other countries imposed similar restrictions. A number of U.S. states based restrictions or bans on leaching rates. The leaching rate of TBT from paints has been considerably improved in recent years by incorporating the TBT in a copolymer matrix.

Because TBT is moderately lipophilic, it would be expected to bioaccumulate. Its ability to bioaccumulate in sediments and various organisms is shown in Figure 16.7. Concentration factors are greatest for organisms that lack efficient degradation systems.

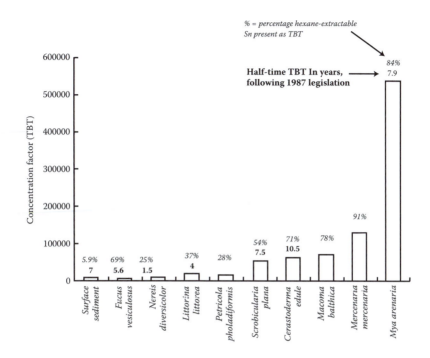

FIGURE 16.7

TBT concentration factors in sediment and organisms from the Itchen Estuary, Southampton (water concentration 67 ng TBT/liter). Figures above bars indicate percent organotin as TBT and half-life following 1987 legislation. Concentration factor = TBT concentration in tissue or sediment (dry weight) divided by TBT in water. (*Source*: Langston, W.J. (1996). *Toxicology and Ecology News* 3, 179–187. With permission.)

TBT is rapidly degraded by fish, birds, and mammals and thus is not considered a risk to top-level consumers.

In the decade after legislation to restrict TBT, imposex indices (Box 11.2) improved in many surviving populations in the UK (Miller et al., 1999). However, populations that were near collapse have shown little evidence of recovery and some have continued to extinction. The situation is made more difficult because Nucella does not have a planktonic larval stage to aid dispersal, and thus recolonization depends on chance incursions of egg-laying adults.

16.4 Forest Spraying in Eastern Canada to Control Spruce Budworm

The spraying program in eastern Canada to control (or, as some of its critics would say, to attempt to control) spruce budworm (*Choristoneura fumigerana*) has been the longest and largest program of its kind in the world. In total, some 17 million kg of pesticides were sprayed on 67 million ha of forest. The total area sprayed is approximately half the area of England and Wales and some areas were sprayed annually or semiannually. Even so, at the peak of the spraying operation (1974 through 1976), well over half the province of New Brunswick was sprayed at least once annually (Figure 16.8). The aerial spraying operation essentially stopped in the early 1990s, and by 1996 only trivial amounts of chemical pesticides were

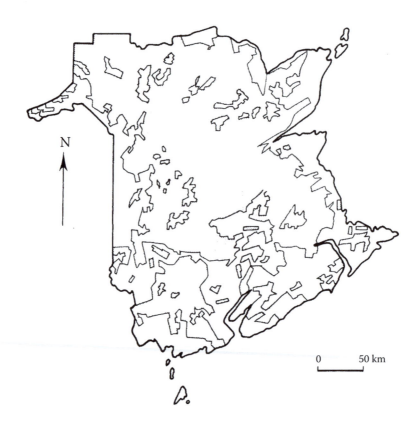

FIGURE 16.8
Map of New Brunswick, Canada, showing the extent of forest spraying (shaded) in 1976. (*Source:* Pearce, P.A. et al. (1976). *Canadian Wildlife Series Progress Notes No. 62.* With permission.)

used; even the use of the *Bacillus thuringiensis* microbial agent was phased out. The only chemical sprayed on the forests of New Brunswick in an appreciable amount (112,000 kg) in 1996 was the herbicide glyphosphate. Table 16.3 shows the total pesticides used.

What ecological consequences resulted from this spraying? Impact studies showed that birds, especially canopy-living birds, were the most vulnerable fauna. Mammals living on the forest floor appeared unaffected; amphibians and reptiles appeared somewhat insensitive. Fish were severely affected by DDT, but not by organophosphorous or carbamate pesticides.

TABLE 16.3

Total Pesticides Used in New Brunswick in
Forest Spray Operations, 1952–1996

Pesticide and Operation Years	Total (kg)
DDT (1952–1964)	5,745,000
Phosphamidon (1963–1977)	771,000
Fenitrothion (1969–1993)	9,696,906
Aminocarb (1972–1992)	551,762
Trichlorofon (1977–1978)	289,000

The impact of the organochlorine DDT was quite different from the impacts of organophosphorous and carbamate pesticides that act by inhibiting cholinesterase. The most important impact from a commercial view was the mortality of young salmon. Entire year classes of salmon were eliminated in some important salmon rivers of maritime Canada. Less important commercially but just as dramatic was the loss from the area of the fish-eating hawk, the osprey.

The DDT used in forest spraying made a significant contribution to the DDT pollution that caused the extinction of the peregrine falcon in eastern North America south of the Arctic. This impact, mediated through reproductive failure caused by eggshell thinning, was discussed earlier in this chapter.

The first pesticide to replace DDT in the spray programs in eastern Canada was the organophosphorous insecticide phosphamidon. It turned out to be a most effective avicide, but acted differently from DDT. DDT accumulates through the food chain, causing reproductive failure to species sensitive to eggshell thinning. It did not cause outright mortality at the dosages used in forest spraying operations. The amounts of DDT that caused widespread mortality of American robins after attempts to stop the spread of Dutch elm disease were much higher (Section 15.2.3). Spraying during these programs dropped some 250 to 550 g of DDT on individual trees.

The main tool for calculating impacts of pesticides on forest birds is the line transect of singing males (Box 16.2). Another useful tool is the determination of cholinesterase levels (Chapter 7). This technique has the advantage that inhibition of cholinesterase is clearly

---

BOX 16.2   LINE TRANSECT FOR MEASURING IMPACTS OF PESTICIDES

Repeated counts of singing male birds were made along roads and trails through forests before and after spray treatment. The routes were 5 km long, took about 3 hours to cover, and the counts were done early in the morning. This method was considered to be more effective than the intensive study of small plots (the numbers recorded were often too small for statistical analysis to be valid). Another problem with the small-plot approach is that the coverage of operational spraying is far from uniform; a small plot may be undersprayed, oversprayed, or even missed completely. The line transect, especially if undertaken at right angles to a line of spray, does not have this weakness. Nevertheless, the approach has some limitations, for example:

1. The counts may underestimate decreases in vocal output because songs are more easily heard as their total number decreases.
2. The mortality of birds from spraying may be masked by immigration from unsprayed areas. The huge blocks of forest sprayed in eastern Canada at the height of the spray program made this less of a problem, although long-distance migration may occur.
3. The method does not prove that the cause of the change was the pesticide, although a marked decline immediately after spraying strongly suggests that the pesticide was the cause. The most likely confounding factor is mortality caused by bad weather.
4. Even if several transits are run and the results averaged, the degree of extrapolation required is large.

related to exposure to organophosphorous or carbamate pesticides. Further experimental studies show that chronic inhibition of 50% or acute inhibition of 80% is associated with mortality. Thus a bird found dead with 80% inhibition of its brain cholinesterase was almost certainly killed by a spray operation. Nevertheless, under field conditions, the approach presents several limitations.

Using the time transect technique, it was estimated that nearly 3 million birds were killed in New Brunswick in 1975, largely because of phosphamidon. The impact of fenitrothion was considered far lower. These calculations, despite vulnerability to criticism, suggest that considerable mortality occurred and ultimately led to the phasing out of phosphamidon in forest-spraying programs.

1. The time course of AChE depression varies between pesticides and probably between species. It is often difficult to collect enough specimens of a species at any one time interval after a spray operation.

2. Sampling may be influenced by birds that enter an area after spraying has occurred, so that their only exposure is secondary.

3. Birds with marked AChE inhibition are inactive, unwilling to fly or be flushed, and are thus less likely to be captured or collected.

All these factors bias the data in favor of underestimating the degree of AChE inhibition. If the application rate is plotted against the degree of AChE inhibition (Figure 16.9), the application rate known (from transient surveys or other population studies) to cause effects produces 35% to 40% inhibition (Mineau and Peakall, 1987).

It is clear that pesticides used in forest spraying can cause mortality, sometimes considerable mortality, at operational dosages. The dose of fenitrothion commonly used in Canadian spray programs (210 g/ha) does not have an appreciable safety margin. Some

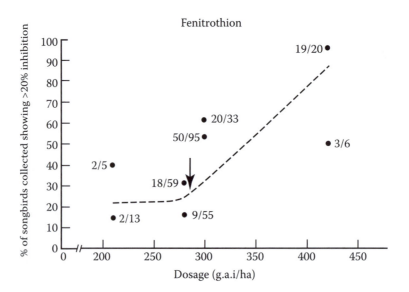

FIGURE 16.9
Relationship between inhibition of cholinesterase in songbirds and the doses of fenitrothion applied to the forest. The arrow marks the dosage above which effects are seen on songbirds as judged by transect analysis; g.a.i./ha = grams active ingredient per hectare.

of the many studies conducted showed effects; others did not. The safety margins for malathion and aminocarb are greater, and several studies with these pesticides have not revealed adverse effects. Fenitrothion was used in New Brunswick for twenty-five years (1969 to 1993) and leads to a fair question: What are the effects of the persistent use of a nonpersistent pesticide? Regrettably, the answer to this interesting question is that we do not know. An assessment on the environmental effects of fenitrothion used in forestry (Ernst et al., 1989) failed to reach any definite conclusion.

The question is indeed difficult to answer. First, the accurate measurement of long-term trends of avian (or other wildlife) populations is difficult and expensive. Even if this effort is made, it is difficult, if not impossible, to assign a cause to changes that have been established. For example, declines of North American songbirds have most frequently been attributed to the destruction of wintering habitats in Latin America.

To collect enough information on populations, it is necessary for many researchers to conduct large numbers of surveys over long periods of time. Virtually the only way to do this is to enlist amateurs (people who do such work for the love of it; use of the term in this context does not indicate their work is second rate). The British Trust for Ornithology (BTO) has followed this practice for years. Some of the current surveys by BTO are shown in Tables 12.3 and 12.4. Specializing in efforts to gather information include the heronries survey, the seabird colonies register, national wildfowl counts, and the estuaries enquiry.

16.5 Summary

Specific examples of linkages of biochemical, physiological, individual, population, and community responses were described. The examples were DDE-induced eggshell thinning in predatory birds, effects of contaminants on fish-eating birds in the Great Lakes of North America, reproductive failures of mollusks caused by tributyl tin, and the forest-spraying programs in eastern Canada. In the first three examples, population changes were well documented; in the final example, although considerable songbird mortality was found, population effects were not demonstrated.

Further Reading

Bryan, G.W. et al. (1986). The first major paper documenting the decline of dog whelks and linkage of the decline to tributyl tin.

Cade, T.J. et al., Eds. (1988). *Peregrine Falcon Populations: Their Management and Recovery*. Detailed account of the recovery of the peregrine and several other birds of prey worldwide.

Gilbertson, M. et al. (1998). *Trends in Levels and Effects of Persistent Toxic Substances in the Great Lakes*. Papers about trends in levels and effects of toxic substances in the Great Lakes.

Hoffman, D. J. et al., Eds. (1995). *Handbook of Ecotoxicology*. Case studies.

IPCS (International Programme on Chemical Safety) (1990). *Tributyl Tin Compounds*. A formal document reviewing analytical methods, environmental levels, and effects on organisms (from microorganisms to humans).

Matthiessen, P. and Gibbs, P.E. (1998). Critical review of the effects of TBT on mollusks.

Mineau, P. and Peakall, D.B. (1987). Evaluation of the methods—transect counts, numbers of singing male birds, and cholinesterase levels—for assessing impacts of OP and carbamate pesticides used in forest spraying.

Peakall, D.B. (1993). Personal account of the discovery of eggshell thinning in the peregrine and its relationship to population changes.

Peakall, D.B. and Bart, J.R. (1983). Review of the amounts of pesticides used, the areas sprayed, and the effects of spraying programs in eastern Canada and the northeastern US.

17

Ecotoxicology:
Looking to the Future

We begin this chapter by outlining in very broad terms certain topics in the history of ecotoxicology that we consider particularly relevant to consideration of future developments. Then we identify several areas where we think it possible there could be technical developments in the near future. This chapter attempts to look ahead and is inevitably less objective than the rest of the book; some parts of it represent the personal views of the authors.

17.1 Changing Patterns of Chemical Pollution

The foregoing text has been concerned with chemicals that are regarded as pollutants—chemicals that can cause environmental harm. The great majority of the chemicals described have been produced by man. However, examples have also been given of naturally occurring chemicals that have caused harm, usually in situations where they have existed at unusually high environmental levels. Examples have included heavy metals such as mercury, lead, and cadmium, and gases such as sulfur dioxide and nitrogen oxides. Organic chemicals of natural origin have also been shown to have adverse effects. Organomercury and organoarsenic compounds are biosynthesized in the environment, and mutagenic polycyclic aromatic hydrocarbons can be produced where there are forest fires. Nevertheless, much of the concern about chemicals has been about those synthesized by man. It follows that the emergence of pollution problems has been related to the growth of the chemical industry.

The chemical industry developed strongly in the early part of the twentieth century, but there was little recognition of pollution problems before the Second World War. After 1945 there was a rapid growth of the chemical industry—notably in the fields of drugs and pesticides. Alarm bells began to ring with the publication of Rachel Carson's book *Silent Spring* (1962). Although hers was not a totally objective scientific text, it served the important function of drawing attention to the fact that some pollutants—such as the persistent organochlorine insecticides—could cause environmental damage. In 1965 a meeting entitled "Pesticides in the Environment and Their Effects on Wildlife" was held at Monk's Wood Experimental Station, UK, which brought together leading researchers from Europe and North America. Particular attention was given to environmental side effects of dichlorodiphenyl-trichloroethane (DDT) and other organochlorine insecticides, as well as those of organomercury fungicides (Moore, 1966). From this time on, there was a growth of interest in the side effects of pesticides, and much work was carried out in this area—particularly in North America and Western Europe. Ecotoxicology came to be recognized as a scientific discipline. Some of this early work is described in the present text.

An important outcome of this activity was the placing of restrictions and outright bans on the use of certain pesticides. Persistent organochlorine insecticides and certain organo-mercury fungicides were targeted. In Western countries, severe restrictions were placed on the agricultural use of the organochlorine insecticides aldrin, dieldrin, heptachlor, and DDT, and on methyl mercury fungicides. In this early stage, particular attention was given to compounds that were highly persistent. However, with time, more restrictions were placed upon certain pesticides that were not very persistent but were judged to be environmentally hazardous because of their high toxicity to mammals and other vertebrates. These included organophosphorous compounds such as disyston, phorate, carbophenothion, and phosdrin, and carbamates such as aldicarb. These bans encouraged the development of less toxic and less persistent (more ecofriendly) pesticides by industry.

With this growing awareness of the undesirable environmental side effects of chemicals and the need for greater control of their marketing and their use, it became necessary to develop better methods for hazard and risk assessment (see Sections 6.5 and 10.5). Unfortunately this is a complex and costly matter, and the best practices for environmental risk assessment are the subject of lively debate. This issue is returned to later in the present chapter.

What has been said up to this stage about the regulation of the use of pesticides applies to much of the developed world. However, the relatively strict regulations that are applied to the marketing and use of pesticides in the developed world do not necessarily operate in the developing world. Pesticides that have been banned in the West are still used in, for example, parts of Africa and parts of the Indian subcontinent. Sometimes, in lands where there is famine—or there are problems with vector-borne diseases such as malaria or trypanosomiasis (sleeping sickness)—regulations of this kind may be seen as a rather expensive luxury. DDT has been used for control of malarial mosquitos in such countries long since a total ban was applied in Western countries.

With the much tighter control of chemical pollution in Western countries, many of the most obvious problems have been addressed—and, as we have seen, certain species severely affected by pollutants have recovered. Examples include the recovery of certain birds of prey in Great Britain and North America following restrictions and bans on the persistent organochlorine insecticides aldrin, dieldrin, and DDT (see Sections 12.7.2 and 16.2) and the recovery of dog whelk populations in British coastal waters and of oysters along the French coast following restrictions placed upon the use of tributyltin (TBT) in antifouling paints used to treat boats (see Sections 13.6.4 and 16.3; Walker, 2009; Chapter 8).

Minor effects of pollutants have been of a growing interest in recent years. Attention has been given to their possible sublethal effects. Included here are (1) putative neurotoxic effects and related behavioral disturbances, and (2) possible endocrine disruption and consequent breeding failure. Regarding the first of these possibilities, there is evidence that sublethal doses of neurotoxic insecticides such as pyrethroids and neonicotinoids can disturb the "waggle dance" of honeybees and thereby disrupt foraging (see Section 8.4.2; Thompson, 2003). This is topical, at the time of writing, because there have been suggestions that pesticides have played a part in the decline of honeybee populations reported from different parts of the world. Regarding the second possibility, there is evidence that EE2, a constituent of the contraceptive pill, can cause the feminization and consequent sterility of male fish in waters near sewage outfalls (see Sections 8.4.3 and 14.3; Walker, 2009; Chapter 15).

Consideration of the sublethal effects of pesticides suggests a link between ecotoxicology and stress ecology (see Van Straalen, 2003). Stress ecology studies how the physiological or reproductive performance or the survival of an organism is affected by the presence of stress factors in the environment, such as regimes of temperature, acidity, humidity, and certain inorganic ions not usually regarded as toxic. Stress factors can be defined as factors causing adverse effects on an individual organism's birth, growth, or mortality rates (Sibly and Hone, 2002), but this is the same as our definition of a pollutant. Thus, in concept, it is logical to set chemical toxicity in the wider context of stress ecology alongside other stress factors. While this proposal has its attractions when pollutants are at low levels and their effects are small, there are still situations—even in the developed world—where effects of chemicals are dominant over those of other stress factors and are sometimes the principal cause of population decline. The decline of vultures in India is such a case (Green et al., 2006). Three species of *Gyps* vulture, once common across the Indian subcontinent, have declined by more than 97% in India since 1992 and are now on the verge of global extinction (Senacha et al., 2008). The decline is due to contamination of their food with diclofenac, a nonsteroidal antiinflammatory drug (NSAID) commonly used as a painkiller for livestock in India (see Section 1.2.16). The use of diclofenac was banned in 2006 but the ban has not completely eradicated its use, and at the date of writing at least one species continues to decline at 18% per year (Cuthbert et al., 2011). This example serves as a reminder that obvious environmental effects of pollutants alone are still evident in some parts of the world—effects that are solely or predominantly attributable to the action of environmental chemicals, without the implication of other factors. Along the same lines, there are in the United States and Canada still sites (e.g., Superfund sites) that are highly polluted with, for example, organomercury or polychlorinated biphenyls (PCBs) and where persistent pollutants are adversely affecting ecosystems outside of effects of other stressors (see, e.g., Section 13.6.5).

There is also the question of possible potentiation of toxicity where there is exposure in the field to complex mixtures of compounds at very low concentrations. Where, for example, surface waters contain very low levels of many different drugs, there may be potentiation of their toxic or pharmacological effects. Also, in the case of declining bee populations mentioned above, questions have been asked about the possible synergistic effects of ergosterol-biosynthesis-inhibiting (EBI) fungicides upon pyrethroid and neonicotinoid insecticides. Bees are sometimes exposed simultaneously to these two different types of pesticides—and exposure to such combinations can lead to potentiation of toxicity (see Sections 9.5 and 9.6).

The developing concern over environmental side effects of pollution has stimulated research into methods of identifying and quantifying such effects. Ecotoxicity tests, biomarker assays, bioassays, macrocosms, quantitative structure–activity relationships (QSARs), and population modeling are all included here. Such methods yield data that can be utilized in environmental risk assessment (see Sections 6.5, 10.5, and 12.8). The very rapid advances made recently in molecular biology, biochemistry, biochemical toxicology, and related disciplines have opened up an embarrassment of riches—a bewildering variety of possibilities. The problem here is that there are only limited resources available—material or human—to study the development of new testing procedures that make use of these technologies. Also, one should not underestimate the complexity of certain testing procedures (e.g., microarrays) and the demands that they make for expensive materials and highly skilled scientists to run them. Often, methods currently used for environmental risk assessment are seen to be woefully inadequate, and progress toward their improvement is very slow. These issues are discussed later in this chapter.

17.2 Environmental Risk Assessment

Since the turn of the century, there has been growing interest in assessing the risks posed by novel chemicals to populations of nontarget organisms. The main scientific association concerned with this area is the Society of Environmental Toxicology and Chemistry (SETAC, http://www.setac.org/). SETAC promotes the advancement and application of scientific research related to contaminants and other stressors in the environment and the use of science in environmental policy and decision making. SETAC has held several workshops and conference sessions considering the problem of assessing the risks posed by novel chemicals to populations of nontarget organisms (see, e.g., Barnthouse et al., 2008; Thorbek et al., 2010). There is also a requirement in European legislation to show that new chemicals will not harm populations of nontarget organisms (EFSA Panel on Plant Protection Products and Their Residues, 2010).

The normal method used in science to show the effects of novel treatments is to carry out a replicated experiment. The attractions of the experimental approach using replicates were outlined in Section 12.7, and the problems that arise when an experiment is not replicated were noted in discussion of the Boxworth experiment in Section 12.7.3. So, how should experiments to show the effects of novel treatments be designed? First, a series of representative landscapes should be selected and each divided into areas to which the chemical treatments of interest could be applied. Replication would be essential to allow distinction of the effects of the chemical from the natural variation that already exists between landscape replicates.

In practice, experiments of this type are rarely feasible. The problems are well illustrated by considering a recent example of a landscape-scale experiment. Known colloquially as the Krebs trial, this was designed to show the effects on bovine TB of two ways of managing badgers, *Meles meles*, which are suspected of carrying the bacteria that cause bovine TB (Donnelly et al., 2007). Three experimental treatments were applied. Two were management methods, and the third was an unmanaged control. Ten UK hotspots of bovine TB were selected and three suitable 100 km² areas were identified within each TB hotspot, and the three experimental treatments were then randomly assigned to the three areas in each hotspot. In this way the experiment was replicated 10 times. The experiment cost over $100 million, was beset by interference from animal rights groups, was disrupted because needed veterinarians had to give priority to a major foot and mouth epidemic, and ran into other practical problems. This experiment illustrates that in general there are major logistical, financial, and ethical problems in carrying out landscape-scale experiments.

Although the experimental method is without doubt the most secure route to advancement of scientific knowledge, it is clear that landscape-scale experiments are not in general feasible for risk assessment of chemicals. An alternative way forward is needed. In SETAC meetings, two possible ways forward have been extensively considered; these are modeling and the use of biomarkers.

17.3 The Use of Models in Population Risk Assessment

The purpose of models in population risk assessment is to integrate all available information to provide the best possible estimate of the likely effects of chemicals. The development of mathematical modeling techniques in the twentieth century identified population growth rate (PGR) as a key variable in population analysis, as described in Chapter 12. PGR is the per capita rate at which a population grows. PGR depends on the age structure of the population, i.e., on the proportion of the population that are juveniles, young adults, and so on, and on immigration and emigration rates. PGR is specific to the landscape in which the population occurs. If the population increases in size, there eventually comes a point at which the population's demand for resources outstrips supply, so that PGR becomes zero or indeed negative if the population declines. Use of PGR methods provides much qualitative and quantitative insight. However, in practice, PGR methods are insufficient on their own for effective population risk assessment. This conclusion emerged from a workshop designed to evaluate their use.

The problem of practical population risk assessment for vertebrates was studied in a workshop in York, UK, in 2004 (Hart and Thompson, 2005). To make the project as realistic as possible, it was decided to evaluate the effects on populations of skylark, *Alauda arvensis*, and wood mice, *Apodemus sylvaticus*, of a fictitious pesticide with realistic properties in traditional laboratory tests. The risk assessment was divided logically into five steps:

1. Assess the information that can be obtained from present-day laboratory toxicity tests on mallard, quail, house mice, and rats. Such tests only go a small part of the way to assessing how life history traits respond to increasing doses of the chemical in the laboratory.

2. Extrapolate to species of interest.

3. Assess exposure of the skylarks and wood mice to the chemical in the field.

4. Estimate effects of the chemical on the life history traits of skylarks and wood mice in the field.

5. Evaluate effects on populations of skylarks and wood mice.

In practice, each of these steps poses major challenges. In step 1, effects are not measured on all the life history traits of mallard and house mice, and it is assumed that the ones that are measured are those most sensitive to the chemical. Since a complete inventory of all effects on the life history is needed for population risk assessment, a method has to be found to estimate effects on traits where measurements are lacking. In step 2, there is only very limited information available on effects on other species, so to be on the safe side a precautionary approach has to be adopted in extrapolating between species. Step 3 inevitably involves extensive computer simulation of the daily lives of individuals and requires much knowledge of the behavioral ecology of skylarks and wood mice. Steps 4 and 5 integrate the results obtained in the earlier steps, and Step 5 requires an additional estimation of how life history traits are affected by population density. This is particularly difficult even for well-studied species using classical methods, and in general can only be achieved by biologically informed guesswork. Given the scale and extent of these challenges, it has seemed to many scientists that a new methodology is needed for steps 3–5. This would involve a single simulation of the whole process, including application of the chemical in the modeled landscape; the behavior through time of individual animals moving through the landscape and in some cases acquiring doses of the chemical; the effects on each individual's life history depending on its individual history of dosing, using estimates from steps 1 and 2; and lastly, the emergent consequences for population numbers.

Computer simulations of populations of individual animals are now widespread in ecology. They are known variously as individual-based models (IBMs) or agent-based models

(ABMs). We use the latter term here. There are more than a thousand published ABMs in the ecological literature, and guidance is available as to how to construct and describe them (Grimm and Railsback, 2005; Grimm et al., 2010; Grimm et al., 2006). An example of an ABM designed for risk assessment was given in Section 12.8. However, at the time of writing, ABMs have not gained general acceptance in environmental risk assessment.

To gain acceptance in risk assessment, ABMs need to convince ecologists and risk managers that (A) they provide realistic models of how individuals' decisions in the field are affected by environmental factors and by the behavior of other individuals; (B) the ABM is correctly coded in the computer program; and (C) the model correctly predicts the dynamics of populations in the field.

Step A is the aim of behavioral ecology and much progress has been made in understanding the factors that affect individuals' decisions about what to do minute by minute throughout the day—for example, which foods to select, where to go to feed, how long to stay there before moving on (see, e.g., Krebs and Davies, 2012). Nevertheless such scientific knowledge is never complete, and it requires judgment to decide whether sufficient information is available to build a robust ABM. Where knowledge is lacking but is needed to complete the ABM, then extrapolation from other species has to be guided by general biological knowledge. Step B would be trivial except that the relevant computer programs are sometimes long and complex. Step C requires collation of all available population field data and objective evaluation of how well such data agree with the predictions of the ABM.

It will be interesting to see whether ABMs become used in population risk assessment. If they can be made reliable, they will be useful for many purposes—not only for risk assessment of environmental chemicals but also for wildlife management, conservation, and the study of fundamental questions in population ecology such as the mechanistic basis of density dependence.

17.4 Technological Advances and New Biomarker Assays

The twentieth century was a period of rapid scientific and technological advance. Biochemical pharmacology and toxicology progressed in leaps and bounds, and this led to corresponding advances in the design of novel drugs and pesticides. In due course, this progress was reinforced by the development of ever more sophisticated computer systems—including computer graphics, which could give three-dimensional images of sites of action for drugs and pesticides and for active centers of enzymes responsible for the metabolism of such compounds. Accordingly, it became easier to design new biologically active compounds for commercial development with structures that can successfully interact at targeted sites of action. Among pesticides, examples of this approach include the development of the EBI fungicides and the neonicotinoid insecticides.

This fundamental, targeted approach to pesticide design can provide the basis for the discovery of more ecofriendly selective compounds. Small alterations in molecular structure can make the pesticide more selective and less harmful to beneficial organisms. In particular, pesticides can be designed that interact with specific forms of their sites of action that exist in pest species and confer resistance. Thus, in principle, insecticides can be designed that will interact with a particular form of a target site (e.g., acetylcholinesterase, a nicotinic receptor, or a sodium channel; see Section 7.4.2) that occurs in a resistant strain of a pest but that is not present, or at low incidence, in populations of beneficial insects existing in the same farmland ecosystem. Problems of resistance of insects to insecticides that are due to "insensitive" target sites can, in principle, be resolved by this approach; new insecticides can be designed that have a high affinity with these insensitive sites. The question of insensitive forms of target sites is discussed in Section 13.6.1.

Another approach to the design of more ecofriendly pesticides is to make them more biodegradable by nontarget species. This means the creation of molecular structures that are susceptible to metabolic detoxication and therefore less persistent than existing pesticides (see Section 5.1.5). Indeed, this follows the principle that was adopted in the 1960s and 1970s when persistent organochlorine insecticides were replaced by already existing organophosphorous and carbamate insecticides that were more biodegradable and less persistent.

A mechanistic approach to ecotoxicology can also give guidance in the development of new biomarker assays, as is discussed in Section 15.1. In principle the "omics" can make an important contribution here. However, this does bring the discussion back to matters of cost. Genomic techniques such as microarray analysis are expensive to run and make considerable demands on highly skilled scientific and technical staff. They can also be difficult to interpret. As yet, "omics" might be described as being "at the research stage." They have not, as yet, yielded useful, user-friendly biomarker assays that can be employed on a regular basis by workers who do not have specialized skills. Hopefully, this situation will change when there is a better appreciation of the longer term financial benefits that can come if there is investment in new technologies of this kind.

Referring back to earlier discussion about the use of mechanistic biomarker assays for ecotoxicity testing and risk assessment in Section 6.7.1, there are a relatively small number of toxic mechanisms important in ecotoxicology in comparison to a very large number of pollutants. This is clearly the case with pesticides—which account for many of the organic pollutants about which there has been serious concern (Walker, 2006). As we have seen, many pesticides are members of quite large families, all members of which share a common mode of action—e.g., organophosphorous, carbamate, neonicotinoid, and pyrethroid insecticides, anticoagulant rodenticides, ergosterol-biosynthesis-inhibitor (EBI) fungicides, etc. It follows from this that there are quite a small number of mechanisms upon which mechanistic biomarker assays may be based. In principle, about five biomarker assays should cover all the important insecticides. Perhaps about 10 such assays would cover environmentally important pesticides more generally. This is far fewer than the number of pesticides under consideration. Similar arguments can be applied to families of pollutants other than pesticides, e.g., coplanar PCBs, polycyclic aromatic hydrocarbons (PAHs), and polychlorinated dibenzodioxins (PCDDs).

If a relatively small number of such biomarker tests were developed, the use of them could, in principle, replace a great deal of highly expensive lethal toxicity testing. Use of such biomarkers would also have the potential for producing data of considerably greater scientific value. Mechanistic biomarker responses could be measured both in the laboratory and field using environmentally realistic levels of exposure. The critical issue here would be to decide whether there were any harmful effects at the population level following realistic levels of exposure in the field. If so, then a threshold biomarker response could be identified above which the population suffered adverse effects. This threshold, once identified, would have value in interpreting laboratory tests. Levels of exposure to a chemical would be regarded as safe if the mechanistic biomarker response fell below threshold. The results would apply for different chemicals of the same type working through the same mechanism of toxicity. The success of this approach would, however, depend on the existence of a single dominant toxic mechanism. In the more complex situation where more than one significant toxic mechanism was involved, the situation would be less straightforward. For example, it might be necessary to utilize more than one biomarker assay to properly assess risk (Walker, 2008).

Along similar lines, more sophisticated QSAR models may be developed by the incorporation of parameters that relate to the operation of mechanisms of toxicity (see Box 6.8). Such models can incorporate terms derived for particular forms of sites of action or detoxifying enzymes. Once again, though, there needs to be investment in the proper characterization of the structure and properties of such macromolecules—in their different forms—before such ambitious modeling can be successfully undertaken.

17.5 A Better Integrated Approach to Environmental Risk Assessment?

Two different approaches have been adopted in ecotoxicology in general and in environmental risk assessment in particular. These can be described as the "top-down" approach and the "bottom-up" approach. Our suggestions for future developments of the top-down approach were given in Section 17.3 and for the bottom-up approach in Section 17.4. Now we consider how these two approaches could be deployed together at field sites. Our proposal incorporates checking for any problems caused by the chemical. At the same time, it checks the validity of the risk assessment process. Through comparison of observations with predictions from the initial environmental risk assessment, the whole process of risk assessment can be evaluated. A feedback cycle is set up that allows improvement of risk assessment methodology.

Our practical suggestion is that where questions are raised about the environmental safety of a chemical—e.g., a pesticide—during the course of statutory risk assessment, further investigations should be undertaken along the following lines. Field trials using the chemical should be monitored. Monitoring should record the dates and other details of application of chemicals, together with numbers and breeding success of key species such as skylarks, field voles, and several invertebrate species. In addition selected biomarkers should be measured in a sample of individuals from each species. The measured biomarkers should be appropriate mechanistic biomarkers as described in Section 17.4. Monitoring should take place at relevant times of year, such as after deployment of the chemical and at the start of the reproductive season.

In conjunction with monitoring of selected field sites, the ABM computer simulations used in the risk assessment, described in Section 17.3, should be rerun for the monitored field sites using information obtained during monitoring. The actual dates and areas of application of chemicals would be known and these should be included in the simulations. The fate of each chemical would be modeled day by day, together with the rate at which it is ingested by individual modeled animals, and the rate at which it is degraded. Thus the chemical load in each individual would be simulated day by day. Since effects of chemicals on life history traits were part of the original risk assessment, these should be incorporated in the simulations to show how the chemicals affect the individuals in the population. The numbers of individuals alive each day can then be calculated, and from this the overall population effects can be obtained. Similarly the effects of the chemicals on the biomarkers in each individual can be calculated. From this, the expected distribution of biomarker responses in the population can be obtained. In this way the data and techniques used in the original risk assessment can be deployed to calculate the predicted effects of the chemical at the monitored field site.

The results of the field monitoring in conjunction with the ABM computer simulation should reveal any unanticipated problems that may be caused by the chemical. There would be cause for concern if abundances of monitored species were lower than predicted or if biomarker responses were significantly higher than predicted. Such discrepancies might result from synergistic or antagonistic interactions with other chemicals, failure of some aspect of the risk assessment, or other causes. Note that our proposed approach also provides a check on the validity of the risk assessment process. By obtaining feedback on the efficacy of the process, improvements can be made for the benefit of future assessments.

If adverse effects at the population level were found in such studies, crucial information would be obtained on the degree of response shown in mechanistic biomarker assays employed in field studies associated with these effects. These studies could then be extended to estimate threshold mechanistic biomarker responses below which no adverse population effect was observed in the field. These data could then be used in future laboratory-based studies or field simulation studies on other similar compounds sharing the same mode of action, as explained earlier in Section 17.4. Measuring, as they do, the operation of the actual mechanism of toxicity, these mechanistic biomarker assays would provide ecotoxicity data with a sounder scientific basis for the purposes of environmental risk assessment than do the standard toxicity data (e.g., LD_{50}, LC_{50}) that is widely employed today (see Section 6.7.1).

17.6 Ethical Issues

The development of better and more cost-effective methods of ecotoxicity testing was discussed earlier (see Section 6.7.1) and has been a recurring theme in the foregoing text. In the first place, such improvements are potentially important in respect to the process of environmental risk assessment, which was discussed earlier in the present chapter. Different countries have different regulatory schemes to control the marketing and sale of chemicals, and these include special considerations about the testing requirements for pesticides and other "environmental chemicals." One example is the REACH legislation that has been adopted by the European Commission, this being an acronym for "Registration, Evaluation, Authorization, and Restriction of Chemicals" (Walker, 2006).With the development of new and improved testing procedures, it should be possible to include them, as appropriate, into the testing requirements of these statutory schemes.

While the main driving force in the quest for alternative testing methods has been toward better science and greater cost effectiveness, another consideration has been of growing importance in the developed world—the issue of animal welfare. There has been considerable public pressure to replace, reduce, and refine testing methods that cause suffering to vertebrate animals, an objective commonly referred to as "the three R's." This pressure has come from contrasting quarters. Unfortunately much publicity has been given to a few extremist groups that have taken militant action and have targeted organizations carrying out animal testing. Actions have included breaking into laboratories, causing damage there and intimidating staff. On the other hand, there are a number of responsible organizations such as FRAME (Fund for the Replacement of Animals in Medical Experiments), ICCVAM (Interagency Coordinating Committee on the Validation of Alternative Methods), and ECVAM (European Centre for the Validation of Alternative Methods), whose purpose has been to validate alternative methods (see Section 6.7.1) that follow the principles of the three R's. The activities of these organizations have received less media attention than has the misbehavior of extremists.

A critical point here is that the objectives of these latter organizations are similar to the aims of those who seek "better science." As explained earlier in the present chapter, there is the potential to develop new testing procedures with a sounder scientific basis, the adoption of which would reduce animal suffering. The greater use of biomarkers in ecotoxicity testing, the use of in vitro assays, the development of QSARs, and the use of population modeling are all examples of developments that could serve this objective. Cooperation between ecotoxicologists and organizations such as those named above should further the cause of improved methods of environmental risk assessment. Shared aims could be a driving force for the improvement of testing methods.

17.7 Summary

Discovery of undesirable side effects of pesticides in the mid-twentieth century led to extensive regulation of the chemical industry, and this largely solved the problem of severe pesticide pollution in the developed world. However, existing methods of environmental risk assessment are crude and there is much room for improvement. We see three areas of technical advance where progress is possible. Comprehensive simulations are possible of populations in realistic landscapes to which pesticides with specified effects are applied. These are agent-based models (ABMs). They would seem to have much potential for use in environmental risk assessment but need ecological credibility before they can gain general acceptance. The second hoped-for area of technical advance is in the development of a new generation of mechanistic biomarkers. Because the number of toxic mechanisms is small— perhaps only 10—assays that cover these mechanisms would yield information relevant to most environmentally important pesticides. Once developed, they could potentially be used to identify chemicals that would not adversely affect populations in the field at realistic levels of application. The success of this approach would depend on each chemical operating through a single dominant toxic mechanism. More complex toxicology would demand more complex models. Lastly we propose systematic monitoring of the effects of the agricultural use of chemicals. The agricultural use of pesticides represents an experiment on the largest scale but little attempt is currently made to monitor effects, if any, on nontarget organisms. Tapping this vast resource would provide valuable feedback on the adequacy (or otherwise) of current processes of environmental risk assessment.

Further Reading

Cuthbert, R., Taggart, M.A., Prakash, V. et al. (2011). Effectiveness of action in India to reduce exposure of Gyps vultures to the toxic veterinary drug diclofenac. *PlosOne* 6(5), e19069.

Forbes, V.E., Calow, P., Grimm, V., Hayashi, T.I., Jager, T., Katholm, A., Palmqvist, A., Pastorok, R., Salvito, D., Sibly, R., Spromberg, J., Stark, J., and Stillman, R.A., 2011. Adding Value to Ecological Risk Assessment with Population Modeling. *Human and Ecological Risk Assessment* 17, 287–299.

Senacha, K. R., Taggart, M. A., Rahmani, A. R. et al. (2008). Diclofenac levels in livestock carcasses in India before the ban. *Journal of the Bombay Natural History Society*, 105: 148–161.

Glossary

AChE: Acetylcholinesterase.

Acid rain: Rain made more acidic by the action of oxides of nitrogen and sulfur (pH below 5.5).

Adduct: Product of the linkage between a xenobiotic and an endogenous molecule, e.g., adduct formed between a benzo(a)pyrene metabolite and DNA.

Ah receptor: Aryl hydrocarbon receptor; receptor located on a cytoplasmic protein to which coplanar molecules such as PAHs, coplanar PCBs, and dioxins bind. Binding initiates induction of cytochrome P_{450} 1A.

Ah receptor-mediated toxicity: Toxic effect produced when dioxins (PCDDs) and coplanar PCBs interact with the Ah receptor. The effect is accompanied by induction of cytochrome P_{450} 1 (see above) and the enzyme activities that it expresses—including the activation of certain coplanar PCBs that are inducers.

ALAD: Aminolaevulinic acid dehydrase.

Allele: One of a pair of genes that occupy the same relative position on homologous chromosomes and separate during meiosis.

Anion: Negatively charged atom or radical.

Antagonism: Toxicity of a mixture is less than the sum of the toxicities of its components.

Anthropogenic: Generated by human activities.

Anthropogenic organic enrichment factor: Measurement of contribution of human activity to the global cycle of a pollutant.

ATPase: Adenosine triphosphatase.

Autotroph: Organism that obtains its carbon from carbon dioxide.

Bioaccumulation factor (BAF): Concentration of a chemical in an animal; concentration of the same chemical in its food.

Bioconcentration factor (BCF): Concentration of a chemical in an organism; concentration of the same chemical in the ambient medium.

Biomarker: Biological response to a chemical at the individual level or below demonstrating a departure from normal status. Usually restricted to responses at the level of organization of a whole organism or below.

Biotransformation: Conversion of a chemical into one or more products by a biological mechanism (nearly always enzyme action).

Birth rate: Number of offspring born to a reproductive female per year.

BOD (biochemical oxygen demand): Amount of dissolved oxygen used by microorganisms to oxidize 1 liter of a sewage sample.

Carboxylesterases: Esterases that hydrolyze organic compounds with carboxyester bonds. In the classification of enzymes by the International Union of Biochemistry, EC3.1.1.1 refers to carboxylesterases that are inhibited by OP compounds.

Cation: Positively charged atom or radical.

Cetaceans: Whales and dolphins.

CFC: Chlorofluorocarbon.

ChE (cholinesterase): General term for esterases that hydrolyze cholinesters.

CMPP: 2-methyl-4-chloro-phenoxy propionic acid.

COD (chemical oxygen demand): Amount of oxygen required to achieve a complete chemical oxidation of 1 liter of a sewage sample.

Colloid: Particle with one dimension in the range of 1×10^{-6} to 10^{-3} mm (i.e., 1×10^{-9} to 1×10^{-6} m).

Congener: Member of a group of structurally related compounds.

Conjugate: In biochemical toxicology, a molecule formed by the combination of a xenobiotic (usually a phase I metabolite) with an endogenous molecule (e.g., glucuronic acid, glutathione, or sulfate).

Crankcase blowby: Leakage around pistons into the crankcase of an internal combustion engine.

Curie: Unit of measurement of radioactivity; 1 curie represents 3.7×10^{10} disintegrations per second.

Cyclodiene insecticide: Organochlorine insecticide such as aldrin, dieldrin, endrin, and heptachlor.

Cytochrome P_{450}: Iron-containing protein that catalyzes many biological oxidations.

2,4-D: 2,4-Dichlorophenoxyacetic acid.

2,4-DB: 2,4-Dichlorophenoxybutyric acid.

***p,p′*-DDT:** *p,p′*-Dichlorodiphenyltrichloroethane.

Density dependence: The phenomenon whereby factors vary in their effects with population density.

Dieldrin (HEOD): Organochlorine insecticide.

DNA adduct: See adduct.

DNOC: Dinitroorthocresol.

***e*:** Base of natural logarithm (2.718).

EBI: Ergosterol biosynthesis inhibitor (fungicide).

EC(D)$_{50}$: Concentration (dose) that affects designated criterion (e.g., behavioral trait) of 50% of a population. Also known as median effect concentration (dose).

ENP: Engineered nanoparticle.

Endogenous: Originating within an organism.

Endoplasmic reticulum: Membranous network of cells that contains many enzymes that metabolize xenobiotics. Microsomes consist mainly of vesicles derived from endoplasmic reticulum.

Epoxide hydrolase: Enzyme that converts epoxides to diols by the addition of water.

Ester: Organic salt that yields an acid and a base when hydrolyzed.

17A-Ethinylestradiol (EE2): Constituent of a contraceptive pill that expresses strong estrogenic activity and is found in some surface waters.

Eukaryote: Organism that contains its DNA within nuclei.

Eutrophication: Stimulation of algal growth in surface waters caused by high levels of nitrates and phosphates. Such pollution may be caused by sewage or runoff from agricultural land treated with fertilizers from industrial waste.

Eyrie: Nest of a bird of prey such as a peregrine or golden eagle.

Fitness: Population growth rate of an allele (note that this is different from the definition of fitness used in population genetics).

Fitness cost: Reduction in the fitness of resistant alleles relative to that of susceptible alleles in unpolluted environments.

Free radicals: Chemical species (atoms, molecules, or ions) containing unpaired electrons; usually highly reactive.

Fugacity: Measurement of escaping tendency of a molecule (see Chapter 3).

GABA: Gamma-aminobutyric acid.

Genomics: Study of how the genome translates into biological function.

Genotoxic: Toxic to the genetic material of an organism.

Glucuronyl transferase: Enzyme that catalyzes the formation of conjugates between glucuronic acid and a foreign compound (usually a phase I metabolite).

Gray (Gy): SI unit of absorbed radiation dose corresponding to an energy absorption of 1 $J kg^{-1}$ of matter (cf. sievert).

Hazard: Potential of a chemical to cause harm.

Hazard assessment: Comparison of ability to cause harm (see hazard) with expected environmental concentration.

HCB: Hexachlorobenzene.

Hemoprotein: Protein containing heme as a prosthetic group, e.g., cytochrome P_{450}.

HEOD: Abbreviation for dieldrin.

Hermaphrodite: Organism having both male and female characteristics.

HMO: Hepatic microsomal monooxygenase (see Chapter 5).

Hydrophobic: Water hating; oils and nonpolar solvents immiscible in water are examples.

Immiscible: Pairs of liquids that are unable to mix, e.g., water and oil.

Imposex: Imposition of male characteristics on females in prosobranch molluscs, principally the dog whelk, *Nucella lapillus*. The level of imposex is the percentage of females within the population that possess penises. Another measure is the size of the female penis relative to that of the male [the relative penis size index (RPSI)]. Within a population, imposex can be calculated as

$$\text{level of imposex (\%)} = \frac{(\text{mean length of female penis mm})^3}{(\text{mean length of male penis mm})^3} \times 100$$

Induction: Increase in enzyme activity due to an increase in cellular concentration; may be a response to a xenobiotic and usually involves an increased rate of synthesis of the enzyme.

K: Biological constant; used in population ecology to indicate carrying capacity.

k-value: Measure of mortality.

K_{ow}: Octanol–water partition coefficient.

λ (Greek lambda): Net reproductive rate; measure of the rate of increase of a population, namely the factor by which population size is multiplied each year. $\lambda = er$.

LC(D)$_{50}$: Concentration (dose) that kills 50% of an observed population; also known as median lethal concentration (dose).

Ligand: Substance that binds specifically.

Lipophilic: Fat loving; lipophilic molecules have high affinities for lipids and tend to move from water into membranes and fat depots.

Lipoprotein: Association of lipids and proteins.

Logistic equation: Simple equation showing how population growth rate may depend on population density.

Logistical regression: Type of statistical regression that estimates the probability of an occurrence based on one or more predictor variables.

Macromolecule: Large molecule such as a protein, DNA, or polysaccharide.

MCPA: 2-methyl 4-chloro-phenoxyacetic acid.

Melanism: Possessing the dark pigment melanin.

Metallothionein: Metal-binding protein.

Microarray technology: Technology that measures the sequential switching on and off of genes in toxicological studies (see Box 15.1).

Microcosm, mesocosm, macrocosm: Small, medium, and large multispecies systems, respectively, in which physical and biological parameters can be altered and subsequent effects monitored. They may be field- or laboratory-based and are thought to mimic responses of organisms in the field more realistically than single-species test systems.

Microsomes, microsomal: When tissue homogenates are subjected to differential centrifugation, the microsomal fraction is separated between about 10,000 and 105,000 g and contains mainly vesicles derived from the endoplasmic reticulum of most vertebrate tissues.

Mitochondrion: Subcellular organelle in which oxidative phosphorylation occurs, leading to the generation of ATP.

Monooxygenase (MO): Enzyme system found in the endoplasmic reticulum; contains cytochrome P_{450} as its active center and catalyzes the oxidation of many lipophilic organic compounds.

Mutualism: Relationship between two species; each benefits from the presence of the other.

Nanoparticle: Particle with one dimension smaller than 1×10^{-4} mm (i.e., 1×10^{-7} m).

NOEC(D): No observed effect concentration (dose).

NOEL: No observed effect level. See NOEC(D).

OC: Organochlorine compound.

OP: Organophosphorous compound.

Oxyradical: Unstable form of oxygen containing an unpaired electron.

Ozonosphere: Layer of the atmosphere in which ozone is concentrated.

PAH: Polycyclic aromatic hydrocarbon.

Parthenogenetic: Relating to virgin birth.

Partition coefficient: K_{ow} is an example of a partition coefficient; see Chapter 3.

PCB: Polychlorinated biphenyl.

PCDD: Polychlorinated dibenzodioxin.

PCDF: Polychlorinated dibenzofuran.

PEC: Predicted environmental concentration.

Photochemical smog: Complex pollution arising from photochemical reactions occurring in certain areas subject to strong solar radiation and high levels of aerial pollutants (e.g., from car exhaust fumes).

Phytotoxic: Toxic to plants.

Pinnipeds: Seals, sea lions, and walruses.

PNEC: Predicted no-effect concentration (see Chapter 6).

Poikilotherms: Organisms that cannot regulate their body temperatures.

Polar: Molecules that carry charges.

Population growth rate (r): Per capita rate of increase of a population.

Porphyrin: Chemical structure of a particular type of protein, e.g., cytochrome P_{450}, hemoglobin.

Potentiation: The toxicity of a combination of compounds exceeds the total toxicities of the individual components.

Probit analysis: Statistical procedure for analyzing data from toxicity tests; responses measured by a test are transformed into probit values.

Prosthetic group: An organic-chemical component of a protein that is distinct from a protein or aminoacid.

Pyrethroids: Synthetic insecticides structurally resembling naturally occurring pyrethrins.

QSAR: Quantitative structure–activity relationships; relationships between parameters describing the structures of molecules and toxicity.

Rain-out: Removal of pollutants from air by incorporation into developing rain droplets of rain clouds.

Reductase: Enzyme that performs reductions.

Resistance: Reduced susceptibility to the toxicity of a chemical that is genetically determined; i.e., characteristic of a resistant strain of animal or plant (see Chapter 13).

Resistant allele: Allele that increases the fitness of its carriers in polluted environments.

Risk assessment: Used in ecotoxicology to mean an assessment of the probability that a chemical will cause harm at the level of population or above.

Selective toxicity: Difference in the toxicity of a chemical among different species, sexes, strains, or age groups; expressed as a selectivity ratio:

$$\frac{LD_{50} \text{ to species A}}{LD_{50} \text{ to species B}}$$

Selectivity: See selective toxicity.

SFG: Scope for growth.

SH: Sulfydryl group.

Sievert (Sv): SI unit. The damage caused by radiation depends on the rate at which it is absorbed. Thus a dose of relatively massive alpha particles of 20 Sv is typically equal to 1 Gy. For less damaging beta particles and gamma rays, 20 Sv = 20 Gy.

Sister chromatid exchange (SCE): Reciprocal exchange of DNA at loci between chromatids during replication of DNA.

Somatic growth rate: Rate of increase of individual body mass.

SR: Synergistic ratio (see Chapter 10).

Standard deviation: Measure of the variation in a sample.

Standard error: Measure of the precision of an estimate.

Stereochemistry: Branch of chemistry concerned with three-dimensional structures of chemicals.

Sublimation: Volatilization of a solid.

Sulfotransferases: Enzymes that catalyze the formation of conjugates between xenobiotics and sulfates.

Survivorship: The proportion of animals surviving between two specified ages.

Survivorship curve: Graph showing how survivorship from birth varies with age.

Susceptible allele: Nonresistant allele.

Synergism: Similar to potentiation. Some authors use the term in a more restricted way, e.g., where one component of a mixture (synergist) would cause no toxicity if applied alone at a stated dose.

2,4,5-T: 2,4,5-Trichlorophenoxyacetic acid.

TBT: Tributyl tin.

TCDD: Tetrachlorodibenzodioxin.

TEF: Toxic equivalency factor; ratio of the toxicity of a compound to that of a reference compound, for example, the ratio of the toxicity of a PCB congener to that of 2,3,7,8, TCDD is used to estimate dioxin equivalents (Section 16.2).

TER: Toxicity exposure ratio; ratio of toxicity to exposure, for example, LC50/PEC

Tolerance: Ability of an organism to withstand adverse effects of pollution.

Toxic: Harmful to living organisms.

Toxic equivalent: Value that expresses the toxicity of a mixture of chemicals relative to toxicity of a compound used as a standard.

Toxicodynamics: Harmful effects of chemicals upon living organisms.

Toxicokinetics: Uptake, metabolism, distribution, and excretion of chemicals that express toxicity; the fates of chemicals in living organisms.

Trade-off: Exchange of one advantageous characteristic for another.

Vitellogenin: Protein found in the yolks of eggs from vertebrates.

Wash-out: Removal of air pollutants by falling rain or snow.

Xenobiotic: Foreign compound; not a component of the normal biochemistry of an organism. A foreign compound to one species may be a normal endogenous compound to another.

Bibliography

Addison, R.F. (1996). The use of biological effects monitoring in studies of marine pollution. *Environmental Reviews* 4, 225–237.

Aguilar, A. and Borrell, A. (1994). Assessment of organochlorine pollutants in Cetaceans by means of skin and hypodermic biopsies. In Fossi, M.C. and Leonzio, C., eds. *Non-Destructive Biomarkers in Vertebrates*. Boca Raton, FL: Lewis.

Ahlborg, U.G., Becking, G.C., Birnbaum, L.S., Brouwer, A. et al. (1994). Toxic equivalency factors for dioxin-like PCBs. *Chemosphere* 28, 1049–1067.

Åhman, B. and Åhman, G. (1994). Radiocesium in Swedish reindeer after the Chernobyl fallout seasonal variations and long-term decline. *Health Physics* 66, 503–512.

Aldridge, W.N. (1953). Serum esterases I and II. *Biochemical Journal* 53, 110–124.

Allen, Y., Scott, A.P., Matthiessen, P., Haworth, S. et al. (1999). Survey of estrogenic activity in United Kingdom estuarine and coastal waters and its effect on gonadal development of the flounder *Platicthys flesus. Environmental Toxicology and Chemistry* 18, 1791–1800.

Alloway, B.J. and Jackson, A.P. (1991). The behaviour of heavy metals in sewage sludge-amended soils. *Science of the Total Environment* 100, 151–176.

Andrewartha, H.G. and Birch, L.C. (1954). *The Distribution and Abundance of Animals*. Chicago: University of Chicago Press.

Anthony, N., Unruh, T., Ganser, D. et al. (1998). Duplication of the Rdl GABA receptor subunit gene in an insecticide resistant aphid *Myzus persicae. Molecular Genetics and Genomics* 260, 165–175.

Applequist, H., Asbirk, S., and Drabaek, I. (1995). Variation in mercury content of guillemot feathers over 150 years. *Marine Pollution Bulletin* 16, 244–248.

Arcand-Hoy, L.D. and Benson, W.H. (1998). Fish reproduction: an ecologically relevant indicator of endocrine disruption. *Environmental Toxicology and Chemistry* 17, 49–57.

Atchison, G.J., Sandheinrich, M.B., and Bryan, M.D. (1996). Effects of environmental stressors on interspecific interactions of aquatic animals. In Newman, M.C. and Jagoe, C.H., eds. *Quantitative Ecotoxicology: A Hierarchical Approach*. Boca Raton, FL: Lewis.

Atienzar, F.A., Conrad, M., Evenden, A.J., Jha, A.N., and Depledge, M.H. (1999). Qualitative assessment of genotoxicity using random amplified polymorphic DNA: comparison of genomic template stability with key fitness parameters in *Daphnia magna* exposed to benzo(a)pyrene. *Environmental Toxicology and Chemistry* 18, 2283–2288.

Atienzar, F.A. and Jha, A.N. (2004). The random amplified polymorphic DNA (RAPD) assay to determine DNA alterations, repair, and transgenerational effects in B(a)P exposed *Daphnia magna. Mutation Research* 552, 125–140.

Axelsen, J., Holst, N., Hamers, T., and Krogh, P.H. (1997). Simulations of the predator–prey interactions in a two species ecotoxicological test system. *Ecological Modelling* 101, 15–25.

Baatrup, E. and Bayley, M. (1993). Effects of the pyrethroid insecticide cypermethrin on the locomotor activity of the wolf spider *Pardosa amentata*: quantitative analysis employing computer-automated video tracking. *Ecotoxicology and Environmental Safety* 26, 138–152.

Bacci, E. (1993). *Ecotoxicology of Organic Contaminants*. Boca Raton, FL: Lewis.

Baird, D.J., Barber, I., and Calow, P. (1990). Clonal variation in general responses of *Daphnia magna* Straus to toxic stress. I. Chronic life-history effects. *Functional Ecology* 4, 399–407.

Baker, A.J.M. (1987). Metal tolerance. *New Phytologist* 106, 93–111.

Baker, A.J.M. and Proctor, J. (1990). The influence of cadmium, copper, lead and zinc on the distribution and evolution of metallophytes in the British Isles. *Plant Systematics and Evolution* 173, 91–108.

Baker, A.J.M. and Walker, P.L. (1989). Physiological responses of plants to heavy metals and the quantification of tolerance and toxicity. *Chemical Speciation and Bioavailability* 1, 7–12.

Balk, F. and Koeman, J.H. (1984). Future hazards from pesticide use. Commission on Ecology Papers No. 6. International Union for the Conservation of Nature and Natural Resources.

Ballantyne, B. and Marrs, T.C. (1992). *Clinical Experimental Toxicology of Organophosphates and Carbamates.* Oxford, Butterwood—Heineman, Ltd.

Balls, M., Bridges, J.M., and Southee, J., eds. (1991). *Animals and Alternatives in Toxicology: Present Status and Future Prospects.* Basingstoke: Macmillan.

Barnthouse, L.W., Munns, W.R., and Sorensen, M.T., eds. (2008). *Population Level Ecological Risk Assessment.* Boca Raton, FL: Taylor & Francis.

Barth, H., ed. (1987). *Reversibility of Acidification.* London: Elsevier.

Bass, C and Field L.M. (2011). Gene amplification and insecticide resistance. *Pest Management Science* 67, 886–890.

Baumann, P.C. and Harshbarger, J. (1998). Long term trends in liver neoplasm epizootics of brown bullhead in the Black River, Ohio. *Environmental Monitoring and Assessment* 53, 213–223.

Bawden, D., Tute, M.S., and Dearden, J.C. (1991). Computer modelling and information technology. In Balls, M., Bridges, J.M., and Southee, J., eds. *Animals and Alternatives in Toxicology: Present Status and Future Prospects*, pp. 253–290. Basingstoke: Macmillan.

Bayley, M., Nielsen, J.R. and Baatrup, E. (1999). Guppy sexual behaviour as an effect biomarker of estrogen mimics. *Ecotoxicology and Environmental Safety* 43, 68–73.

Beauvais, S.L., Jones, S.B., Brewer, S.K., and Little, E.E. (2000). Physiological measures of neurotoxicity of diazinon and malathion and their correlation with behavioural measures. *Environmental Toxicology and Chemistry* 19, 1875–1880.

Beeby, A. (1991). Toxic metal uptake and essential metal regulation in terrestrial invertebrates: a review. In Newman, M.C. and McIntosh, A.W., eds. *Metal Ecotoxicology: Concepts and Applications*, pp. 65–89. Boca Raton, FL: Lewis.

Begon, M., Mortimer, M., and Thompson, D.J. (1996). *Population Ecology: A Unified Study of Animals and Plants*, 3rd ed. Oxford: Blackwell Scientific.

Beitinger, T.L. (1990). Behavioral reactions for the assessment of stress on fishes. *Journal of Great Lakes Research* 16, 495–528.

Bello, S.M., Franks, D.G., Stegemann, J.J. et al. (2001). Aquired resistance to Ah receptor agonists in a population of Atlantic killifish (*Fundulus heteroclitus*) inhabiting a Superfund site; *in vivo* and *in vitro* studies on the inducibility of xenobiotic metabolising enzymes. *Toxicological Sciences* 60, 77–91.

Bengtsson, G., Gunnarsson, T., and Rundgren, S. (1983). Growth changes caused by metal uptake in a population of *Onychiurus armatus* (Collembola) feeding on metal polluted fungi. *Oikos* 40, 216–225.

Bengtsson, G., Gunnarsson, T., and Rundgren, S. (1985). Influence of metals on reproduction, mortality and population growth in *Onychiurus armatus* (Collembola) *Journal of Applied Ecology* 22, 967–978.

Benn, F.R. and McAuliffe, C.A., eds. (1975). *Chemistry and Pollution.* London: Macmillan.

Berggren, D., Bergvist, B., Falkengren-Grerup, U., Folkeson, L., and Tyler, G. (1990). Metal solubility and pathways in acidified forest ecosystems of South Sweden. *Science of the Total Environment* 96, 103–114.

Bergman, A., Bergstrand, A. and Bignert, A. (2001). Renal lesions in Baltic grey seals (*Halichoerus grypus*) and ringed seals (*Phoca hispida botnica*). *Ambio* 30, 397–409.

Bergman, A., Olsson, M., and Reiland, S. (1992). Skull bone lesion in the Baltic grey seal (*Haliochoerus grypus*). *Ambio* 21, 517–519.

Bernard, R.F. (1966). DDT residues in avian tissues. *Journal of Applied Ecology* 3, 193–198.

Berrow, S.D., Long, S.C., McGarry, A.T., Pollard, D. et al. (1998). Radionuclides (137 Cs and 10K) in harbour porpoises *Phocoena phocoena* from British and Irish coastal waters. *Marine Pollution Bulletin* 36, 569–576.

Besselink, H.T., Van Santen, E., Vorstman, W., Vethaak, A.D. et al. (1997). High induction of cyto-chrome $P_{450}1A$ activity without changes in retinoid and thyroid hormone levels in flounder (*Platichthys flesus*) exposed to 2,3,7,8-tetrachlorodibenzo-p-dioxin (TCDD). *Environmental Toxicology and Chemistry* 16, 816–823.

Bettiin, C., Oehlmann, J., and Stroben, E. (1996). TBT-induced imposex in marine neogastropods is mediated by an increasing androgen level *Helgolander Meeresuntersuchungen* 50, 217–299.

Bickham, J.W., Mazet, J.A., Blake, J., Smolen, M.J. et al. (1998). Flow cytometric determination of genotoxic effects of exposure to petroleum in mink and sea otters. *Ecotoxicology* 7, 191–199.

Bishop, J.A. (1981). A neoDarwinian approach to resistance: examples from mammals. In Bishop, J.A. and Cook, L.M., eds. *Genetic Consequences of Man-Made Change*, pp. 37–51. London: Academic Press.

Bishop, J.A. and Cook, L.M. (1981). *The Genetic Consequences of Man-Made Change*. London: Academic Press.

Bloxham, M.J., Worsfold, P.J., and Depledge, M.H. (1999). Integrated biological and chemical moni-toring: *in situ* physiological responses of freshwater crayfish to fluctuations in environmental ammonia concentrations. *Ecotoxicology* 8, 225–237.

Bongers, M., Rusch, B. and Van Gestel, C.A.M. (2004). The effect of counterion and percolation on the toxicity of lead for the springtail *Folsomia candida* in soil. *Environmental Toxicology and Chemistry* 23, 195–199.

Boon, J.P., Eijgenraam, F., Everaarts, J.M., and Duinker, J.C. (1989). A structure–activity relationship (SAR) approach toward metabolism of PCBs in marine animals from different trophic levels. *Marine Environmental Research* 27, 159–176.

Borg, H., Andersson, P., and Johansson, K. (1989). Influence of acidification on metal fluxes in Swedish forest lakes. *Science of the Total Environment* 87/88, 241–253.

Borio, R., Chiocchini, S., Cicioni, R., Esposti, P.D. et al. (1991). Uptake of radiocesium by mushrooms. *Science of the Total Environment* 106, 183–190.

Bosveld, A.T.C. and van den Berg, M. (1994). Effects of polychlorinated biphenyls, dibenzo-p-dioxins and dibenzofurans on fish-eating birds. *Environmental Reviews* 2, 147–166.

Bower, J.S., Broughton, G.F.J., Stedman, J.R., and Williams, M.L. (1994). A winter NO_2 smog episode in the UK. *Atmospheric Environment* 28, 461–475.

Brakefield, P.M. (1987). Industrial melanism: Do we have the answers? *Trends in Ecology and Evolution* 2, 117–122.

British Ecological Society (1990). River water quality. *Ecological Issues No. 1*, Field Studies Council, Preston Montford.

Brooks, G.T. (1974). *Chlorinated Insecticides*. Boca Raton, FL: CRC Press.

Brouwer, A. and van den Berg, K.J. (1986). Binding of a metabolite of 3,4,3′,4′-tetrachlorobiphe-nyl to transthyretin reduces serum vitamin A transport by inhibiting the formation of the protein complex carrying both retinol and thyroxin. *Toxicology and Applied Pharmacology* 85, 301–312.

Brouwer, A., Murk, A.J., and Koeman, J.H. (1990). Biochemical and physiological approaches in eco-toxicology. *Functional Ecology* 4, 275–281.

Brouwer, A., Reijnders, P.J.H., and Koeman, J.H. (1989). Polychlorinated biphenyl (PCB)-contaminated fish induces vitamin A and thyroid hormone deficiency in the common seal (*Phoca vitulina*). *Aquatic Toxicology* 15, 99–105.

Bryan, G.W. and Langston, W.J. (1992). Bioavailability, accumulation and effects of heavy metals in sediments with special reference to United Kingdom estuaries: a review. *Environmental Pollution* 76, 89–131.

Bryan, G.W., Gibbs, P.E., and Burt, G.R. (1988). A comparison of the effectiveness of tri-n-butyltin chloride and five other organotin compounds in promoting the development of imposex in the dog whelk *Nucella lapillus*. *Journal of the Marine Biological Association of the United Kingdom* 68, 733–744.

Bryan, G.W., Gibbs, P.E., Hummerstone, L.G., and Burt, G.R. (1986). The decline of the gastropod *Nucella lapillus* around southwest England: evidence for the effect of tributyltin from antifoul-ing paints. *Journal of the Marine Biological Association of the United Kingdom* 66, 140–611.

Bunce, N. (1991). *Environmental Chemistry*. Winnipeg: Wuerz.

Busby, D.G., White, L.M., and Pearce, P.A. (1990). Effects of aerial spraying of fenitrothion on breeding white-throated sparrows. *Journal of Applied Ecology* 27, 743–755.

Butler, J.D. (1979). *Air Pollution Chemistry*. London: Academic Press.

Cade, T.J., Enderson, J.H., Thelander, C.G., and White, C.M. eds. (1988). *Peregrine Populations: Their Management and Recovery*. Boise, ID: Peregrine Fund.

Caffrey, P.B. and Keating, K.I. (1997). Results of zinc deprivation in Daphnid culture. *Environmental Toxicology and Chemistry* 16, 572–575.

Calabrese, E.J. (1999). Evidence that hormesis represents an "overcompensation" response to a disruption in homeostasis. *Ecotoxicology and Environmental Safety* 42, 135–137.

Calabrese, E.J. and Baldwin, L.A. (2003). Toxicology rethinks its central belief. *Nature* 421, 691–692.

Calamari, D. and Vighi, M.F. (1992). Role of evaluative models to assess exposure to pesticides. In Tardiff, R.G., ed. *Methods to Assess Adverse Organisms (SCOPE)*, pp. 119–132. Chichester: John Wiley.

Callahan, C.A., Menzie, C.A., Burmaster, D.E. et al. (1991). On-site methods for assessing chemical impact on the soil environment using earthworms: a case study at the Baird and McGuire superfund site. *Environmental Toxicology and Chemistry* 10, 817–826.

Calow, P. (1989). Ecotoxicology? *Journal of Zoology* 218, 701–704.

Calow, P., ed. (1993). *Handbook of Ecotoxicology*, Vol. 1. Oxford: Blackwell Science.

Calow, P., ed. (1994). *Handbook of Ecotoxicology*, Vol. 2. Oxford: Blackwell Science.

Calow, P., Sibly, R.M., and Forbes, V. (1997). Risk assessment on the basis of simplified life-history scenarios. *Environmental Toxicology and Chemistry* 16, 1983–1989.

Campbell, P.M. and Hutchinson, T.H. (1998). Wildlife and endocrine disruptors: requirements for hazard identification. *Environmental Toxicology and Chemistry* 17, 127–135.

Camplani, A., Saino, N., and Miller, A.P. (1999). Carotenids, sexual signals and immune function in barn swallows from Chernobyl. *Proceedings of the Royal Society of London* 266B, 1111–1116.

Caquet, T., Lagadic, L., and Sheffield, S.R. (2000). Mesocosms in ecotoxicology [1]: outdoor aquatic systems. *Reviews Environmental Contamination and Toxicology* 165, 1–38.

Carson, R. (1962). *The Silent Spring*. Boston: Houghton Mifflin.

Carter, L.J. (1976). Michigan's PBB incident: chemical mix-up leads to disaster. *Science* 192, 240–243.

Castellini, S., Savva, D., Renzoni, A., and Mattei, E.N. (1996). Biomarkers of genotoxicity in two species of bivalves molluscs from coastal ecosystems of north Adriatic Sea. *Proceedings of the Italian Society of Ecology* 17, 399–402.

Caswell, H. (2001). *Matrix Population Models*, 2nd ed. Sunderland, MA: Sinauer.

Cattani, O., Fabbri, D., Salvati, M., Trombini, C., and Vassura, I. (1999). Biomonitoring of mercury pollution in a wetland near Ravenna, Italy, by translocated bivalves (*Mytilus galloprovincialis*). *Environmental Toxicology and Chemistry* 18, 1801–1805.

Causton, H.C., Ren, B., Koh, S.S. et al. (2001). Remodelling of yeast genome expression in response to environmental changes. *Molecular Biology of the Cell* 12, 323–337.

Champ, M.A. (1999). The need for the formation of an Independent, International Marine Coatings Board. *Marine Pollution Bulletin* 38, 239–246.

Chen, D.R., Toone, W.M., Mata, J. et al. (2003). Global transcriptional responses of fission yeast to environmental stress. *Molecular Biology of the Cell* 14, 214–229.

Clark, R.B. (1992). *Marine Pollution*, 3rd ed. Oxford: Clarendon Press.

Clarke, B. (1975). The contribution of ecological genetics to evolutionary theory: detecting the direct effects of natural selection on particular polymorphic loci. *Genetics* 79, 101–108.

Clarke, K.R. and Warwick, R.M. (1998). A taxonomic distinctness index and its statistical properties. *Journal of Applied Ecology* 35, 523–531.

Coale, K.H., Johnson, K.S., Fitzwater, S.E. et al. (1996). A massive phytoplankton bloom induced by an ecosystem-scale iron fertilization experiment in the equatorial Pacific Ocean. *Nature* 383, 495–501.

Coleman, J.E. (1967). Metal ion dependent binding of sulphonamide to carbonic anhydrase. *Nature* 214, 193–194.

Combes, R.D., Dandrea, J., and Balls, M. (2003). FRAME and the Royal Commission on Environmental Pollution: common recommendations for assessing risks posed by chemicals under the EU REACH system. *ATLA* 31, 529–535.

Cook, L.M., Dennis, R.L.H., and Mani, G.S. (1999). Melanic morph frequency in the peppered moth in the Manchester area. *Proceedings of the Royal Society of London* 266B, 293–297.

Craig, P.J., ed. (1986). *Organometallic Compounds in the Environment: Principles and Reactions*. London: Longman S.

Cresswell, J.E. (2011). Meta analysis of experiments testing the effects of a neonicotinoid insecticide (imidacloprid) on honeybees. *Ecotoxicology* 20, 149–157.

Crick, H.Q.P., Baillie, S.R., Balmer, D.E., Bashford, R.I. et al. (1998). Breeding birds in the wider countryside; their conservation status. Research Report 198. British Trust for Ornithology. Thetford.

Crommentuijn, T., Brils, J., and Van Straalen, N.M. (1993). Influence of cadmium on life-history characteristics of *Folsomia candida* (Willem) in an artificial soil substrate. *Ecotoxicology and Environmental Safety* 26, 216–227.

Crosby, D.G. (1998). *Environmental Toxicology and Chemistry*. Oxford: Oxford University Press.

Crossland, N.O. (1988). A method for evaluating effects of toxic chemicals on fish growth rates. In Adams, J.A., Chapman, G.A., and Landis, W.G., eds. *Aquatic Toxicology and Hazard Assessment*. Philadelphia: American Society for Testing and Materials.

Crossland, N.O. (1994). Extrapolating from mesocosms to the real world. *Toxicology and Ecotoxicology News* 1, 15–22.

Crouau, Y., Chenon, P., and Gislard, C. (1999). The use of *Folsomia candida* (Collembola: Isotomidae) for the bioassay of xenobiotic substances and soil pollutants. *Applied Soil Ecology* 12, 103–111.

Crout, N.M.J., Beresford, H.A., and Howard, B.J. (1991). The radiological consequences for lowland pastures used to fatten upland sheep contaminated with radiocaesium. *Science of the Total Environment* 103, 73–88.

Crump, R.G., Morley, H.S., and Williams, A.D. (1999). West Angle Bay, a case study. Littoral monitoring of permanent quadrats before and after the Sea Empress oil spill. *Field Studies* 9, 497–511.

Crump, R.G., Williams, A.D. and Crothers, J.H. (2003). West Angle Bay: a case study. the fate of limpets. *Field Studies* 10, 579–599.

Culbard, E.B., Thornton, I., Watt, J. et al. (1988). Metal contamination in British urban dusts and soils. *Journal of Environmental Quality* 17, 226–234.

Curtis, C.F., Cook, L.M., and Wood, R.J. (1978). Selection for and against insecticide resistance and possible methods of inhibiting the evolution of resistance in mosquitoes. *Ecological Entomology* 3, 273–287.

Cuthbert, R., Taggart, M.A., Prakash, V. et al. (2011). Effectiveness of action in India to reduce exposure of gyps vultures to the toxic veterinary drug diclofenac. *PlosOne*, 6, e19069.

Dallinger, R. (1993). Strategies of metal detoxification in terrestrial invertebrates. In Dallinger, R. and Rainbow, P.S. eds. *Ecotoxicology of Metals in Invertebrates*, pp. 245–289. Boca Raton, FL: Lewis.

Dallinger, R., Berger, B., Hunziker, P., and Kagi, J.H.R. (1997). Metallothionein in snail Cd and Cu metabolism. *Nature* 388, 237.

Dallinger, R. and Rainbow, P.S., eds. (1993). *Ecotoxicology of Metals in Invertebrates*. Boca Raton, FL: Lewis.

Daniels, R.E. and Allan, S.D. (1981). Life table evaluation of chronic exposure to a pesticide. *Canadian Journal of Fisheries and Aquatic Sciences* 38, 485–494.

Davies, N.B., Krebs, J.R., and West, S.A., 2012. *An Introduction to Behavioural Ecology*. Oxford: Wiley-Blackwell.

Deininger, R.A. (1987). Survival of Father Rhine. *Journal of American Water Works Association* 79, 78–93.

Dell'Omo, G. and Shore, R.F. (1996a). Behavioral and physiological effects of acute sublethal exposure to dimethoate on wood mice, *Apodemus sylvaticus*. I. Laboratory studies. *Archives of Environmental Contamination and Toxicology* 31, 91–97.

Dell'Omo, G. and Shore, R.F. (1996b). Behavioral and physiological effects of acute sublethal exposure to dimethoate on wood mice, *Apodemus sylvaticus*. II. Field studies on radiotagged mice in a cereal ecosystem. *Archives of Environmental Contamination and Toxicology* 31, 538–542.

Dell'Omo, G., ed. (2000). *Behaviour in Ecotoxicology.* Chichester: Wiley.

Dempster, J.P. (1975). Effects of organochlorine insecticides on animal populations. In Moriarty, F., ed. *Organochlorine Insecticides: Persistent Organic Pollutants,* pp. 231–248. London: Academic Press.

Depledge, M.H. (1994). The rational base for the use of biomarkers as ecotoxicological tools. In Fossi, C.M. and Leonzio, C., eds. *Nondestructive Biomarkers in Vertebrates,* pp. 271–295. Boca Raton, FL: Lewis.

Depledge, M.H. and Andersen, B.B. (1990). A computer-aided physiological monitoring system for the continuous, long-term recording of cardiac activity in selected invertebrates. *Comparative Biochemistry and Physiology* 96A, 473–477.

Depledge, M.H., Amaral-Mendel, J. J., Daniel, B. et al. (1993). The conceptual basis of the biomarker approach. In Peakall, D. B. and Shugart, R.L., eds. *Biomarkers: Research and Application in the Assessment of Environmental Health,* pp. 15–29. Berlin: Springer-Verlag.

Desbrow, C., Routledge, E.J., Brighty, J.P. et al. (1998). Identification of estrogenic chemicals in STW effluent, 1. Chemical fractionation and in vitro biological screening. *Environmental Science and Technology* 32, 1549–1558.

Devonshire, A.L. and Sawicki, R.M. (1979). Insecticide-resistant *Myzus persicae* as an example of evolution by gene duplication. *Nature* 280, 140–141.

Dix, H.M. (1981). *Environmental Pollution: Atmosphere, Land, Water and Noise.* Chichester, John Wiley.

Donker, M.H., Eijsackers, H., and Heimbach, F., eds. (1994). *Ecotoxicology of Soil Organisms.* Boca Raton, FL: Lewis.

Donnelly, C.A., Wei, G., Johnston, W.T. et al. (2007). Impacts of widespread badger culling on cattle tuberculosis: concluding analyses from a large-scale field trial. *International Journal of Infectious Diseases* 11, 300–308.

Drobne, D. (1997). Terrestrial isopods — a good choice for toxicity testing of pollutants in the terrestrial environment. *Environmental Toxicology and Chemistry* 16, 1159–1164.

Drobne, D. and Hopkin, S.P. (1994). Ecotoxicological laboratory test for assessing the effects of chemicals on terrestrial isopods. *Bulletin of Environmental Contamination and Toxicology* 53, 390–397.

Edwards, C.A. (1976). *Persistent Pesticides in the Environment,* 2nd ed. Boca Raton, FL, CRC Press.

Edwards, T. (1994). Chernobyl. *National Geographic* 186, 100–115.

Eeva, T., Lehikoinen, E., and Rönkä, M. (1998). Air pollution fades the plumage of the great tit. *Functional Ecology* 12, 607–612.

EFSA Panel on Plant Protection Products and their Residues. (2010). Scientific opinion on the development of specific protection goal options for environmental risk assessment of pesticides, in particular in relation to the revision of the guidance documents on aquatic and terrestrial ecotoxicology (SANCO/3268/2001 and SANCO/10329/2002). *EFSA Journal,* 8, 1821. doi:10.2903/j. efsa.2010.1821. www.efsa.europa.eu/efsajournal.htm

Eldefrawi, M.E. and Eldefrawi, A.T. (1990). Nervous system-based insecticides. In Hodgson, E. and Kuhr, R.J., eds. (1990). *Safer Insecticides: Development and Use.* New York: Marcel Dekker.

Elliott, J.E., Harris, M.L., Wilson, L.K., Whitehead, P.E. and Norstrom, R.J. (2001). Monitoring temporal and spatial trends in polychlorinated dibenzo-p-dioxins (PCDDs) and dibenzofurans (PCDFs) in eggs of Great Blue Heron (*Ardea herodias*) on the coast of British Columbia, Canada, 1983–1998. *Ambio* 30, 416–428.

Elliott, J.M. (1993). A 25-year study of production of juvenile sea-trout, *Salmo trutta,* in an English Lake District stream. *Canadian Special Publication of Fisheries and Aquatic Sciences* 118, 109–122.

Enderson, J.H., Temple, S.A., and Swartz, L.G. (1972). Time-lapse photographic records of nesting Peregrine falcons. *Living Bird* 11, 113–128.

Environmental Health Criteria No. 101 (1990). *Methylmercury.* Geneva, WHO.

Environmental Health Criteria No. 116 (1990). *Tributyl Tin Compounds.* Geneva, WHO.

Environmental Health Criteria No. 63 (1986). *Organophosphorus Insecticides: A General Introduction.* Geneva, WHO.

Environmental Health Criteria No. 64 (1986). *Carbamate Pesticides: A General Introduction.* Geneva, WHO.

Environmental Health Criteria No. 86 (1989). *Mercury: Environmental Aspects.* Geneva, WHO.

Environmental Health Criteria No. 9 (1979). *DDT and Its Derivatives.* Geneva, WHO.

Ernst, W. (1976). Physiological and biochemical aspects of metal tolerance. In Mansfield, T.A., ed. *Effects of Air Pollutants on Plants.* Cambridge: Cambridge University Press.

Ernst, W.H.O. and Peterson, P.J. (1994). The role of biomarkers in environmental assessment. (4). Terrestrial plants. *Ecotoxicology* 3, 180–192.

Ernst, W.H.O., Verkleij, J.A.C., and Schat, H. (1992). Metal tolerance in plants. *Acta Botanica Neerlandica* 41, 229–248.

Ernst, W.R., Pearce, P.A., and Pollock, T.L. (1989). Environmental effects of fenitrothion use in forestry. *Environment Canada,* Atlantic Region Report.

Eto, M. (1974). *Organophosphorus Insecticides: Organic and Biological Chemistry.* Boca Raton, FL, CRC Press.

Evans, P.R. (1990). Population dynamics in relation to pesticide use, with particular reference to birds and mammals. In Somerville, L. and Walker, C.H., eds. *Pesticide Effects on Terrestrial Wildlife,* pp. 307–317. London: Taylor & Francis.

Everaarts, J.M., Den Besten, P.J., Hillebrand, M.T.J., Halbrook, R.S., and Shugart, L.R. (1998). DNA strand breaks, cytochrome P_{450}-dependent monooxygenase system activity and level of chlorinated biphenyl congeners in the pyloric caeca of the seastar (*Asterias rubens*) from the North Sea. *Ecotoxicology* 7, 69–80.

Everaarts, J.M., Shugart, L.R., Gustin, M.K., Hawkins, W.E., and Walker, W.W. (1993). Biological markers in fish: DNA integrity, hematological parameters and liver somatic index. *Marine Environmental Research* 35, 101–107.

Fairbrother, A. (1994). Clinical biochemistry. In Fossi, M.C. and Leonzio, C., eds. *Nondestructive Biomarkers in Vertebrates,* pp. 63–89. Boca Raton, FL: Lewis.

Fairbrother, A., Marden, B.T., Bennett, J.K., and Hooper, M.J. (1991). Methods used in determination of cholinesterase activity. In Mineau, P., ed. *Cholinesterase-Inhibiting Insecticides: Their Impact on Wildlife and the Environment,* pp. 35–71. Amsterdam: Elsevier.

Feder, M.E., Bennett, A.F., Burggren, W.W., and Huey, R.B. (1987). *New Directions in Ecological Physiology.* Cambridge: Cambridge University Press.

Fest, C. and Schmidt, K.J. (1982). *Chemistry of Organophosphorus Pesticides,* 2nd ed. Berlin, Springer-Verlag.

Fischer, H. (1989). Cadmium in seawater recorded by mussels: regional decline established. *Marine Ecology Progress Series* 55, 159–169.

Forbes, V.E., ed. (1999). *Genetics and Ecotoxicology.* Boca Raton, FL: Taylor & Francis.

Forbes, V.E. and Calow, P. (1999). Is the per capita rate of increase a good measure of population-level effects on ecotoxicology? *Environmental Toxicology and Chemistry* 17, 1544–1556.

Forrester, N.W., Cahill, M., Bird, L.J., and Layland, J.K. (1993). Management of pyrethroid and endosulfan resistance in *Helicoverpa armigera* (Lepidoptera: Noctuidae) in Australia. *Bulletin of Entomological Research* l, 1–132.

Fossi, M.C. and Leonzio, C., eds. (1994). *Nondestructive Biomarkers in Vertebrates.* Boca Raton, FL, Lewis.

Fountain, M. and Hopkin, S.P. (2004a). Biodiversity of Collembola in urban soils and the use of *Folsomia candida* to assess soil quality. *Ecotoxicology* 13, 555–572.

Fountain, M. and Hopkin, S.P. (2004b). A comparative study of the effects of metal contamination in Collembola in the field and in the laboratory. *Ecotoxicology* 13, 573–587.

Fountain, M.T. and Hopkin, S.P. (2001). Continuous monitoring of *Folsomia candida* (Insecta: Collembola) in a metal exposure test. *Ecotoxicology and Environmental Safety* 48, 275–286.

Fountain, M.T. and Hopkin, S.P. (2005). *Folsomia candida* (Collembola): a 'standard' soil arthropod. *Annual Review of Entomology* 50, 201–222.

Fox, G.A., Kennedy, S.W., Norstrom, R.J., and Wingfield, D.C. (1988). Porphyria in herring gulls: a biochemical response to chemical contamination of Great Lake food chains. *Environmental Toxicology and Chemistry* 7, 831–839.

Frank, P.W., Boll, C.D., and Kelly, R.W. (1957). Vital statistics of laboratory cultures of *Daphnia pulex* DeGeer as related to density. *Physiological Zoology* 30, 287–305.

Freedman, B. (1989). *Environmental Ecology: The Impacts of Pollution and Other Stresses on Ecosystem Structure and Function.* San Diego: Academic Press.

Freedman, B. (1995). *Environmental Ecology: The Environmental Effects of Pollution Disturbance and Other Stresses,* 2nd ed. San Diego, Academic Press.

Fryday, S.L., Hart, A.D.M., and Dennis, N.J. (1994). Effects of exposure to an organophosphate on the seed-handling efficiency of the house sparrow. *Bulletin of Environmental Contamination and Toxicology* 53, 869–876.

Futuyma, D.J. (1986). *Evolutionary Biology,* 2nd ed. Sunderland, MA: Sinauer.

Gdula-Argasinska, J., Appleton, J., Sawicka-Kapusta, K. et al. (2004). Further investigation of the heavy metal content of the teeth of the bank vole as an exposure indicator of environmental polllution in Poland. *Environmental Pollution* 131, 71–79.

Gibbs, P.E. (1993). A male genital defect in the dog-whelk, *Nucella lapillus* (Neogastropoda), favouring survival in a TBT-polluted area. *Journal of the Marine Biological Association of the United Kingdom* 73, 667–678.

Gibbs, P.E. and Bryan, G.W. (1986). Reproductive failure in populations of the dog whelk *Nucella lapillus*, caused by imposex induced by tributyltin from antifouling paints. *Journal of the Marine Biological Association of the United Kingdom* 66, 767–777.

Gibbs, P.E., Pascoe, P.L., and Burt, G.R. (1988). Sex change in the female dog whelk, *Nucella lapillus*, induced by tributyl tin from antifouling paints. *Journal of the Marine Biological Association of the United Kingdom* 68, 715–731.

Gibson, G. and Skett, P. (1986). *Introduction to Drug Metabolism.* London: Chapman & Hall.

Gibson, R., Smith, M.D., Spary, C.J. et al. (2005). Mixtures of estrogenic contaminants in bile of fish exposed to wastewater treatment works effluents. *Environmental Science and Technology.* 29, 2461–2471.

Giddings, J.M., Solomon, K.R., and Maund, S.J. (2001). Probabilistic risk assessment of cotton pyrethroids. II. Aquatic mesocosm and field studies. *Environmental Toxicology and Chemistry* 20, 660–668.

Gilbertson, M., Fox, G.A., and Bowerman, W.W. (1998). *Trends in Levels and Effects of Persistent Toxic Substances in the Great Lakes.* Dordrecht: Kluwer.

Gough, J.J., McIndoe, E.C., and Lewis, G.B. (1994). The use of dimethoate as a reference compound in laboratory acute toxicity tests on honey bees (*Apis mellifera* L.) 1981–1992. *Journal of Apicultural Research* 33, 119–125.

Grand, T., Heinz, S.K., Huse, G. et al. 2006. A standard protocol for describing individual-based and agent-based models. *Ecological Modelling* 198, 115–126.

Grant, A. and Middleton, R. (1990). An assessment of metal contamination in the Humber Estuary, U.K. *Estuarine, Coastal and Shelf Science* 31, 71–85.

Gray, J.S. (1981). *The Ecology of Marine Sediments.* Cambridge: Cambridge University Press.

Green, R.E., Taggart, M.A, Das, D. et al. (2006). Collapse of Asian vulture populations: risk of mortality from the veterinary drug diclofenac in carcasses of treated cattle *Journal of Applied Ecology* 43 949–956.

Greig-Smith, P.W. and Hardy, A.R. (1992). Design and management of the Boxworth project. In Greig-Smith, P.W., Frampton, G., and Hardy, T., eds. *Pesticides, Cereal Farming and the Environment: The Boxworth Project.* London: HMSO.

Greig-Smith, P.W., Becker, H., Edwards, P.J. et al., eds. (1992a). *Ecotoxicology of Earthworms.* Andover, Intercept.

Greig-Smith, P.W., Frampton, G., and Hardy, T. (1992b). *Pesticides, Cereal Farming and the Environment. The Boxworth Project.* London: HMSO.

Grimm, V. and Railsback, S.F. (2005). *Individual-Based Modeling and Ecology.* Princeton, NJ: Princeton University Press.

Grimm, V., Berger, U., Bastiansen, F., Eliassen, S., Ginot, V., Giske, J., Goss-Custard, J., Grand, T., Heniz, S.K., Huse, G., et al. (2006). A standard protocol for describing individual-based and agent-based models. *Ecological Modelling* 198, 115–126.

Grimm, V., Berger, U., Bastiansen, F. et al. (2010). The ODD protocol: a review and first update. *Ecological Modelling* 221, 2760–2768.

Grue, C.E., Fleming, W.J., Busby, D.G., and Hill, E.F. (1983). Assessing hazards of organophosphate pesticides to wildlife. *Transactions North American Wildlife Conference* 48, 200–220.

Grue, C.E., Hart, A.D.M., and Mineau, P. (1991). Biological consequences of depressed brain cholinesterase activity in wildlife. In Mineau, P., ed. *Cholinesterase-Inhibiting Insecticides: Their Impact on Wildlife and the Environment*, pp. 151–210. Amsterdam: Elsevier.

Grue, C.E., Hoffman, D.J., Beyer, W.N., and Franson, L.P. (1986). Lead concentrations and reproductive success in European starlings *Sturnus vulgaris* nesting within highway roadside verges. *Environmental Pollution* 42A, 157–182.

Guilhermino, L., Barros, P., Silva, M.C., and Soares, A.M.V.M. (1998). Should the use of inhibition of cholinesterases as a specific biomarker for organophosphate and carbamate pesticides be questioned? *Biomarkers* 3, 157–163.

Gunther, A.J., Davis, J.A., Hardin, D.D., Gold, J. et al. (1999). Long-term bioaccumulation monitoring with transplanted bivalves in the San Francisco Estuary. *Marine Pollution Bulletin* 38, 170–181.

Gupta, S.K. and Sundararaman, V. (1991). Correlation between burrowing capability and AChE activity in the earthworm, *Pheretima posthuma*, on exposure to carbaryl. *Bulletin of Environmental Contamination and Toxicology* 46, 859–865.

Guthrie, F.E. and Perry, J.J., eds. (1980). *Introduction to Environmental Toxicology*. New York: Elsevier.

Hågvar, S. and Abrahamsen, G. (1990). Microarthropoda and Enchytraeidae (Oligochaeta) in naturally lead-contaminated soil: a gradient study. *Environmental Entomology* 19, 1263–1277.

Hahn, M.E. (1998). Mechanisms of innate and acquired resistance to dioxin-like compounds. *Reviews in Toxicology* 2, 395–443.

Haimi, J. and Siira-Pietikainen, A. (1996). Decomposer animal communities in forest soil along heavy metal pollution gradient. *Fresenius Journal of Analytical Chemistry* 354, 672–675.

Hamer, D.M. (1986). Metallothionein. *Annual Review of Biochemistry* 55, 913–951.

Hamers, T. and Krogh, P.H. (1997). Predator–prey relationships in a two species toxicity test system. *Ecotoxicology and Environmental Safety* 37, 203–212.

Hamilton, G.A., Hunter, K., Ritchie, A.S. et al. (1976). Poisoning of wild geese by carbophenothion treated winter wheat. *Pesticide Science* 7, 175–183.

Handy, R.D. and Depledge, M.H. (1999). Physiological responses; their measurement and use as environmental biomarkers in ecotoxicology. *Ecotoxicology* 8, 329–349.

Harborne, J.B. (1993). *Introduction to Ecological Biochemistry*, 4th ed. London: Academic Press.

Hario, M., Hirvi, J.P., Hollmen, T. and Rudback, E. (2004). Organochlorine concentrations in diseased vs. healthy gull chicks from the northern Baltic. *Environmental Pollution* 127, 411–423.

Harries, J.E., Janbakhsh, A., Jobling, S., Matthiessen, P. et al. (1999). Estrogenic potency of effluent from two sewage treatment works in the United Kingdom. *Environmental Toxicology and Chemistry* 18, 932–937.

Harries, J.E., Sheahan, D.A., Jobling, S., Matthiessen, P. et al. (1997). Estrogenic activity in five United Kingdom rivers detected by measurement of vitellogenesis in caged male trout. *Environmental Toxicology and Chemistry* 16, 534–542.

Hart, A.D.M. (1993). Relationships between behaviour and the inhibition of acetylcholinesterase in birds exposed to organophosphorus pesticides. *Environmental Toxicology and Chemistry* 12, 321–336.

Hart, A.D.M. and Thompson, H.M. (2005). Improved approaches to assessing long-term risks to birds and mammals. *Ecotoxicology* 14.

Hart, L.G., Shultice, R.W., and Fouts, J.R. (1963). Stimulatory effects of chlordane on hepatic microsomal drug metabolism in the rat. *Toxicology and Applied Pharmacology* 5, 371–386.

Hartl, D.L. and Clark, A.G. (1989). *Principles of Population Genetics*. Sunderland, MA: Sinauer.

Hartwell, S.I., Cherry, D.S., and Cairns, J., Jr. (1987). Field validation of avoidance of elevated metals by fathead minnows (*Pimephales promelas*) following *in situ* acclimation. *Environmental Toxicology and Chemistry* 6, 189–200.

Hassall, K.A. (1990). *The Biochemistry and Uses of Pesticides*, 2nd ed. London: Macmillan.

Hayes, W.J. and Laws, E.R. (1991). *Handbook of Pesticide Toxicology, Vol. 2. Classes of Pesticides*. San Diego: Academic Press.

Hedgecott, S. (1994). Prioritization and standards for hazardous chemicals. In Calow, P., ed. *Handbook of Ecotoxicology*, pp. 368–393. Oxford: Blackwell.

Hegdal, P.L. and Blaskiewicz, R.W. (1984). Evaluation of the potential hazard to barn owls of talon (brodifacoum bait) used to control rats and house mice. *Environmental Toxicology and Chemistry* 3, 167–179.

Heggestad, H.E. (1991). Origin of Bel-W3, Bel-C and Bel-B tobacco varieties and their use as indicators of ozone. *Environmental Pollution* 74, 264–291.

Heliövaara, K., Väisänen, R., Braunschweiler, H., and Lodenius, M. (1987). Heavy metal levels in two biennial pine insects with sap-sucking and gall-forming life styles. *Environmental Pollution* 48, 13–23.

Hensbergen, P.J., Donker, M.H., Van Velzen, M.J.M., Roelofs, D. et al. (1999). Primary structure of a cadmium-induced metallothionein from the insect *Orchesella cincta* (Collembola). *European Journal of Biochemistry* 259, 197–203.

Herbert, I.N., Svendsen, C., Hankard, P.K. and Spurgeon, D.J. (2004). Comparison of instantaneous rate of population increase and critical-effect estimates in *Folsomia candida* exposed to four toxicants. *Ecotoxicology and Environmental Safety* 57, 175–183.

Hickey, J. J. (1969). *The Peregrine Falcon Populations: Their Biology and Decline.* Madison, WI: University of Wisconsin Press.

Hickey, J.J., Keith, J.A., and Coon, F.B. (1966). An exploration of pesticides in a Lake Michigan ecosystem. *Journal of Applied Ecology* 3, 141–154.

Hill, E.F. and Fleming, W.J. (1982). Anticholinesterase poisoning of birds: field monitoring and diagnosis of acute poisoning. *Environmental Toxicology and Chemistry* 1, 27–38.

Hites, R.A., Foran, J.A., Carpenter, D.O., Hamilton, M.C., Knuth, B.A. and Schwager, S.J. (2003). Global assessment of organic contaminants in farmed salmon. *Science* 303, 226–229.

Hobbs, K.E., Muir, D.C.G., Michaud, R. et al. (2003). PCBs and organochlorine pesticides in blubber biopsies from free-ranging St. Lawrence River estuary beluga whales (*Delphinapterus leucas*), 1994–1998. *Environmental Pollution* 122, 291–302.

Hodgson, E. and Kuhr, R.J., eds. (1990). *Safer Insecticides: Development and Use.* New York, Marcel Dekker.

Hodgson, E. and Levi, P. (1994). *Introduction to Biochemical Toxicology,* 2nd ed. Norwalk, CT: Appleton and Lange.

Hoffman, A.A. and Parsons, P.A. (1991). *Evolutionary Genetics and Environmental Stress.* Oxford: Oxford University Press.

Hoffman, D.J., Rattner, B.A., Burton, G.A., Jr., et al., eds. (1995). *Handbook of Ecotoxicology.* Boca Raton, FL: Lewis.

Holloway, G.J., Sibly, R.M., and Povey, S.R. (1990). Evolution in toxin-stressed environments. *Functional Ecology* 4, 289–294.

Holmes, S.B. and Boag, P.T. (1990). Inhibition of brain and plasma cholinesterase activity in zebra finches orally dosed with fenitrothion. *Environmental Toxicology and Chemistry* 9, 323–334.

Hopkin, S.P. (1989). *Ecophysiology of Metals in Terrestrial Invertebrates.* Barking, UK: Elsevier Applied Science.

Hopkin, S.P. (1990). Critical concentrations, pathways of detoxification and cellular ecotoxicology of metals in terrestrial arthropods. *Functional Ecology* 4, 321–327.

Hopkin, S.P. (1993a). *In situ* biological monitoring of pollution in terrestrial and aquatic ecosystems. In Calow, P., ed. *Handbook of Ecotoxicology,* Vol. 1, pp. 397–427. Oxford: Blackwell.

Hopkin, S.P. (1993b). Deficiency and excess of copper in terrestrial isopods. In Dallinger, R. and Rainbow, P.S., eds. *Ecotoxicology of Metals in Invertebrates,* pp. 359–382. Boca Raton, FL: Lewis.

Hopkin, S.P. (1993c). Ecological implications of 95% protection levels for metals in soils. *Oikos* 66, 137–141.

Hopkin, S.P. (1995). Deficiency and excess of essential and non-essential metals in terrestrial insects. *Symposia of the Royal Entomological Society of London* 17, 251–270.

Hopkin, S.P. (1997). *Biology of the Springtails (Insecta: Collembola).* Oxford: Oxford University Press.

Hopkin, S.P. and Martin, M.H. (1984). Heavy metals in woodlice. *Symposia of the Zoological Society of London* 53, 143–166.

Hopkin, S.P. and Spurgeon, D.J. (2000). Forecasting the environmental effects of zinc, the metal of benign neglect in soil ecotoxicology. In Rainbow, P.S., Hopkin, S.P., and Crane, M., eds. *Forecasting the Fate and Effects of Toxic Chemicals*. Chichester: John Wiley.

Hopkin, S.P., Hames, C.A.C., and Dray, A. (1989). X-ray microanalytical mapping of the intracellular distribution of pollutant metals. *Microscopy and Analysis* 14, 23–27.

Hopkin, S.P., Hardisty, G.N., and Martin, M.H. (1986). The woodlouse *Porcellio scaber* as a biological indicator of zinc, cadmium lead and copper pollution. *Environmental Pollution* (Series B) 11, 271–290.

Hove, K., Pederson, O., Garmo, T.H. et al. (1990). Fungi; a major source of radiocesium contamination of grazing ruminants in Norway. *Health Physics* 59, 189–192.

Huckle, K.R., Warburton, P.A., Forbes, S., and Logan, C.J. (1989). Studies on the fate of flucoumafen in the Japanese quail (*Coturnix coturnix japonica*). *Xenobiotica* 18, 51–62.

Hudson, R.H., Tucker, R.K., and Haegele, M.A. (1984). *Handbook of Toxicity of Pesticides to Wildlife*, p. 153. U.S. Fish and Wildlife Service Resource Publishers.

Huet, M., Paulet, Y.M., and Le Pennec, M. (1996). Survival of *Nucella lapillus* in a tributyltin-polluted area in West Brittany: a further example of a male genital defect (Dumpton syndrome) favouring survival. *Marine Biology* 125, 543–549.

Huggett, D.B., Brooks, B.W., Peterson, B. et al. (2002). Toxicity of select beta adrenergic receptor blocking pharmaceuticals (β blockers) on aquatic organisms. *Archives Environmental Contamination and Toxicology* 41, 229–235.

Huggett, R.J., Kimerle, R.H., Mehrle, P.M., Jr., et al., eds. (1992). *Biomarkers. Biochemical, Physiological, and Histological Markers of Anthropogenic Stress*. Boca Raton, FL: Lewis.

Hutton, M. (1980). Metal contamination of feral pigeons *Columba livia* from the London area. Part 2. Biological effects of lead exposure. *Environmental Pollution* 22A, 281–293.

Hynes, H.B.N. (1960). *The Biology of Polluted Waters*. Liverpool: Liverpool University Press.

Inoue, K., Osakabe, M., Ashihard, W., and Hamamura, T. (1986). Factors affecting abundance of the Kanzawa spider mite *Tetranychus kanzawai* on grapevine in a glasshouse. *Bulletin of the Fruit Tree Research Station* Series E, 103–116.

Iwasa, T., Motoyama, N., Ambrose, J.T. et al. (2004). Mechanism for the differential toxicity of neonicotinoid insecticides in the honeybee. *Crop Protection* 23, 371–378.

Jaeger, K. (1970). *Aldrin, Dieldrin, Endrin and Telodrin*. Amsterdam: Elsevier.

Jagoe, C.H., Dallas, C.E., Chesser, R.K., Smith, M.H. et al. (1998). Contamination near Chernobyl: radiocaesium, lead and mercury in fish and sediment radiocaesium from waters within the 10 km zone. *Ecotoxicology* 7, 201–209.

Jak, R.G., Ceulemans, M., Scholten, M.C.T., and Van Straalen, N.M. (1998). Effects of tributyltin on a coastal North Sea plankton community in enclosures. *Environmental Toxicology and Chemistry* 17, 1840–1847.

Janssen, P.A.H., Faber, J.H., and Bosveld, A.T.C. (1998). (Fe)male? *IBN Science Contribution* 13.

Janssen, R.P.T., Posthuma, L., Baerselman, R., Den Hollander, H.A. et al. (1997). Equilibrium partitioning of heavy metals in Dutch field soils. II. Prediction of metal accumulation in earthworms. *Environmental Toxicology and Chemistry* 16, 2479–2488.

Jensen, C.S., Garsdal, L., and Baatrup, E. (1997). Acetylcholinesterase inhibition and altered locomotor behavior in the carabid beetle *Pterostichus cupreus*. A linkage between biomarkers at two levels of biological complexity. *Environmental Toxicology and Chemistry* 16, 1727–1732.

Jeschke, P. and Nauen, R. (2008). Neonicotinoids–from zero to hero in insecticide chemistry. *Pest Management Science* 64, 1082–1098.

Jobling, S., Beresford, N., Nolan, M. et al. (2002a). Altered sexual maturation and gamete production in wild roach living in rivers that receive treated sewage effluents. *Biology of Reproduction* 66, 272–281.

Jobling, S., Coey, S., Whitmore, J.G. et al. (2002b). Wild intersex roach have reduced fertility. *Biology of Reproduction* 67, 515–524.

Johansen, P., Asmund, G. and Riget, F. (2004). High human exposure to lead through consumption of birds hunted with lead shot. *Environmental Pollution* 127, 125–129.

Johnston, G.O. (1995). The study of interactive effects of pesticides in birds — a biomarker approach. *Aspects of Applied Biology* 41, 25–31.

Jones, K.C., Symon, C., Taylor, P.J. et al. (1991). Evidence for a decline in rural herbage lead levels in the U.K. *Atmospheric Environment* 25A, 361–369.

Jorgensen, S.E., ed. (1991). *Modelling in Ecotoxicology.* Amsterdam: Elsevier.

Jukes, T. (1985). Selenium not for dumping. *Nature* 349, 438–440.

Kannan, K., Kajiwara, N., Le Boeuf, B.J. and Tanabe, S. (2004). Organochlorine pesticides and poly-chlorinated biphenyls in Californian sea lions. *Environmental Pollution* 131, 425–434.

Kazakov, V.S., Demidchik, E.P., and Astakhova, L.N. (1992). Thyroid cancer after Chernobyl. *Nature* 359, 21.

Kedwards, T.J., Maund, S.J., and Chapman, P.F. (1999). Community analysis of ecotoxicological field studies: II. Replicated design studies. *Environmental Toxicology and Chemistry* 18, 158–166.

Kelce, W.R., Stone, C.R., Laws, S.C. et al. (1995). Persistent DDT metabolite p,p'DDE is a potent androgen receptor antagonist. *Nature* 375, 581–585.

Kennedy, S.W. and James, C.A. (1993). Improved method to extract and concentrate porphyrins from liver tissue for analysis by high-performance liquid chromatography. *Journal of Chromatography* 619, 127–132.

Kettlewell, B. (1973). *The Evolution of Melanism.* Oxford: Clarendon Press.

Kidd, K.A., Blanchfield, P.A., Hines, K.H. et al. (2007). Collapse of a fish population after exposure to a synthetic estrogen *Proceedings of National Academy of Science of USA* 104, 8897–8901.

Kinter, W.B., Merkins, L.S., Janicki, R.H., and Guarino, A.M. (1972). Studies on the mechanism of toxicity of DDT and polychlorinated biphenyls (PCBs): disruption of osmoregulation in marine fish. *Environmental Health Perspectives* 1, 169–173.

Koeman, J.H. and Van Genderen, H. (1972). Tissue levels in animals and effects caused by chlori-nated hydrocarbons, chlorinated biphenyls and mercury in the marine environment. In *Marine Pollution and Sea Life*, pp. 1–8. West Byfleet, Fishing New Books.

Konemann, H. (1981). QSAR relationships in fish toxicity studies. I: Relationships of 50 industrial pollutants. *Toxicology* 19, 209–221.

Korsloot, A., Van Gestel, C.A.M., and Van Straalen, N.M (2004). *Environmemntal Stress and Cellular Response in Arthropods.* Boca Raton, FL: CRC Press.

Koss, G., Schuler, E., Arndt, B., Siedel, J. et al. (1986). A comparative toxicological study on pike (*Esox lucius* L.) from the River Rhine and River Lahn. *Aquatic Toxicology* 8, 1–9.

Krebs, C.J. (1999). *Ecological Methodology,* 2nd ed. Menlo Park, CA: Benjamin/Cummings.

Kruckeberg, A.L. and Wu, L. (1992). Copper tolerance and copper accumulation of herbaceous plants colonizing inactive Californian copper mines. *Ecotoxicology and Environmental Safety* 23, 307–319.

Kudo, A., Miyahara, S., and Miller, D.R. (1980). Movement of mercury from Minimata Bay into Yatsushiro Sea. *Progress in Water Technology* 12, 509–524.

Kuhr, R. and Dorough, W. (1977). *Carbamate Insecticides.* Boca Raton, FL: CRC Press.

Lacerda, L.D., Pfeiffer, W.C., Ott, A.T., and Silveira, E.G. (1989). Mercury contamination in the Madeira River, Amazon — Hg inputs to the environment. *Biotropica* 21, 91–93.

Lammenga, J. and Laskowski, R., eds. (2000). *Demography in Ecotoxicology.* Chichester: John Wiley.

Lange, R., Hutchinson, T.H., Croudace, C.P. et al. (2001). Effects of the synthetic estrogen 17 alpha-ethinylestradiol on the life cycle of the fathead minnow. *Environmental Toxicology and Chemistry* 20, 1216–1227.

Langston, W.J. (1996). Recent developments in TBT ecotoxicology. *Toxicology and Ecotoxicology News* 3, 179–187.

Langston, W.J. and Bebianno, M.J., eds. (1998). *Metal Metabolism in Aquatic Environments.* London: Chapman & Hall.

Larsson, D.J.G., Adolfsson-Erici, M., Parkkonen, J. et al. (1999). Ethinyloestradiol: an undesired fish contraceptive? *Aquatic Toxicology* 45, 91–97.

Leahey, J.P. (1985). *The Pyrethroid Insecticides.* London: Taylor & Francis.

Leivestad, H., Hendry, G., Muniz, I.P., and Snekvik, E. (1976). Effects of acid precipitation on freshwater organisms. In Brakke, F.H., ed. *Impact of Acid Precipitation on Forest and Freshwater Ecosystems in Norway*, pp. 87–111. Research Report FR 6/76. Oslo, SNSF Project.

Lepp, N.W. and Dickinson, N.M. (1994). Fungicide-derived copper in tropical plantation crops. In Ross, S.M., ed. *Toxic Metals in Soil–Plant Systems*, pp. 367–393. Chichester: John Wiley.

Levin, L., Caswell, H., Bridges, T., Dibacco, C. et al. (1996). Demographic responses of estuarine polychaetes to sewage, algal and hydrocarbon additions: life table response experiments. *Ecological Application* 6, 1295–1313.

Levinton, J.S., Suatoni, E., Wallace, W. et al. (2003). Rapid loss of genetically based resistance to metals after clean up of a Superfund site. *Proceedings of the National Academy of Science of USA* 100, 9889–9891.

Lewis, D.F.V (1996). *Cytochromes P450: Structure, Function, and Mechanism*. London: Taylor & Francis.

Lewis, M.A. (1990). Are laboratory-derived toxicity data for freshwater algae worth the effort? *Environmental Toxicology and Chemistry* 9, 1279–1284.

Lima, S.L. and Dill, L.M. (1990). Behavioural decisions made under the risk of predation: a review and prospectus. *Canadian Journal of Zoology* 68, 619–640.

Linke-Gamenick, I., Forbes, V.E., and Sibly, R.M. (1999). Density-dependent effects of a toxicant on life history traits and population dynamics of a capitellid polychaete. *Marine Ecology Progress Series* 184, 139–148.

Little, E.J., McCaffery, A.R., Walker, C.H., and Parker, T. (1989). Evidence for an enhanced metabolism of cypermethrin by a monooxygenase in a pyrethroid-resistant strain of the tobacco budworm (*Heliothis virescens* F.). *Pesticide Biochemistry and Physiology* 34, 58–68.

Livingstone, D.R. and Stegeman, J.J., eds. (1998). Forms and functions of cytochrome P_{450}. *Comparative Biochemistry and Physiology* 121C(suppl).

Lizotte, R.E., Dorn, P.B., Wade S.R. et al. (2002). Ecological effects of an anionic alkylethoxysulphate surfactant in outdoor stream mesocosms. *Environmental Toxicology and Chemistry* 21, 2742–2751.

Loganathan, B.G., Tanabe, S., Tanaka, H., Watanabe, S. et al. (1990). Comparison of organochlorine residue levels in the striped dolphin from Western Northern Pacific, 1978–1979 and 1986. *Marine Pollution Bulletin* 21, 435–439.

Løkke, H. and Van Gestel, C.A.M., eds. (1998). *Handbook of Soil Invertebrate Toxicity Tests*. Chichester: John Wiley.

Losey, J.E., Rayor, L.S., and Carter, M.E. (1999). Transgenic pollen harms monarch larvae. *Nature* 399, 214.

Lovelock, J. (1989). *The Ages of Gaia: A Biography of Our Living Planet*. Oxford, Oxford University Press.

Lowe, V.P.W. (1991). Radionuclides and the birds at Ravenglass. *Environmental Pollution* 70, 1–26.

Lukashev, V.K. (1993). Some geochemical and environmental aspects of the Chernobyl nuclear accident. *Applied Geochemistry* 8, 419–436.

Lundholm, C.E. (1997). DDE-induced eggshell thinning in birds: effects of p,p'-DDE on the calcium and prostaglandin metabolism of the eggshell gland. *Comparative Biochemistry and Physiology* 118C, 113–128.

Lutgens, F.K. and Tarbuck, E.J. (1992). *The Atmosphere: An Introduction to Meteorology*. Hemel Hempstead: Prentice-Hall.

Mackay, D. (1991). *Multimedia Environmental Models: The Fugacity Approach*. Boca Raton, FL: Lewis.

Mackenzie, A.B. and Scott, R.D. (1993). Sellafield waste radionuclides in Irish Sea intertidal and salt marsh sediments. *Environmental Geochemistry and Health* 15, 173–184.

Mackenzie, A.B., Scott, R.D., Allan, R.L., Ben Shaban, Y.A. et al. (1994). Sediment radionuclide profiles: implications for mechanisms of Sellafield waste dispersal in the Irish Sea. *Journal of Environmental Radioactivity* 23, 36–69.

Macnair, M.R. (1981). Tolerance of higher plants to toxic materials. In Bishop, J.A. and Cook, L.S., eds. *Genetic Consequences of Man-Made Changes*. London: Academic Press.

Macnair, M.R. (1987). Heavy metal tolerance in plants: a model evolutionary system. *Trends in Evolution and Ecology* 2, 354–359.

Main, A.R. (1980). (A) Toxicant–receptor interactions: fundamental principles, pp. 180–192. (B) Cholinesterase inhibitors, pp. 193–223. In Hodgson, E. and Guthrie, F.E., eds. *Introduction to Biochemical Toxicology*. New York: Elsevier.

Majerus, M. (1998). *Melanism: Evolution in Action.* Oxford: Oxford University Press.

Malins, D.C. and Collier, T.K. (1981). Xenobiotic interactions in aquatic organisms: effects on biological systems. *Aquatic Toxicology* 1, 257–268.

Mallet, J. (1989). The evolution of insecticide resistance: have the insects won? *Trends in Evolution and Ecology* 4, 336–340.

Maltby, L., Kedwards, T.K., Forbes, V.E. et al. (2001). Linking individual-level responses and population-level consequences. In Baird, D.J. and Burton, G.A., Jr., eds. *Ecological Variability: Separating Natural from Anthropogenic Causes of Ecosystem Impairment*, pp. 27–82. Pensacola, FL: SETAC.

Maltby, L. and Naylor, C. (1990). Preliminary observations on the ecological relevance of the Gammarus "scope for growth" assay: effect of zinc on reproduction. *Functional Ecology* 4, 393–397.

Maltby, L., Naylor, C., and Calow, P. (1990a). Effect of stress on a freshwater benthic detrivore: scope for growth in *Gammarus pulex. Ecotoxicology and Environmental Safety* 19, 285–291.

Maltby, L., Naylor, C., and Calow, P. (1990b). Field deployment of a scope for growth assay *Gammarus pulex* a freshwater benthic detrivore. *Ecotoxicology and Environmental Safety* 19, 292–300.

Manahan, S.E. (1994). *Environmental Chemistry*, 6th ed. Boca Raton, FL: Lewis.

Markert, B. and Weckert, V. (1994). Higher lead concentrations in the environment of former West Germany after the fall of the Berlin Wall. *Science of the Total Environment* 158, 93–96.

Marshall, J. (1992). Weeds. In Grieg-Smith, P., Frampton, G., and Hardy, T., eds. *Pesticides, Cereal Farming and the Environment. The Boxworth Project.* London: HMSO.

Martikainen, E.A.T. and Krogh, P.H. (1999). Effects of soil organic matter content and temperature on toxicity of dimethoate to *Folsomia fimetaria* (Collembola: Isotomidae). *Environmental Toxicology and Chemistry* 18, 865–872.

Martin, M.H. and Bullock, R.J. (1994). The impact and fate of heavy metals in an oak woodland ecosystem. In Ross, S.M., ed. *Toxic Metals in Soil–Plant Systems*, pp. 327–365. Chichester: John Wiley.

Martineau, D., Lagace, A., Beland, P., Higgins, C.R. et al. (1988). Pathology of stranded beluga whales (*Delphinapterus leucas*) from the St. Lawrence estuary (Quebec, Canada). *Journal of Comparative Pathology* 38, 287–308.

Martinson, L., Lamersdorf, N. and Warfvinge, P. (2005). The Solling roof revisited: slow recovery from acidification observed and modelled despite a decade of 'clean rain' treatment. *Environmental Pollution* 135, 293–302.

Matthiessen, P. and Gibbs, P.E. (1998). Critical appraisal of the evidence for tributyl-mediated endocrine disruption in molluscs. *Environmental Toxicology and Chemistry* 17, 37–43.

Maund, S.J., Chapman, P., Kedwards, T. et al. (1999). Application of multivariate statistics to ecotoxicological field studies. *Environmental Toxicology and Chemistry* 18, 111–112.

McCarthy, J.F. and Shugart, L.R. (1990). Biological markers of environmental contamination. In McCarthy, J.F. and Shugart, L.R., eds. *Biomarkers of Environmental Contamination*, pp. 3–14. Boca Raton, FL: Lewis.

McNeilly, T. (1968). Evolution in closely adjacent plant populations. III. *Agrostis tenvis* on a small copper mine. *Heredity* 23, 99–108.

Mehlhorn, H., O'Shea, J.M., and Wellburn, A.R. (1991). Atmospheric ozone interacts with stress ethylene formation by plants to cause visible plant injury *Journal of Experimental Botany* 42, 17–24.

Mellanby, K. (1967). *Pesticides and Pollution.* London: Collins.

Merian, E., ed. (1991). *Metals and Their Compounds in the Environment.* Weinheim: VCH.

Meyer, J.N., Nacci, D.E., and Di Giulio, R.T. (2002). Cytochrome P4501A in killifish (*Fundulus heteroclitus*). Heritability of altered expression and relationship to survival in contaminated sediments. *Toxicological Sciences* 68, 69–81.

Miller, K.L., Fernandes, T.F., and Read, P.A. (1999). The recovery of populations of dog whelks suffering from imposex in the Firth of Forth 1987–1997/98. *Environmental Pollution* 106, 183–192.

Mineau, P., ed. (1991). *Cholinesterase-Inhibiting Insecticides: Their Impact on Wildlife and the Environment.* Amsterdam: Elsevier.

Mineau, P. and Peakall, D.B. (1987). An evaluation of avian impact assessment techniques following broad-scale forest insecticide spraying. *Environmental Toxicology and Chemistry* 6, 781–791.

Moore, N.W. and Walker, C.H. (1964). Organic chlorine insecticide residues in wild birds. *Nature* 201, 1072–1073.

Moran, P.J. and Grant, T.R. (1991). Transference of marine fouling communities between polluted and unpolluted sites; impact on structure. *Environmental Pollution* 72, 89–102.

Moriarty, F., ed. (1975). *Organochlorine Insecticides: Persistent Organic Pollutants.* London: Academic Press.

Moriarty, F. (1999). *Ecotoxicology,* 3rd ed. London: Academic Press.

Moriarty, F. and Walker, C.H. (1987). Bioaccumulation in food chains: a rational approach. *Ecotoxicology and Environmental Safety* 13, 208–215.

Mose, D.G., Mushrush, G.W., and Chrosniak, C.E. (1992). A two-year study of seasonal indoor radon variations in southern Maryland. *Environmental Pollution* 76, 195–199.

Mouw, O., Kalitis, K., Anver, M., Schwartz, J. et al. (1975). Lead. Possible toxicity in urban vs. rural rats. *Archives Environmental Health* 30, 276–280.

Muller www.es.ucsb.edu/faculty/muller/toxicokineticses120

Murk, A.J., Leonards, P.E.G., Bulder, A.S., Jonas, A. et al. (1997). The CALUX (chemical-activated luciferase expression) assay, a sensitive *in vitro* bioassay for measuring TCDD-equivalents in blood plasma. *Environmental Toxicology and Chemistry* 16, 1583–1589.

Murton, R.K. (1965). *The Wood Pigeon.* London, Collins New Naturalist Series No. 20.

Nacci, D.E., Champlin, D., Coiro, L. et al. (2002). Predicting the occurence of genetic adaptation to dioxin-like compounds in populations of the estuarine fish *Fundulus heteroclitus. Environmental Toxicology and Chemistry* 21, 1525–1532.

Nacci, D.E., Coiro, L., Champlin, D. et al. (1999). Adaptations of wild populations of the estuarine fish *Fundulus heteroclitus* to persistent environmental contaminants. *Marine Biology* 134, 9–17.

Nayak, B.N. and Petras, M.L. (1985). Environmental monitoring for genotoxicity: *in vivo* sister chromatid exchange in the house mouse (*Mus musculus*). *Canadian Journal of Genetics and Cytology* 27, 351–356.

Naylor, C., Maltby, L., and Calow, P. (1989). Scope for growth in *Gammarus pulex,* a freshwater benthic detrivore. *Hydrobiologia* 188/189, 517–523.

Nebert, D.W. and Gonzalez, F.J. (1987). P_{450} genes: structure, evolution and regulation. *Annual Review of Biochemistry* 56, 945–993.

Nelson, D.R. (1998). Metazoan P_{450} evolution. In Livingstone, D.R. and Stegeman, J.J., eds. *Forms and Functions of Cytochrome P_{450},* pp. 15–22. *Comparative Biochemistry and Physiology* 121C.

Nelson, R.W. (1976). Behavioural aspects of egg breakage in peregrine falcons. *Canadian Field Naturalist* 90, 320–329.

Newman, M.C. (2010). *Fundamentals of Ecotoxicology.* CRC Press: Boca Raton, FL.

Newman, M.C. and Unger, M.A. (2003). *Fundamentals of Ecotoxicology,* 2nd ed. Boca Raton, FL: Lewis.

Newton, I. (1986). *The Sparrowhawk.* Calton, Poyser.

Newton, I. (1988). Determination of critical pollutant levels in wild populations, with examples from organochlorine insecticides in birds of prey. *Environmental Pollution* 55, 29–40.

Newton, I. and Haas, M.B. (1984). The return of the sparrowhawk. *British Birds* 77, 47–70.

Newton, I. and Wyllie, I. (1992). Recovery of a sparrowhawk population in relation to declining pesticide contamination. *Journal of Applied Ecology* 29, 476–484.

Nieboer, E. and Richardson, D.H.S. (1980). The replacement of the nondescript term "heavy metals" by a biologically and chemically significant classification of metal ions. *Environmental Pollution* 1B, 3–26.

Nosengo, N. (2003). Fertilized to death. *Nature* 425, 894–895.

NRC (National Research Council) (1985). *Oil in the Sea: Inputs Fates and Effects.* Washington: National Academy Press.

Oberdorster, G. (2001). Pulmonary effects of inhaled ultrafine particles. *International Archives of Occupational and Environmental Health* 74, 1–8.

OECD (1991). *State of the Environment.* Paris: OECD.

Ollson, M., Anderson, O., Bergman, A., Blomkist, G. et al. (1992). Contaminants and diseases in seals from Swedish waters. *Ambio* 21, 561–562.

344 Bibliography

Oppenoorth, F.J. (1985). Biochemistry and genetics of insecticide resistance. In Kerkut, G.A. and Gilbert, L.I., eds. *Comprehensive Insect Physiology Biochemistry and Pharmacology* 12. Oxford: Pergamon Press.

Paasivirta, J. (1991). *Chemical Ecotoxicology*. Chelsea, MI: Lewis.

Pain, D. and Pienkowski, M., eds. (1997). *Farming and Birds in Europe*. New York: Academic Press.

Pankhurst, C.E., Doube, B.M., and Gupta, V.V.S.R., eds. (1997). *Biological Indicators of Soil Health*. Wallingford: CAB International.

Pastor, N., Baos, R., Lopez-Lazaro, M., Jovani, R., Tella, J.L., Hajji, N., Hiraldo, F. and Cortes, F. (2004). A 4 year follow-up analysis of genotoxic damage in birds of the Donana area (southwest Spain) in the wake of the 1998 mining waste spill. *Mutagenesis* 19, 61–65.

Payne, J.F., Fancey, L.L., Rahimtula, A.D., and Porter, E.L. (1987). Review and perspective on the use of mixed-function oxygenase enzymes in biological monitoring. *Comparative Pharmacology and Physiology* 86C, 233–245.

Payne, J.F., Mathieu, A., Melvin, W., and Fancey, L.L. (1996). Acetylcholinesterase, an old biomarker with a new future? Field trials in association with two urban rivers and a paper mill in Newfoundland. *Marine Pollution Bulletin* 32, 225–231.

Peakall, D.B. (1985). Behavioural responses of birds to pesticides and other contaminants. *Residue Reviews* 96, 45–77.

Peakall, D.B. (1992). *Animal Biomarkers as Pollution Indicators*. London: Chapman & Hall.

Peakall, D.B. (1993). DDE-induced eggshell thinning: an environmental detective story. *Environmental Reviews* 1, 13–20.

Peakall, D.B. and Bart, J.R. (1983). Impacts of aerial application of insecticides on forest birds. *Critical Reviews in Environmental Control* 13, 117–165.

Peakall, D.B. and Carter, N. (1997). Decreases in farmland birds and agricultural practices: a huge ecotoxicological experiment. *Toxicology and Ecotoxicology News* 4, 162–163.

Peakall, D.B. and Shugart, L.R., eds. (1993). *Biomarkers. Research and Application in the Assessment of Environmental Health*. Berlin: Springer-Verlag.

Peakall, D.B., Walker, C.H., and Migula, P. (1999). *Biomarkers: A Pragmatic Basis for Remediation of Severe Pollution in Eastern Europe*. NATO Science Series, Vol. 54. NATO. Dordrecht, Kluwer.

Pearce, P.A., Peakall, D.B., and Erskine, A.J. (1976). Impact on forest birds of the 1975 spruce budworm spray operations in New Brunswick. Canadian Wildlife Series Progress Notes No. 62.

Perkins, D.F. and Millar, R.O. (1987). Effects of airborne fluoride emissions near an aluminium works in Wales. 1. Corticolous lichens growing on broad-leaved trees. *Environmental Pollution* 47, 63–78.

Persoone, G., Janssen, C., and de Coen, W. (2000). *New Microbiotests for Routine Toxicity Screening*. New York: Kluwer.

Pfeiffer, W.C., De Lacerda, L.D., Malm, O., Souza, C.M.M. et al. (1989). Mercury concentrations in inland waters of gold-mining areas in Rondonia, Brazil. *Science of the Total Environment* 87/88, 233–240.

Pfeiffer, W.C., De Larcerda, L.D., Salomons, W., and Malm, O. (1993). Environmental fate of mercury from gold mining in the Brazilian Amazon. *Environmental Research* 1, 26–37.

Pickering, Q.C., Henderson, C., and Lemke, A.E. (1962). The toxicity of organic phosphorus insecticides to different species of warmwater fishes. *Transactions of the American Fisheries Society* 91, 175–184.

Pilling, E.D., Bromley-Challenor, K.A.C., Walker, C.H., and Jepson, P.C. (1995). Mechanisms of synergism between the pyrethroid insecticide t-cyhalothrin and the imidazole fungicide prochloraz in the honeybee (*Apismellifera*). *Pesticide Biochemistry and Physiology* 51, 1–11.

Plenderleith, R.W. and Bell, L.C. (1990). Tolerance of the twelve tropical grasses to high soil concentrations of copper. *Tropical Grasslands* 24, 103–110.

Porteous, A. (1992). *Dictionary of Environmental Science and Technology*. Chichester: John Wiley.

Postel, S. (1984). Air pollution, acid rain and the future of forests. Worldwatch Paper 58.

Posthuma, L. (1997). Effects of toxicants on population and community parameters in field conditions, and their potential use in the validation of risk assessment models. In Van Straalen, N.M. and Lokke, H., eds. *Ecological Risk Assessment of Contaminants in Soil*, pp. 85–123. London: Chapman & Hall.

Posthuma, L. and Van Straalen, N.M. (1993). Heavy-metal adaptation in terrestrial invertebrates; a review of occurrence, genetics, physiology and ecological consequences. *Comparative Biochemistry and Physiology* 106C, 11–38.

Potts, G.R. (1986). *The Partridge.* London: HarperCollins.

Potts, G.R. (2000). The grey partridge. In Pain, D. and Dixon, J., eds. *Bird Conservation and Farming Policy in the European Union.* London: Academic Press.

Prein, A.E., Thie, G.M., Koeman, J.H., and Poels, C.L.M. (1978). Cytogenic changes in fish exposed to water of the river Rhine. *Science of the Total Environment* 9, 287–291.

Purdom, C.E., Hardiman, P.A., Bye, V.J., Eno, N.C. et al. (1994). Estrogenic effects of effluents from sewage treatment works. *Chemistry Ecology* 8, 275–285.

Pyza, E., Mak, P., Kramarz, P., and Laskowski, R. (1997). Heat shock proteins (HSP70) as biomarkers in ecotoxicological studies. *Ecotoxicology and Environmental Safety* 38, 244–251.

Ramade, F. (1992). *Precis dí Ecotoxicologie.* Paris: Masson.

Ratcliffe, D. (1993). *The Peregrine Falcon,* 2nd ed. London: Carlton.

Ratcliffe, D.A. (1967). Decrease in eggshell weight in certain birds of prey. *Nature* 215, 208–210.

Read, H.J., Martin, M.H., and Rayner, J.M.V. (1998). Invertebrates in woodlands polluted by heavy metals: an evaluation using canonical correspondence analysis. *Water, Air and Soil Pollution* 106, 17–42.

Richter, C.A., Drake, J.B., Giesy, J.P., and Harrison, R.O. (1994). Immunoassay monitoring of polychlorinated biphenyls (PCBs) in the Great Lakes. *Environmental Science and Pollution Research* 1, 69–74.

Risebrough, R.W. and Peakall, D.B. (1988). The relative importance of the several organochlorines in the decline of peregrine falcon populations. In Cade, T.J., Enderson, J.H., Thelander, C.G., and White, C.M., eds. *Peregrine Falcon Populations: Their Management and Recovery*, pp. 449–468. Boise, ID, Peregrine Fund.

Robertson, L.W. and Hansen, L.G., eds (2001). *PCBs: Recent Advances in Environmental Toxicology and Health Effects.* Lexington: University of Kentucky Press.

Robinson, J., Richardson, A., Crabtree, A.N., Coulson, J.C., and Potts, G.R. (1967). Organochlorine residues in marine organisms. *Nature* 214, 1307–1311.

Roefer, P.S., Snyder, R.E., Zegers, D.J. et al. (2000). Endocrine disrupting chemicals in source water. *Journal of American Water Association* 92, 52–58.

Ronday, R. and Houx, N.W.H. (1996). Suitability of seven species of soil-inhabiting invertebrates for testing toxicity of pesticides in soil pore water. *Pedobiologia* 40, 106–112.

Ronis, M.J.J. and Walker, C.H. (1989). The microsomal monooxygenases of birds. *Reviews in Biochemical Toxicology* 10, 310–384.

Rosman, K.J.R., Chisholm, W., Boutron, C.F., Candlestone, J.P., and Gorlach, U. (1993). Isotopic evidence for the source of lead in Greenland snows since the late 1960s. *Nature* 362, 333–335.

Ross, S.M., ed. (1994). *Toxic Metals in Soil–Plant System.* Chichester: John Wiley.

Roush, R.T. and Daly, D.C. (1990). The role of population genetics in resistance research and management. In Roush, R.T. and Tabashnik, B.E., eds. *Pesticide Resistance in Arthropods.* New York: Chapman & Hall.

Routledge, E.J., Sheahan, D., Desbrow, C. et al. (1998). Identification of estrogenic chemicals in STW effluent 11. In vitro responses in trout and roach. *Environmental Science and Technology* 32, 1559–1565.

Ruhling, A. and Tyler, G. (2004). Changes in the atmospheric deposition of minor and rare elements between 1975 and 2000 in south Sweden, as measured by moss analysis. *Environmental Pollution*, 131, 417–423.

Safe, S. (1990). Polychlorinated biphenyls (PCBs), dibenzo-p-dioxins (PCDDs), dibenzofurans (PCDFs), and related compounds: environmental and mechanistic considerations which support the development of toxic equivalency factors (TEFs). *Critical Reviews in Toxicology* 21, 51–88.

Safe, S. (2001). PCBs as aryl hydrocarbon receptor agonists: implications for risk assessment. In Robertson, L.W. and Hansen, L.G., eds. *PCBs: Recent Advances in Environmnetal Toxicology and Health Effects*, pp. 171–177. Lexington: University of Kentucky Press.

Salminen, J. and Haimi, J. (1997). Effects of pentachlorophenol on soil organisms and decomposition in forest soil. *Journal of Applied Ecology* 34, 101–110.

Salomons, W., Bayne, B.L., Duursma, E.K., and Förstner, V., eds. (1988). *Pollution of the North Sea: an Assessment*. Berlin: Springer-Verlag.

Sandahl, J.F., Baldwin, D.H., Jenkins, J.J. et al. (2005). Comparative thresholds for AChase inhibition and behavioural impairment in Coho salmon exposed to chlorpyriphos. *Environmental Toxicology and Chemistry* 24, 136–145.

Sanders, B.M. (1993). The cellular stress response. *Environs* 16, 3–6.

Savva, D. (1998). Use of DNA fingerprinting to detect genotoxic effects. *Ecotoxicology and Environmental Safety* 41, 103–106.

Sawicki, R.M. and Rice, A.D. (1978). Response of susceptible and resistant peach-potato aphid *Myzus persicae* (Sulz.) to insecticides in leaf-dip assays. *Pesticide Science* 9, 513–516.

Schat, H. and Bookum, W.M.T. (1992). Genetic control of copper tolerance in *Silene vulgari. Heredity* 68, 219–229.

Scheuhammer, A.M. (1989). Monitoring wild bird populations for lead exposure. *Journal of Wildlife Management* 53, 759–765.

Scheuhammer, A.M. and Templeton, D.M. (1998). Use of stable isotope ratios to distinguish sources of lead exposure in wild birds. *Ecotoxicology* 7, 37–42.

Schindler, D.W. (1996). Ecosystems and ecotoxicology: a personal perspective. In Newman, M. and Jagoe, C., eds. *Quantitative Ecotoxicology: A Hierarchical Approach*. Boca Raton, FL: Lewis.

Schindler, D.W., Frost, T.M., Mills, K.H., Chang, P.S.S. et al. (1991). Comparisons between experimentally and atmospherically-acidified lakes during stress and recovery. *Proceedings of the Royal Society of Edinburgh* 97B, 193–226.

Schmitt, C.J. and Brumbaugh, W.G. (1990). National contaminant biomonitoring program: concentrations of arsenic, cadmium copper, lead, mercury, selenium and zinc in US freshwater fish, 1976–1984. *Archives of Environmental Contamination and Toxicology* 19, 731–747.

Schmuck, R., Schoning, R., Stonk, A., et al. (2000). Risk posed to honey bees by an imidacloprid seed dressing of sunflowers. *Pest Management Science* 57, 225–238.

Schmuck, R., Stadler, T., and Schmitt, H.-W. (2003). Field reference of a synergistic effect observed in the laboratory between an EBI fungicide and a chloronicethyl insecticide in the honeybee *Pest Management Science* 59, 279–286.

Schüürmann, G. and Markert, B., eds. (1998). *Ecotoxicology*. Chichester: John Wiley.

Schwarzenbach, R.P., Gschwend, P.M., and Imboden, D.M. (1993). *Environmental Organic Chemistry*. New York: John Wiley.

Scown, T.M., van Aele, R., and Tyler, C.R. (2010). Do engineered nanoparticles pose a significant threat in the aquatic environment? *Critical Review in Toxicology* 10, 653–670.

Seager, J. and Maltby, L. (1989). Assessing the impact of episodic pollution. *Hydrobiologia* 188/189, 633–640.

Senacha, K.R., Taggart, M.A., Rahmani, A.R. et al. (2008). Diclofenac levels in livestock carcasses in India before the ban. *Journal of the Bombay Natural History Society* 105, 148–161.

Shaw, A.J. (1999). The evolution of heavy metal tolerance in plants: adaptations, limits, and costs. In Forbes, V. E., ed. *Genetics and Ecotoxicology*. Boca Raton, FL: Taylor & Francis.

Shaw, A.J., ed. (1989). *Heavy Metal Tolerance in Plants: Evolutionary Aspects*. Boca Raton, FL: CRC Press.

Sheppard, S., Bembridge, J., Holmstrup, M., and Posthuma, L., eds. (1998). *Advances in Earthworm Ecotoxicology*. Pensacola, FL: SETAC.

Shore, R.F. and Rattner, B.A., eds. (2001). *Ecotoxicology of Wild Mammals*. Chichester: John Wiley.

Shugart, L. (1994). Genotoxic responses in blood. In Fossi, M.C. and Leonzio, C., eds. *Nondestructive Biomarkers in Vertebrates*, pp. 131–145. Boca Raton, FL: Lewis.

Sibly, R.M. (1996). Effects of pollutants on individual life histories and population growth rates. In Newman, M.C. and Jagge, C., eds. *Quantitative Ecotoxicology: A Hierarchical Approach.* Boca Raton, FL: Lewis.

Sibly, R.M. (1999). Efficient experimental designs for studying stress and population density in animal populations. *Ecological Applications* 9, 496–503.

Sibly, R.M. and Antonovics, J. (1992). Life-history evolution. In Berry, R.J., Crawford, T.J., and Hewitt, G.M., eds. *Genes in Ecology.* Oxford: Blackwell Scientific Publications.

Sibly, R.M. and Calow, P. (1989). A life cycle theory of responses to stress. *Biological Journal of the Linnean Society* 37, 110–116.

Sibly, R.M. and Hone, J. (2002). Population growth rate and its determinants: an overview. *Philosophical Transactions of the Royal Society London B* 357, 1153–1170.

Sibly, R.M., Akcakaya, H.R., Topping, C. et al. (2005). Population level assessment of risks of pesticides to birds and mammals in the U.K. *Ecotoxicology* 14, 863–876.

Sibly, R.M., Newton, I., and Walker, C.H. (2000). Effects of dieldrin on population growth rates of sparrowhawks 1963–1986. *Journal of Applied Ecology* 37, 540–546.

Sibly, R.M., Williams, T.D., and Jones, M.B. (2000). How environmental stress affects density dependence and carrying capacity in a marine copepod. *Journal of Applied Ecology* 37, 388–397.

Sibuet, M., Calmet, D., and Auffret, G. (1985). Reconnaissance photographique de contencurs en place dans la zone d'immersion des dechets faiblement radioactifs de l'Atlantique Nord-Est. *Compte Rendu Hebdomadaire des Seances de l'Academie des Sciences Paris*, Serie 111, 301, 497–550.

Simkiss, K. (1993). Radiocaesium in natural systems: a UK coordinated study. *Journal of Environmental Radioactivity* 18, 133–149.

Sinclair, A.R.E. (1989). Population regulation in animals. In Cherrett, J.M., ed. *Ecological Concepts.* Oxford: Blackwell Scientific Publications.

Sinclair, A.R.E. (1996). Mammal populations: fluctuation, regulation, life history theory and their implications for conservation. In Floyd, R.B., Sheppard, A.W., and De Barro, P.J., eds. *Frontiers of Population Ecology*, pp. 127–154. Melbourne: CSIRO Publishing.

Smit, C.E. and Van Gestel, C.A.M. (1998). Effects of soil type, prepercolation, and ageing on bioaccumulation and toxicity of zinc for the springtail *Folsomia candida. Environmental Toxicology and Chemistry* 17, 1132–1141.

Smit, C.E., Stam, E.D., Baas, N., Hollander, R. and Van Gestel, C.A.M. (2004). Effects of dietary zinc exposure on the life history of the parthenogenetic springtail *Folsomia candida* (Collembola: Isotomidae). *Environmental Toxicology and Chemistry* 23, 1719–1724.

Smith, P.J. (1996). Selective decline in imposex levels in the dog whelk *Lepsiella scobina* following a ban on the use of TBT anti-foulants in New Zealand. *Marine Pollution Bulletin* 32, 362–365.

Snape, J.R., Maund, S.J., Pickford, D.B. et al. (2004). Ecotoxicogenomics: the challenge of integrating genomics into aquatic and terrestrial ecotoxicology. *Aquatic Toxicology* 67, 143–154.

Snell, T.W., Brogdon, S.E., and Morgan, M.B. (2003). Gene expression profiling in ecotoxicology. *Ecotoxicology* 12, 475–483.

Södergren, A. (1991). Environmental fate and effects of bleached pulp mill effluents. Swedish Environment Protection Agency Report 4031.

Solomon, K.R., Giddings, J.M., and Maund, S.J. (2001). Probabilistic risk assessment of cotton pyrethroids: I. Distributional analyses of laboratory aquatic toxicity data. *Environmental Toxicology and Chemistry* 20, 652–659.

Solomons, W.G. (1986). *Organic Chemistry*, 2nd ed. New York: John Wiley.

Solonen, T. and Lodenius, M. (1990). Feathers of birds of prey as indicators of mercury contamination in Southern Finland. *Holarctic Ecology* 13, 229–237.

Somerville, L. and Greaves, M.P. (1987). *Pesticide Effects on Soil Microflora.* London: Taylor & Francis.

Somerville, L. and Walker, C.H., eds. (1990). *Pesticide Effects on Terrestrial Wildlife.* London: Taylor & Francis.

Sørensen, F.F., Weeks, J.M., and Baatrup, E. (1997). Altered locomotory behaviour in woodlice *Oniscus asellus* (L.) collected at a polluted site. *Environmental Toxicology and Chemistry* 16, 685–690.

Sorensen, T.S. and Holmstrup, M. (2005). A comparative analysis of the toxicity of eight common soil contaminants and their effects on drought tolerance in the collembolan *Folsomia candida*. *Ecotoxicology and Environmental Safety* 60, 132–139.

Spear, P.A., Moon, T.W., and Peakall, D.B. (1986). Liver retinoid concentrations in natural populations of herring gulls (*Larus argentatus*) contaminated by 2,3,7,8-tetrachlorodibenzo-p-dioxin and in ring doves (*Streptopelia risoria*) injected with a dioxin analogue. *Canadian Journal of Zoology* 64, 204–208.

Spurgeon, D.J. and Hopkin, S.P. (1996). Effects of variations of the organic matter content and pH of soils on the availability and toxicity to zinc of the earthworm *Eisenia fetida*. *Pedobiologia* 40, 80–96.

Spurgeon, D.J. and Hopkin, S.P. (1999). Seasonal variation in the abundance, biomass and biodiversity of earthworms in soils contaminated with metal emissions from a primary smelting works. *Journal of Applied Ecology* 36, 173–183.

Spurgeon, D.J. and Hopkin, S.P. (2000). The development of genetically inherited resistance to zinc in laboratory selected generations of the earthworm *Eisenia fetida*. *Environmental Pollution* 109, 193–201.

Spurgeon, D.J., Hopkin, S.P., and Jones, D.T. (1994). Effects of cadmium, copper, lead and zinc on growth, reproduction and survival of the earthworm *Eisenia fetida* (Savigny): assessing the environmental impact of point-source metal contamination in terrestrial ecosystems. *Environmental Pollution* 84, 123–130.

Stanley, T.R., Jr., Smith, G.J., Hoffman, D.D.J. et al. (1996). Effects of boron and selenium on mallard reproduction and duckling growth and survival. *Environmental Toxicology and Chemistry* 15, 1124–1132.

Stark, J.E. and Banken, J.A.O. (1999). Importance of population structure at the time of toxicant exposure. *Ecotoxicology and Environmental Safety* 42, 282–287.

Stein, J.E., Collier, T.K., Reichert, W.L. et al. (1992). Bioindicators of contaminant exposure and sublethal effects: studies with benthic fish in Puget Sound, Washington. *Environmental Toxicology and Chemistry* 11, 701–714.

Stellman, J.M., Stellman, S.D., Christian, R. et al. (2003). The extent and patterns of usage of Agent Orange and other herbicides in Vietnam. *Nature* 422, 681–687.

Suter, G.W., ed. (1993). *Ecological Risk Assessment*. Boca Raton, FL: Lewis.

Sutherland, W.J., ed. (1996). *Ecological Census Techniques*. Cambridge: Cambridge University Press.

Svendsen, C., Meharg, A.A., Freestone, P., and Weeks, J. (1996). Use of an earthworm lysosomal biomarker for the ecological assessment of pollution from an industrial plastics fire. *Applied Soil Ecology* 3, 99–107.

Tanford, C. (1980). *The Hydrophobic Effect: Formation of Micelles and Biological Membranes*, 2nd ed. New York: John Wiley.

Tardiff, R., ed. (1992). *Methods to Assess Adverse Effects of Pesticides on Non-Target Organisms*. Chichester: John Wiley.

Taylor, C.E. (1986). Genetics and evolution of resistance to insecticides. *Biological Journal of the Linnean Society* 27, 103–112.

Ternes, T.A., Stumpf, M., Mueller, K. et al. (1999). Behaviour and occurrence of estrogens in municipal sewage treatment plants 1. Investigations in Germany, Canada and Brazil. *Science of the Total Environment* 225, 81–90.

Terriere, L.C. (1984). Induction of detoxification enzymes in insects. *Annual Review of Entomology* 29, 71–88.

Theodorakis, C.W., Blaylock, B.G., and Shugart, L.R. (1997). Genetic ecotoxicology. I. DNA integrity and reproduction in mosquitofish exposed *in situ* to radionuclides. *Ecotoxicology* 6, 205–218.

Thompson, H.M. (2003). Behavioural effects of pesticides in bees — their potential for use in risk assessment. *Ecotoxicology* 12, 307–316.

Thompson, H.M. and Hunt, L.V. (1999). Extrapolating from honeybees to bumblebees in pesticide risk assessment. *Ecotoxicology* 8, 147–166.

Thorbek, P., Forbes, V.E., Heimbach, F., et al., eds. (2010). *Ecological Models for Regulatory Risk Assessments of Pesticides*. Boca Raton, FL: CRC Press.

Timbrell, J.A. (1995). *Introduction to Toxicology*, 2nd ed. London: Taylor & Francis.

Timbrell, J.A. (1999). *Principles of Biochemical Toxicology*, 3rd ed. London: Taylor & Francis.

Tolba, M.K. (1992). *Saving Our Planet*. London: Chapman & Hall.

Topping, C., Sibly,R.M., Akcakaya, H.R et al (2005) Risk assessment of UK skylark populations using life history and individual-based landscape models. *Ecotoxicology* 14, 925–936.

Trewavas, A. (2004). Are we oversensitive to risks from toxins? *Chemistry and Industry*, pp. 14–15.

Turekian, K.K. (1976). *Oceans*, 2nd ed. Hemel Hempstead: Prentice-Hall.

Turle, R., Norstrom, R.J., and Collins, B. (1991). Comparison of PCB quantitation methods: re-analysis of archived specimens of herring gull eggs from the Great Lakes. *Chemosphere* 22, 201–213.

Turtle, E.E., Taylor, A., Wright, E.N., Thearle, R.J.P. et al. (1963). The effects on birds of certain chlorinated insecticides used as seed dressings. *Journal of Science in Food and Agriculture* 8, 567–576.

Tyler, C.R. and Jobling, S. (2005). *The Ecological Relevance of Endocrine Disrption in Wildlife Environmental Health Perspectives*. In press.

Tyler, C.R., Jobling, S., and Sumpter, J.P. (1998). Endocrine disruption in wildlife: a critical review of evidence. *Critical Reviews in Toxicology* 28, 319–361.

Ueno, D. et al. (2004). Global pollution monitoring of butyltin compounds using skipjack tuna as a bioindicator. *Environmental Pollution* 127, 1–12.

UNEP (1993). *Environmental Data Report 1993–94*. Oxford: Blackwell.

van der Gaag, M.A., van der Kerkhof, J.F.J., van der Klift, H.W., and Poels, C.L.M. (1983). Toxicological assessment of river quality in bioassays with fish. *Environmental Monitoring and Assessment* 3, 247–255.

Van Gestel, C.A.M. and Hensbergen, P.J. (1997). Interaction of Cd and Zn toxicity for *Folsomia candida* Willem (Collembola: Isotomidae) in relation to bioavailability in soil. *Environmental Toxicology and Chemistry* 16, 1177–1186.

Van Gestel, C.A.M. and Koolhaas, J.E. (2004). Water-extractability, free ion activity, and pH explain cadmium sorption and toxicity to *Folsomia candida* (Collembola) in seven soil–pH combinations. *Environmental Toxicology and Chemistry* 23, 1822–1833.

Van Straalen, N.M. (1993). Soil and sediment quality criteria derived from invertebrate toxicity data. In Dallinger, R. and Rainbow, P.S., eds. *Ecotoxicology of Metals in Invertebrates*, pp. 427–441. Boca Raton, FL: Lewis.

Van Straalen, N.M. (1996). Critical body concentrations: their use in bioindication. In Van Straalen, N.M. and Krivolutsky, D.A., eds. *Bioindicator Systems for Soil Pollution*, pp. 5–16. Dordrecht: Kluwer.

Van Straalen, N.M. (1997). Community structure of soil arthropods as a bioindicator of soil health. In Pankhurst, C.E., Doube, B.M., and Gupta, V.V.S.R., eds. *Biological Indicators of Soil Health*, pp. 235–264. Wallingford, CAB International.

Van Straalen, N.M. (1998). Evaluation of bioindicator systems derived from soil arthropod communities. *Applied Soil Ecology* 9, 429–437.

Van Straalen, N.M. (2003). Ecotoxicology becomes stress. *Environmental Science and Technology*, pp. 324–330.

Van Straalen, N.M. (2004). The use of soil invertebrates in ecological surveys of contaminated soils. In *Vital Soil: Function Values and Properties*, P. Doelman and H.J.P. Eijsackers, eds., pp. 159–195. Elsevier: Amsterdam.

Van Straalen, N.M. and Kammenga, J.E. (1998). Assessment of ecotoxicity at the population level using demographic parameters. In Schüürmann, G. and Markert, B., eds. *Ecotoxicology*, pp. 621–644. Chichester: John Wiley.

Van Straalen, N.M. and Løkke, H., eds. (1997). *Ecological Risk Assessment of Contaminants in Soil*. London: Chapman & Hall.

Van Straalen, N.M. and Verhoef, H.A. (1997). The development of a bioindicator system for soil acidity based on arthropod pH preferences. *Journal of Applied Ecology* 34, 217–232.

Van Zutphen, L.F.M. and Balls, M., eds. (1997). *Animal Alternatives, Welfare and Ethics*. Amsterdam: Elsevier.

Van't Hof, A.E., Edmonds, N., Dalikova, M., Marec, F., and Saccheri, I.J. (2011). Industrial melanism in British Peppered Moths has a singular and recent mutational origin. *Science* 332, 958–960.

Varanasi, U., Reichart, W.L., and Stein, J.E. (1989). 32P-Postlabelling analysis of DNA adducts in liver of wild English sole (*Parophys vetulus*) and winter flounder (*Pseudopleuronectes americanus*). *Cancer Research* 49, 1171–1177.

Vasquez, M.D., Poschenrieder, C., Bercelo, J., Baker, A.J.M. et al. (1994). Compartmentation of zinc in roots and leaves of the zinc hyperaccumulator *Thlaspi caerulescens* J and C Presl. *Botanica Acta* 107, 243–250.

Vasseur, P., and Leguille-Cossu, C. (2006). Linking molecular interactions to consequent effects upon populations. *Chemosphere* 62, 1033–1042.

Vijver, M.G., Van Gestel, C.A.M., Lanno, R.P., Van Straalen, N.M. and Peijnenburg, W.J.G.M. (2004). Internal metal sequestration and its ecotoxicological relevance: a review. *Environmental Science and Technology* 38, 4705–4712.

Waid, J.S., ed. (1985–1987). *PCBs in the Environment*, Vols I–III. Boca Raton, FL: CRC Press.

Walker, C.H. (1975). *Environmental Pollution by Chemicals*, 2nd ed. London: Hutchinson.

Walker, C.H. (1980). Species differences in some hepatic microsomal enzymes that metabolize xenobiotics. *Progress in Drug Metabolism* 5, 118–164.

Walker, C.H. (1983). Pesticides and birds: mechanisms of selective toxicity. *Agricultural Ecosystems and Environment* 9, 211–226.

Walker, C.H. (1987). Kinetic models for predicting bioaccumulation of pollutants in ecosystems. *Environmental Pollution* 44, 227–240.

Walker, C.H. (1990a). Kinetic models to predict bioaccumulation of pollutants. *Functional Ecology* 4, 295–301.

Walker, C.H. (1990b). Persistent pollutants in fish-eating sea birds: bioaccumulation metabolism and effects. *Aquatic Toxicology* 17, 293–324.

Walker, C.H. (1992). Biochemical responses as indicators of toxic effects of chemicals in ecosystems. *Toxicology Letters* 64/65, 527–532.

Walker, C.H. (1994). Comparative toxicology. In Hodgson, E. and Levi, P., eds. *Introduction to Biochemical Toxicology*, pp. 193–218. Norwalk, CT: Appleton and Lange.

Walker, C.H. (1998a). Avian forms of cytochrome P450. In Livingstone, D.R. and Stegeman, J.J., eds. *Forms and Functions of Cytochrome P_{450}: Comparative Biochemistry and Physiology*.

Walker, C.H. (1998b). Alternative approaches and tests in ecotoxicology: a review of the present position and prospects for change taking into account ECVAMs duties, topic selection, and test criteria. *Alternatives to Laboratory Animals* 26, 649–677.

Walker, C.H. (2003). Neurotoxic pesticides and behavioural effects upon birds. *Ecotoxicology* 12, 307–316.

Walker, C.H. (2006). Ecotoxicity testing of chemicals with particular reference to pesticides. *Pest Management Science* 62, 571–583.

Walker, C.H. (2009). *Organic Pollutants: An Ecotoxicological Perspective*. London: Taylor & Francis.

Walker, C.H. and Livingstone, D.R. (1992). *Persistent Pollutants in Marine Ecosystems*. Oxford: Pergamon Press.

Walker, C.H. and Newton, I. (1999). Effects of cyclodiene insecticides on raptors in Britain: correction and updating of an earlier paper. *Ecotoxicology* 8, 185–189.

Walker, C.H. and Thompson, H.M. (1991). Phylogenetic distribution of cholinesterases and related esterases. In Mineau, P., ed. *Cholinesterase-Inhibiting Insecticides: Impact on Wildlife and the Environment*, pp. 1–17. Amsterdam: Elsevier.

Walker, C.H., Greig-Smith, P.W., Crossland, N.O., and Brown, R. (1991a). Ecotoxicology. In Balls, M., Bridges, J., and Southee, J., eds. *Proceedings of FRAME Meeting, Animals and Alternatives in Toxicology*. pp. 223–252. London: Macmillan.

Walker, C.H., Johnston, G.O., and Dawson, A. (1993). Enhancement of toxicity due to the interaction of pollutants at the toxicokinetic level in birds. *Science of the Total Environment* 134(s), 525–531.

Walker, C.H., Kaiser, K., Klein, W., Lagadic, L. et al. (1998). Alternative testing methodologies for ecotoxicity. *Environmental Health Perspectives* 106, 441–451.

Walker, C.H., Mackness, M.I., Brearley, C.J., and Johnston, G.O. (1991b). Toxicity of pesticides to birds; the enzymic factor. *Biochemical Society Transactions* 19, 741–745.

Walton, K.C. (1986). Fluoride in moles, shrews and earthworms near an aluminium reduction plant. *Environmental Pollution* 42A, 361–371.

Ward, D.V., Howes, B.L., and Ludwig, D.F. (1976). Interactive effects of predation pressure and insecticide (Temefos) toxicity on populations of the marsh fiddler crab *Uca pugnax*. *Marine Biology* 5, 119–126.

Warren, C.E. and Davis, G.L. (1967). Laboratory studies on the feeding of fishes. In Gerking, S.D., ed. *The Biological Basis of Freshwater Fish Production*, pp. 175–214. Oxford: Blackwell Scientific Publications.

Warwick, R.M. and Clarke, K.R. (1998). Taxonomic distinctness and environmental assessment. *Journal of Applied Ecology* 35, 532–543.

Waterman, M.R. and Johnston, E.F., eds. (1991). *Methods in Enzymology*, p. 206. San Diego: Academic Press.

Watt, W.D., Scott, C.D., and White, W.J. (1983). Evidence of acidification of some Novia Scotia rivers and its impact on Atlantic salmon, *Salmo salar*. *Canadian Journal of Fisheries and Aquatic Sciences* 40, 462–473.

Wayne, R.P. (1991). *Chemistry of Atmospheres*, 2nd ed. Oxford: Clarendon Press.

Weeks, J.M. and Svendsen, C. (1996). Neutral red retention by lysosomes from earthworms (*Lumbricus rubellus*) coelomocytes: a simple biomarker of exposure to soil copper. *Environmental Toxicology and Chemistry* 15, 1801–1805.

Weseloh, D.V., Teeple, S.M., and Gilbertson, M. (1983). Double-crested cormorants of the Great Lakes: egg-laying parameters, reproductive failure and contaminant residues in eggs, Lake Huron 1972–1973. *Canadian Journal of Zoology* 61, 427–436.

Whittier, J.B. and McBee, K. (1999). Use of flow cytometry to detect genetic damage in mallard dosed with mutagens. *Environmental Toxicology and Chemistry* 18, 1557–1563.

Widdows, J. and Donkin, P. (1991). Role of physiological energetics in ecotoxicology. *Comparative Biochemistry and Physiology* 100C, 69–75.

Widdows, J. and Donkin, P. (1992). Mussels and environmental contaminants: bioaccumulation and physiological aspects. In Gosling, E., ed. *The Mussel Mytilus: Ecology, Physiology, Genetics and Culture*. Amsterdam: Elsevier.

Widdows, J., Burns, K.A., Menon, K.R. et al. (1990). Measurements of physiological energetics (scope for growth). and chemical contaminants in mussels (*Arca zebra*) transplanted along a contamination gradient in Bermuda. *Journal of Experimental Marine Biology and Ecology* 138, 99–117.

Wigfield, D.C., Wright, S.C., Chakrabarti, C.L., and Karwowska, R. (1986). Evaluation of the relationship between chemical and biological monitoring of low level lead poisoning. *Journal of Applied Toxicology* 6, 231–235.

Wiles, J.A. and Frampton, G.K. (1996). A field bioassay approach to assess the toxicity of insecticide residues on soil to Collembola. *Pesticide Science* 47, 273–285.

Wilkinson, C.F. (1976). Insecticide interactions. In Wilkinson, C.F., ed. *Insecticide Biochemistry and Physiology*. New York: Plenum Press.

Williams, R.J.P. (1981). Natural selection of the chemical elements. *Proceedings of the Royal Society of London* 213B, 361–397.

Willis, K.J., Van den Brink, P., and Green, J.D. (2004). Seasonal variations in plankton community responses of mesocosms dosed with pentachlorophenol. *Ecotoxicology* 13, 707–720.

Wirgin, I. and Waldman, J.R. (2004). Resistance to contaminants in North American fish populations. *Mutation Research* 552, 73–100.

Wolfe, M.F., Schwarzbach, S., and Sulaiman, R.A. (1998). Effects of mercury on wildlife: a comprehensive review. *Environmental Toxicology and Chemistry* 17, 146–160.

Wong, S., Fourneir, M., Coderre, D., Banska, M., and Krzystyniak, K. (1992). Environmental immunotoxicology. In Peakall, D.B., ed. *Animal Biomarkers as Pollution Indicators*, pp. 166–189. London: Chapman & Hall.

Wood, M. (1995). *Environmental Soil Biology*, 2nd ed. Glasgow: Blackie.

Wood, R.J. and Bishop, J.A. (1981). Insecticide resistance: populations and evolution. In Bishop, J.A. and Cook, L.M., eds. *Genetic Consequences of Man-Made Change*, pp. 97–127. London: Academic Press.

Woodford, J.E., Karasov, W.H., Meyer, M.W., and Chambers, L. (1998). Impact of 2,3,7,8-TCDD on survival, growth, and behavior of ospreys breeding in Wisconsin. *Environmental Toxicology and Chemistry* 17, 1323–1331.

Wright, J.F., Furse, M.T., and Armitage, P.D. (1993). RIVPACS — a technique for evaluating the biological quality of rivers in the UK. *European Water Pollution Control* 3, 15–25.

Yang, K., Lin, D., and Xing, B. (2009). Interaction of humic acid with nanonised inorganic oxides. *Langmuir* 25, 3571–3576.

Yokoi, K., Kimura, M., and Irokawa, Y. (1990). Effect of dietary tin deficiency on growth and mineral status in rats. *Biological Trace Element Research* 24, 223–231.

Index